SCIENCE IN AMERICA

THE CHICAGO HISTORY
OF SCIENCE AND MEDICINE

Allen G. Debus, editor

Edited, Selected, and with an Introduction by

NATHAN REINGOLD
and
IDA H. REINGOLD

SCIENCE IN AMERICA
A Documentary History 1900–1939

THE UNIVERSITY OF CHICAGO PRESS | Chicago & London

THE UNIVERSITY OF CHICAGO PRESS, CHICAGO 60637
THE UNIVERSITY OF CHICAGO PRESS, LTD., LONDON

Printed in the United States of America

88 87 86 85 84 83 82 81   5 4 3 2 1

NATHAN REINGOLD is editor of the Joseph Henry Papers at the Smithsonian Institution. IDA H. REINGOLD is a mathematician who has authored numerous technical papers.

For our fathers who struggled for us in the promised land

LIBRARY OF CONGRESS CATALOGING IN PUBLICATION DATA

Main entry under title:

Science in America, a documentary history, 1900–1939.

(History of science and medicine)
Includes index.
1. Science—History. 2. Science—United States—History. I. Reingold, Nathan, 1927–
II. Reingold, Ida H. III. Series.
Q125.S43433   500   81-2584
ISBN 0-226-70946-9   AACR2

Contents

Genius develops in quiet places, character out of the full current of human life.

Preface

A traditional way of regarding science and its history is to envisage a core consisting of concepts, data, and techniques, an account of which constitutes the history of science. Human beings and their institutions are outside the core; if they are sometimes included within the definition of "core," the tendency is to define them in terms of importance to the content of science. In fact, to this day both the ideology of science and some histories have a tendency, explicitly or implicitly, to identify the scientist and the scientific organization with the body of knowledge. Ohm becomes Ohm's Law; Darwin is natural selection; the early Royal Society is Newtonian science. The core exists independently of humans; humans exist for the core.

However one defines the core, somewhere there is an outside—the rest of history—where science, the scientific community, and the scientific way of life do not exist. A further tacit assumption is that processes occur at an interface between the core and the outside. This is where science and society meet and interact. Just how this occurs is a matter of considerable debate—whether certain functions go from in to out and others from out to in, or perhaps they go both ways. This is a convenient model for many purposes, not the least of which is to further particular ideologies about science, public policy, the nature of a modern society.

A quite different set of assumptions, however, support this book. Simply stated, there is no core—everything is part of an interface. And if the core with its weighty contents has the aura of permanence, the interface is like a thin film, almost a soap bubble. Ideas, people, institutions of all sorts—not merely the scientific—exist in the interface. It is too much for such fragility, and tensions are created which produce additional film. The core image necessitates a kind of causal language; the film interface assumes temporal relationships which may or may not be causal in some simple mechanistic manner. It does presuppose that the temporal relationships are interesting and meaningful; it regards history or a society as having a kind of unity or wholeness which is violated by abstracting knowledge and its possessors from the rest of humanity.

From this persective, the concept of a core of scientific ideas, perhaps with associated men and institutions, appears in a new light—an ethereal ideal separate from the generality of mankind and daily concerns. At bottom, it is a belief—impassioned and sincere—in the transcendent value of abstract knowledge of the material world, of truth divorced from mundane realities. It diminishes everything outside the core; when individuals and institutions are admitted, it is reluctantly and with every tacit intention of blending them into intellectual formulations. In essence, it is the core—seemingly solid and weighty even amid change—which turns out to be weightless. The interface, a mere surface, is packed, heavy with contents continually interacting and in tension. Those who believe in a core regard knowledge as deserving explanation. Those who believe in the interface regard human interactions—intellectual, social, economic, political, institutional, etc.—as the problem of history; they are under no illusions—declared or tacit—of practicing either science or philosophy.

One reason tensions exist and the thin film changes has to do with scientists and their labors. Almost invariably, scientists organize themselves and their data in ways which reflect older conceptions. Even when they are being most adventurous intellectually, scientists are usually confined by older ideals which they seek to impose upon ambiguous findings. When this attempt invariably breaks down, it takes the form of outcomes—as, for example, a further extension of the film, a new soap bubble—which is genuinely surprising to the initial presuppositions of most of the actors in the scientific drama. The course of events which we denote as "history" simply does not live up to the prior expectations of humans or to their illusions about rationality, which is best defined as the "comfortably familiar."

Describing and analyzing thin films does not seem a very promising agenda for historians. In fact, however, historians do just that by gathering surviving evidences of the past—the debris from the successive breakdowns of films—and creating an artificial core for the past in the form of an article, an analytic monograph, or even a sweeping historical narrative. We think of what follows as a mosaic composed of bits of reality, manuscripts from past environments. The histories of these documents necessarily removed them from their original contexts of curiosity, pride, greed, ambition, self-sacrifice, bungling, nobility, passion, and the like. Then, they were further removed from the presence of related papers in collections by an interloper from quite a different soap bubble who arranged them in new patterns. The pieces of the mosaic are affixed in the new patterns by the stated and unstated intellectual purposes of the historian who believes, hopefully, that what he has produced intersects past contexts. Somehow, the creative eye of the historian should give a valid sense of the past, even as his product displays strengths and weaknesses of his concepts, techniques, and imagination. Because of that, this preface has an element of self-definition as well as being an imperfect road map for the text to follow.

Histories of science, like all histories, are artifacts of their authors and, consequently, of particular niches in time and space. They are also hemmed in by their sources. We do not want to do a history running from great ideas to great men to great institutions. Our intents are quite different. Despite that, most of the documents included here are from the unpublished papers of near-great and great scientific figures. But there are justifications: (1) the leading scientists are important for the generation of new ideas and data and for setting standards of scientific life; both influence the lesser members of the various scientific disciplines; (2) many of the concerns, intellectual and social, of the leading figures turn out to be shared by others in the community; (3) many of the documents refer to individuals now lost to fame; (4) quite deliberately, a number of the documents convey a sense of the routine, of the day-by-day life in the laboratory, university, and observatory. There is no massive rift in the intellectual landscape between the great creators and their lesser contemporaries. Many of the latter turn out to be classmates, students, colleagues, and friends of the giants.

What does not appear here explicitly, except in a small number of documents, are the many outside the scientific community. That is a great loss, because they too are on the interface. But the study of that aspect of the history of science has barely begun. Even in this collection of documents, the larger society's presence is tangible, a silent, near-invisible Greek chorus. Again and again in the documents, the problem of the position of science and of the scientists on the national scene recurs, a unifying theme, sometimes loud, sometimes muted.

In what follows—the mosaic of bits of the past—the documents themselves and the structure of the entire work exhibit this blend of scientific and nonscientific concerns. Or to use conventional terms in the historiography of science, the internal and external aspects are both given. Gluing them together is a kind of mixture ideology. Our initial presupposition, many years ago, was to expect a juxtaposition, not a rather complex organic linkage. Like almost all historians, we prefer to think we found the ideology in the sources. Perhaps this viewpoint arose because we first encountered the history of science as a research problem in informal sources—unpublished letters and the like. Unlike printed articles and books, the evidences of a complex past were not squeezed out. Instead of a core, neat and solid, there was a glorious jungle of ideas, men, and events. No doubt, we were even then innoculated against the tendency to idealize the knowledge at the expense of personal and social realities. The effect is strange, perhaps, to those accustomed to devout analysis of the growth of knowledge or to respectful curtsying before great ones.

In no way, shape, or form should what follows be considered as demeaning science. There is much that can be said in a muckraking vein, but this is not our purpose or our belief. Unlike some members of the scientific community who prefer to think of themselves and their predecessors as abstract carriers of concepts, unlike some internalist historians of science who prefer to think of concepts and data as growing in a vacuum, unlike some philosophers of science who are unconcerned with the actual persons who generated science, we and many of our colleagues are greatly interested in human beings and their environments. Having spent a good many years reading through a wide range of personal correspondence and other records of scientists from the late eighteenth into the mid-twentieth centuries, we are not surprised to find that scientists can be petty, nasty, vain, quarrelsome, foolish, greedy, vindictive, or show any degree of less than admirable qualities. The lay public or the professional brought up on the idea of science as an abstract search for truth unconnected with the realities of human life may have been startled and shocked to discover the conflicts disclosed in Watson's *The Double Helix*. But something like this is familiar to any person who has done any studying in the history of science based upon original sources, especially manuscript sources. Although we are in the position of the valet who knows many of the secrets of his employer and cannot therefore be too impressed by his employer's pretensions, we have to confess that all this very intimate knowledge has not in the least shattered our high respect and admiration both for science as an endeavor and for many scientists. What impresses us is not that the scientists are flawless people engaged in an abstract search for truth, but that the ideal seems to be approximated in a surprising number of cases, given the ample evidence of mankind's frailties.

The external structure of this book is pyramidal. The detail is greater in the early years and gets less so as the text enters the start of World War II. We hope the reader will sense that the details are all leading, more or less, to the same point, just as in a pyramid. There is a moral here: how the best efforts of a number of very talented individuals came to an ironical conclusion. From World War II came a different order of things at variance with much that was earnestly sought for in the prior decades.

The internal structure derives from the mixture ideology and is best described as a "wraparound." An opening institutional chapter on the Carnegie Institution of Washington introduces principal actors and animating issues intermixed with scientific content. Next follow a number of disciplinary chapters with the physical sciences preceding the biological. Scientific ideas intermix with actors and issues into the year 1915, then a kind of central core around which all the chapters wrap. In a short chapter on the reform of the National Academy of Sciences, George Ellery Hale, one of the great promoters and mythmakers of science, briefly struts across the historical stage. The massive part of the core is a chapter on World War I as a test of the ideals professed by scientists in both warring camps and as an opportunity for furthering Hale's reforms. Advancement of knowledge and of social ideologies are juxtaposed in many documents, perhaps insensibly to their creators. The disciplinary theme is picked up again, roughly from where it was left off earlier, and carried toward the end of the period. Biology now precedes the physical sciences whose chapters point to the ultimate unit of the book. This is an account of the early years of the Institute for Advanced Study at Princeton seen as the culmination of several trends in early portions of the book.

The documents are given with a decent minimum of editing. Each has a brief identification, including the date, in boldface. Headings, addresses, etc., have been omitted. Only in a few cases are there silent corrections such as obvious errors in typing (e.g., "hte" for the definite article) or rectification of confusing usages. Some items (such as retained carbon copies) lack signatures at the end or identification of the author. These are supplied in brackets which also are used in the text to indicate the rare insertions by the editors. At the end of each document, the location is noted in italics at the right-hand margin.

The purpose of the array of documents is to convey a general sense of what being a scientist meant in a given time and place, not the particulars of every scientific development involved. Documents were selected with three criteria in mind: (1) to show the scientist at work or reflecting on the work; (2) to show interactions between members of specific research communities or between members of different research communities; (3) to show the scientists consciously responding to their environments or, at least, acting against a backdrop of events and institutions outside the bounds of the scientific disciplines. Rarely do documents meet all three criteria, but the cumulative effect should communicate a texture of the past as experienced by a selected group of scientists.

Footnotes are a problem in a work of this nature. Although many of the documents are lucid statements of their authors' intentions—some spectacularly so—they are embedded in contexts not well known to the general public. The chapter introductions should clarify these contexts as well as a number of specifics. Footnotes are not intended to be exhaustive, only to be helpful. Many of the documents deliberately contain considerable detail regarding the scientific work, and we believe that the educated layman can understand what is involved in most, if not all, instances. This book is not intended as a work of science or as an example of the internal history of science. But the technical content is there as a necessary part of the historical landscape.

Colleagues at the Joseph Henry Papers provided encouragement and assistance for which we are grateful. The staff of the Smithsonian Library, especially Jack Marquardt, helped in the hunt for secondary literature. Philip Pauly called our attention to one of the documents. Matthew Reingold helped prepare the manuscript. Dr. Brigitta Dassler helped with the German translations. We are indebted to many in the libraries and archives containing unpublished materials in this area. A large body was collected over the years, perhaps enough for three or four volumes of equal interest to this one. Space considerations, however, forced us into a carefully limited mold. Yet the experience of looking at other collections was very valuable, and we want to thank all those individuals in so many institutions who answered our letters and responded to queries in person.

We want to thank the following for their courtesy and cooperation in permitting the publication of documents: Air Force Geophysics Laboratory; American Philosophical Society; Dr. Garrett Birkhoff; Library, University of California, Berkeley; Millikan Library, California Institute of Technology; Carnegie Institution of Washington; Regenstein Library, University of Chicago; Mrs. Grace Richards Conant; Countway Library, Harvard University; Albert Einstein Estate; Embryological Laboratory, Carnegie Institution of Washington; Miss Elizabeth Davisson; Dr. P. H.

Fowler; Franklin Delano Roosevelt Library; Hale Observatories; Harvard University Archives; Houghton Library, Harvard University; Huntington Library; Archives, University of Illinois at Urbana-Champaign; Mr. Alexander R. James; Mrs. Jane Bridgman Koopman; Mrs. Mary B. Lawrence; Mrs. Jean Flexner Lewinson; Library of Congress; Institutional Archives and Special Collections, Massachusetts Institute of Technology; Mrs. Isabel M. Mountain; Archives, National Academy of Sciences; New York Public Library; Mrs. Agnes F. Peterson; Firestone Library, Princeton University; Dr. William J. Robbins; Rockefeller Archives Center; Smithsonian Archives; Dr. William C. Sturtevant; Humanities Research Center, University of Texas; VanPelt Library, University of Pennsylvania; Library, Washington University; Medical Library, Washington University; Dr. Warren Weaver; Dr. Marina Whitman; Sterling Library, Yale University; Historical Medical Library, Yale University.

We want to thank, in no particular order, a few of the individuals who helped us answer troublesome questions: Dorothy Hanks, Arthur Norberg, Judy Goodstein, Amy Levin, Joan Warnow, Ronald Wilkinson, Ellin Wolfe, Spencer Weart, Paul Forman, N. H. Robinson, Clark Elliott, Helen Slotkin, Joseph Ernst, William Hess.

Introduction

By 1900 the community of scientists in the United States had grown and matured. From a position of provinciality in the colonial era and early days of the Republic, the community had achieved metropolitan status. In both quality and quantity, the United States was now one of the scientific powers, not yet an equal of the imperial Germans and Britons but clearly a rising presence on a par with some proud, older scientific communities.

Of course, there is no such thing as "the scientific community." There were and are various groupings by discipline, specialty or subspecialty, institutional location, etc. These were held together loosely by the feeling of their members that a community of science should exist. Often too there was a sense of grievance. Facing colleagues overseas, many were self-conscious about rankings, individual and national. Some felt a sense of receiving only belated, begrudging, and inadequate recognition for their work, and they were ambivalent about their European colleagues. A perceived lack of appreciation—a sense of inadequate deference to the research products—was the bond holding diverse elements together.

Lack of international recognition hurt, because several generations of hard work, shrewd politicking, and adroit proselytizing had paid off. Taking advantage of the expansion of the U.S. economy and the resulting urbanization, the scientific community had somehow convinced key individuals and groups to place a small portion of the nation's surplus wealth into the sciences as a cultural adornment, as a symbol of national prestige, and (in some cases) as a means of knowing the handiwork of the Deity. With the expansion of the economy came an expansion of government. By the end of the last century the idea of trained expertise was decidedly evolving in both. In a very American way, that often came out as using "science" or "scientists." Despite the visible signs of increased scientific research, despite the vocalization of scientific ideologies, and despite the growth of research in industry, medicine, and agriculture, there are ample evidences in print and in manuscript of uneasiness or even distress among a significant number of scientists, including many of note. In ways both gross and subtle, the ideologies and expectations of many in the scientific disciplines did not match the concepts and realities widely present in the American nation. For a number of scientists, that produced another, unifying grievance—a domestic lack of deference to science.

Even in 1900 scientists in the United States had to reconcile an "official ideology" of their calling with the realities of their daily lives. Take the idea of science, a term subject to all manner of problems of definition. In the United States an older definition persisted and could legitimately claim hegemony over competitors. Not specifically Anglo-American, this view loosely thought of any knowledge of the natural world as "science," avoiding any distinction between the pure and applied. By the end of the century most educated Europeans thought of science as something ideally theoretical, even verging on the philosophical. Applications were assumed but regarded as of a lesser intellectual order. Many U.S. scientists agreed in varying degrees. There is ample evidence, however, for both Europe and the United States that this set of beliefs did not wholly correspond to reality. In fact, there were scientists on both sides of the Atlantic with avid concerns for applied topics, or whose labors straddled both theory and practice. Nor should we disregard the clear implication that much of the published research results were inevitably modest in scope, far from the empyrean level of grand theory.

But the old definition in the United States resulted in a great readiness to regard almost anything as "science" that worked or was even merely favorably regarded. The general public and the great institutions of the society had this tendency. Relatively few laymen consciously placed abstruse scientific work in a ranking above, for example, the skillfull labors of talented tinkerers. What was at issue was not the ratio between the support of pure or basic research and what is now known as applied research and development. Implicitly, the nation had a definition of science quite different from the Europeans and from many U.S. scientists. It assumed that everyone involved in occupations related to the scientific disciplines and the applied professions (medicine, engineering) was somehow a scientist, that all their activities to some extent were "science." In contrast, the professional scientists followed European precedents and defined the community as one of researchers adding to knowledge, even sometimes assuming that only significant investigators made the grade. Underlying these beliefs was an unspoken identification of the researchers with the knowledge they produced. That identification had little relevance to the reality they were a part of—a growing mass scientific community. In this conglomeration of institutions, disciplines, and individuals, mass, utility, and higher education had emerged in time, cunningly intertwined.

To further the progress of society and of knowledge, scientists in the United States tacitly adopted two coexisting strategies which persist to this very day. The strategies enabled them to assimilate lesser scientists and practitioners of other fields while generating the pure science ideally needed for attaining utilitarian ends. First, applied goals were defined in such a manner as to require, necessarily, the conduct of abstruse research. The federal government provided ample precedents in the great surveys—the Geological and the Coast and Geodetic—in the Naval Observatory, and, most significantly, in agriculture. Scientific medicine was another example represented in this volume. The documents on or from the rising world of industrial research illustrate this strategy. Second, wherever possible, sheltered enclaves for basic investigations came into being. The research installations of the Carnegie Institution of Washington, the Rockefeller Institute, and the Institute for Advanced Study are presented here because they represent this enclave strategy. In the first strategy, the interaction of data from theory and practice supposedly generated an intellectual excitement producing both theoretical and practical progress. In the second is an assumption of a quiet solitude supporting great talent, even genius, away from any momentary distractions. By the early years of this century, both strategies were being actively pursued.

To man all the research missions and all the enclaves required a very large supply of trained individuals. Some came from overseas. Most were products of the nation's school system. By the turn of the century a significant expansion of secondary schooling and higher education was underway; modest by present standards, the slightly more than 4% of the 18–21-year-old population in institutions of higher education was impressive for 1900.

Although many scientists before and after 1900 thought the government bureau would serve as the loci and model for research, the period 1900–1939 marked a high point in the role of the private sector in the support and conduct of research. The influx of funds from private philanthropy reinforced the conservative inclinations of a considerable number of influential scientists prior to World War II. In their view, the advancement of scientific theory required an autonomy not at the mercy of governmental purposes, popular passions, or utilitarian needs. Most of the influential leaders of the various scientific disciplines neither foresaw nor wanted what developed after 1940.

The universities increasingly served as the loci and model for research after Johns Hopkins University placed graduate education on a firm footing. The problems and limitations of the university arena are a recurrent theme in what follows. Despite continuing strenuous efforts in this century, universities, even the most elite, tended to American patterns, not European ones. This meant open or tacit roles in training a large body of men and women for all the tasks required by a complex modern society. The universities were forced to become mass institutions. Abraham Flexner, the founder of the Institute for Advanced Study, considered the entire trend simply abhorrent. Part of his reaction stemmed from an idealization of the German university, stressing its research and conveniently downplaying its utilitarian mission to train for leadership and the professions and to maintain a specific national culture. Further, the American universities took up the roles both as sites for enclaves of pure research and as conductors of broad missions for the nation. Increasingly, the heirs of cruder empiricisms succumbed to the lure of the genteel tradition of the colleges and universities; many engineers and inventors now aspired to the supposed benefits of "higher" culture. There is a historical irony here—institutions reflecting or deriving from a genteel culture producing the personnel for a mass society which acquires aspects of gentility while remaining resolutely vernacular.

The universities became the ideal site for research, even though most investigations were not carried out at such institutions. What mattered is that pure research was conceived as best conducted in institutions of higher learning. An idealized image of the university influenced the patterns in government and in industry. In part, it was a misreading of history. The view also downplayed the extensive role of the university in applied research and in development.

Impelled by such beliefs, mass higher education was further strengthened, making possible a mass scientific community. An amorphous body, this community has blurred boundaries between the top hierarchs and the rest of its members (i.e., between who may and may not do research). Knowledge-generating and knowledge-using activities are often so intertwined as to defy sharp differentiation. More important, the general public is omnipresent, often conscious of knowing something because of schooling and usually interested in the implications of research. Like it or not, American scientists are more likely to be sensitive to the presence of the general public than scientists in most other industrialized nations. Another blurred boundary exists here, one making it hazardous to differentiate the attitudes of the American scientists from the lay public.

That blurred boundary only aggravated the problem of deference for many American scientists who felt their society lacked a proper structure hierarchy recognizing the high place of science and its creators. This will show again and again in the documents that follow; the feeling persists even into this generation. It may seem odd to those not familiar with, or unsympathetic to, the strong motivations of many scientists who believe in the greatness of their calling and in its importance for humanity. This belief is at bottom an emotional faith, even though often expressed with great display of reasoning and cogent example. Then and now, there is still this concern for proper deference despite the historical data showing the relatively high socioeconomic status of scientists, the willingness of other Americans to support research, the high influence of scientists in the United States compared to many other nations. It is not simply a matter of money and power but of perceived prestige. Given the strong feelings about the essentiality of basic science, no amount of support could ever totally assuage the feelings of inadequate recognition. A democratic society within the historic traditions of this nation could not go beyond certain limits.

Lacking a traditional structured society, deference often takes strange forms in this country. Perhaps popular stardom is a crude way of envisioning it. Politics has that attribute; sometimes the law. From time to time sheer wealth gets the public attention, generating deference even out of proportion to economic power. Religion clearly had that quality once in American life and occasionally still does. The mass media and sports now have the deference due popular stardom.

Prior to 1900 the sciences almost never had such a role. Popular lecturing by men like the elder Silliman and Agassiz briefly attracted mass attention. After Darwin, there was a brief flurry of general interest based on hopes and fears. By and large, the scientists were a small community treated with kindly tolerance but hardly viewed as essential. The idea of utility engendered another flurry of public interest which yielded a modest return in deference but not popular stardom. From 1890 to World War II scientists, engineers, physicians were valued for their expertise but not enough to change essentials. None created the sensation Albert Einstein did in the years before he migrated here. Without consciously trying, he achieved a species of popular stardom. The Langmuirs of the world never made it, even though they combined theory and practice. A Thomas Hunt Morgan simply did not enter the picture, nor did a George Ellery Hale, perhaps the scientific equivalent of a J. P. Morgan.

World War II changed all that. The atom bomb and radar endowed the scientific community and some investigators with star quality. Public policies now were defined consciously around the aspirations of the scientists. But such deference had serious flaws. Within a few years the atom bomb engendered thoughtful concerns over environmental pollution. National defense cast scientists and engineers as protectors of the nation; Vietnam harshly raised the relation between knowledge and death in a setting of national defeat. Even the great biomedical research effort of post–World War II, which had attracted some physical scientists fleeing the bomb, proved a false source of deference. As some health problems proved intractable, the question arose whether the funds benefitted the research professions and organized medicine more than the public. If proper deference for science required popular stardom or mass impact, only Einsten had succeeded as an individual in this century.

All other successes, however temporary, stemmed from big projects—the bomb, the moon walk, the war on cancer. Thoughtful individuals in the research community were not against big projects as such. Like the general public, the scientists lived in a world in which scientific ideas and their tangible products were ubiquitous, a kind of anthropological museum of past Western Civilization. What differentiated them at times from their fellow citizens was another set of beliefs about the nature of the scientific community and the relation of knowledge to utility. Although not all scientists shared the ideals in every particular, an impressive number did, including many of the leaders of the community. The big projects distorted these ideals. Progress occurred, some believed, because of science, generated from its generalities in significant jumps in contrast to the small step-by-step advances from empirical practice. A half-truth at best, this viewpoint had cogency to many in Western Europe where the social structure of the societies tend to segregate the savants from the lesser men of practice. When circumstances permitted, U.S. scientists glibly and sincerely talked about practical results, sometimes obscuring the essentiality of the hard steps between theory and practice; sometimes implying too much in an eagerness to expand the support of research. The result was a general public conditioned to conceive of science solving all problems and inclining to a faith in big projects. But those would only work when basic knowledge had advanced to the point of containing an implicit solution. Big projects could also deflect support from needed, less spectacular research. Somehow the scientists never quite conveyed that point in their ideology, eventually producing a strong skepticism about the claims of science. But that came well after World War II.

1
The Carnegie Institution of Washington

Near the end of 1901 Andrew Carnegie, the great steel maker of the United States, announced a gift to found an institution to somehow encourage what we now call basic research. A $10,000,000 endowment (later enlarged) was an enormous sum by the standards of the time, giving rise to high expectations. Influencing Carnegie were two factors. First, and most decisive, was the turn-of-the-century literature reminding Americans of their humble place in the world of science.[1] While invention and applications thrived, pure science was largely derivative from the great centers in Western Europe. Carnegie responded to the call for the United States to repay the Old World for its great gift of science to modern civilization and to take a place in the world of research befitting a great nation. Internationalism and nationalism mingled easily among the motivations of the founder and his associates in the early years of the Carnegie Institution of Washington (CIW).

If not immediately decisive, the second factor, the role of higher education, was a more fundamental issue in the United States. Despite the writings on American indifference to basic or pure research, the sciences had experienced a very high rate of growth in the previous three decades. Viewed in light of the heavy human and material U.S. commitment to the sciences by the turn of the century, Carnegie's gift was a spectacular instance, not a unique event. At the same time Rockefeller moneys were founding an institute in New York City for fundamental investigations bearing on medical problems.[2] Even though research was on the rise, both the scientific community and the small concerned lay public were dissatisfied with the uneven growth and the inevitable lag in tangible successes.

Nowhere in the United States was there more awareness of the great European centers than in the universities; nowhere were internationalistic and nationalistic hopes greater; nowhere did uneven growth and unfulfilled expectations produce such acute frustrations and anxieties. A few institutions (Johns Hopkins and the University of Chicago) had solid research traditions; at many others, some with eminent names, science was honored but tepidly supported. At most colleges and universities, the overwhelming emphasis on teaching made original investigation an infrequent and hazardous occurrence. In retrospect, knowing that the universities in the United States became the principal site for the conduct of abstruse research, the historian's eye can see precursors and precedents. To many faculty and students around 1900, the increasing quality and quantity of publications only added to their frustrations. How much more could come with better support? Andrew Carnegie was an obvious target.

Initially, the idea of a national university was dangled before him by John W. Hoyt.[3] To ardent proponents like Hoyt, there was an obvious need for a great institution that would provide the apex of the sprawling educational system and be on a par with the great universities of the world. Most leading university presidents intensely disliked this idea; they felt it downgraded their achievements and implied a permanent limit to their potentials. But ANDREW D. WHITE (1832–1918), former president of Cornell, and DANIEL COIT GILMAN (1831–1908), president of Johns Hopkins University, supported the idea. Gilman would serve as the first president of the Carnegie Institution of Washington.[4]

The national university idea also brought on the scene CHARLES D. WALCOTT (1850–1927), the second of the four key men in the early history of the CIW, who was its first secretary. Then the director of the U.S. Geological Survey, Walcott was also secretary of the Smithsonian Institution from 1906 until his death, and president of the National Academy of Sciences from 1917 to 1923. Walcott was the successor to A. D. Bache, Joseph Henry, and John Wesley Powell, the three great hierarchs of federal science in the last century. A great paleontologist, he is probably the last significant American scientist without formal training in a college or university.[5]

His involvement with the idea for a national university arose from a need to do something about the relative decline of governmental science. As head of the Washington Academy of Sciences, he sought an alliance with the university world by opening government facilities to students and faculty. In 1901 the Washington Academy joined with the George Washington Memorial Association to incorporate a "Washington Memorial Institute." An organization of women who wanted to honor the first president by establishing a Washington Memorial University in the District of Columbia, the association joined hands with Walcott and his allies in academe; the institute would bring students to Washington. Its proposed handsome edifice would serve as the headquarters for the national university to emerge from this nucleus. Walcott and his female allies turned to Carnegie for support.

In November 1901 Carnegie called a third man into consultation with him and Gilman, JOHN SHAW BILLINGS (1839–1913).[6] A physician known best for developing two great libraries, the New York Public Library and the Surgeon General's Library (now the National Library of Medicine), Billings was a strong, bright, blunt-spoken individual. Arguing against Gilman and against the idea of a national university, Billings prevailed. The early formal statements of the CIW largely reflect his views. He early became chairman of the Executive Committee of the Trustees of the Institution and, up to his death, chairman of the Board of Trustees for nearly 10 years.

By far the most influential shaper of policy was ROBERT S. WOODWARD (1849–1924), Gilman's successor, who served from 1904 to 1920. A civil engineer with a talent for mathematics, he became an applied mathematician or classical physicist. After serving as an astronomer with various federal surveys, Woodward joined Columbia University in 1892, becoming its dean of pure science in 1895. With great skill and tenacity, he fashioned the Carnegie Institution of Washington along his lines of reasoning. Although differing markedly on one point—the role of the universities—he consciously articulated and embodied many of the positions which would become clichés in the policies of the leaders of the U.S. scientific community in this century.[7]

Even before the formal launching of the institution, the trustees were grappling with the question of how to aid basic research. Carnegie's trust deed (see below) stressed aid to "exceptional men." How are such individuals identified? Although the scope could include "every department of study," somehow a scale of priorities had to emerge. The endowment could not hope to cover every field. What were the forms by which the exceptional men and the favored fields would get support? And most important, where would the research occur? Very early, signs appeared of a calculated intention to keep aloof from the university world.

The trustees quickly established advisory committees to review the needs of different fields and their opportunities. Woodward chaired the committees on physics and geophysics. To aid exceptional men, until the committees reported, two forms of action were adopted: "research assistantships" became available to promising young investigators ("fellowships" in our present terminology), and limited "minor grants" for what we now designate as "research projects."

With the reports in hand, a third form of action became prominent, eventually dominant. This was the formation of research laboratories, departments, and bureaus within the Carnegie Institution of Washington itself. Walcott strongly favored this; Billings was ambiguous. The abortive attempt to merge with the Woods Hole Biological Laboratory[8] and the awkwardness often accompanying relations with university people strengthened an existing bias for in-house research.

Enter the former dean of pure science filled with ideas for cooperation in science and somewhat leery of the founder's "happy phrase," the exceptional man (the characterization is Woodward's). He first went along with Carnegie in supporting a very exceptional but difficult individual, Luther Burbank.[9] Very soon, Woodward declared war against the idea of seeking out and developing exceptional men. In place of the minor grants and research assistantships, he proposed a system of research associateships, long-term aid to proven investigators. Although the trustees first balked, eventually Woodward prevailed. The exceptional men now became proven, full-time trained specialists. Finding and nurturing such individuals was delegated to the universities.

In addition to the idea of research associates—notable scientists like Thomas Hunt Morgan—Woodward furthered the thrust for in-house research. It was all in accordance with his theory of research. The experience with the Department of Economics and Sociology, given below, reinforced his views which worked against funding university people. The institution was not to be merely a "disbursing agency" but a real participant in research.

As to choice of fields, Woodward sympathized with Gilman's desire to aid the humanities and social sciences, in contrast to Billings and Walcott. But his fiscal caution and a belief in the historical and logical primacy of the sciences prevailed. Woodward was willing to give limited support to exceptional men and worthy investigations for their own sake and to further the development of these fields into true disciplines where trained specialists could work with a clearly agreed-upon consensus.

In the period 1902–1920, during the administrations of Gilman and Woodward, the Carnegie Institution of Washington provided a notable stimulus to research in the United States, both in terms of its own units and in consequence of the support of outside investigators, both proven and unproven. Even more significant was the effect on the universities. The CIW was one of a number of ventures throughout the world in the years 1890–1910 to establish research institutions divorced from teaching. Its founding, for example, was a stimulus to the organization of the Kaiser Wilhelm Gesellschaft[10] in Imperial Germany with its own research institutes. Unlike their critics—such as James McKean Cattell[11] and Theodore W. Richards[12]—Billings, Walcott, and Woodward did not see the university as necessarily the site for basic research. Having followed earlier career routes before Gilman's success with graduate education at Hopkins, they generalized from the experiences in the federal establishment. What they did not see was the pervasiveness of research—in practice and as an ideal—in the American definition of higher education. It was too late to call for specialization in research. With great aplomb, the major American universities responded to the threat of the nonuniversity research institute by incorporating more and more aspects of these institutions. Perhaps as important, universities with pretentions to excellence scrambled down that same path.

1. A more detailed account of the early years of the Carnegie Institution of Washington (CIW) is in Nathan Reingold, "National Science Policy in a Private Foundation: The Carnegie Institution of Washington, 1902–1920," in *The Organization of Knowledge in Modern America, 1860–1920,* ed. A. Oleson and J. Voss (Baltimore, 1979), pp. 313–341. See also Nathan Reingold, "American Indifference to Basic Research: A Reappraisal," in *Nineteenth-Century American Science: A Reappraisal,* ed. George H. Daniels (Evanston, Ill., 1972), pp. 38–62. The best account of the founding of CIW is in chap. 9 of Howard S. Miller, *Dollars for Research: Science and Its Patrons in Nineteenth Century America* (Seattle, 1970).

2. George W. Corner, *A History of the Rockefeller Institute, 1901–1953* (New York, 1964).

3. JOHN W. HOYT (1831–1962) was a former governor of Wyoming Territory and the first head of its state university. See David Madsen, *The National University, Enduring Dream of the U.S.A.* (Detroit, 1967).

4. For the general story of universities, see Lawrence R. Veysey, *The Emergence of the American University,* (Chicago, 1965). For Gilman, see Hugh Hawkins, *Pioneer: A History of Johns Hopkins University, 1874–1889* (Ithaca, N.Y., 1960).

5. For a sympathetic account, see Ellis L. Yochelson, "Charles Doolittle Walcott, 1850–1927," in vol. 29 of the National Academy of Sciences *Biographical Memoirs* (New York, 1967).

6. None of the biographic accounts do justice to Billings. His personal papers at the New York Public Library still await effective use. Billings was close to Carnegie because of the latter's program of support for public libraries.

7. See the entry on Woodward in the *Dictionary of Scientific Biography.*

8. CHARLES OTIS WHITMAN (1842–1914) of the University of Chicago headed the Woods Hole Marine Biological Laboratory. He is a very important figure in a number of respects but virtually unstudied. F. R. Lillie's *The Woods Hole Marine Biological Laboratory* (Chicago, 1944) is useful but deliberately vague on the disputes of the early years.

9. See Peter Dreyer, *A Gardner Touched with Genius: The Life of Luther Burbank* (New York, 1975).

10. Now the Max Planck Gesellschaft.

11. JAMES MCKEAN CATTELL (1860–1944) was a psychologist then at Columbia. Besides his great importance as a force in the development of his field in the United States, Cattell was an editor and promoter of scientific periodicals. He had strong views, expressed with great vigor. Cattell was a founder of the American Association of University Professors.

12. THEODORE WILLIAM RICHARDS (1868–1928) of Harvard was a physical chemist. He was the first U.S. citizen to receive the Nobel Prize in chemistry (1914) in recognition of his work on atomic weights.

William Henry Welch[1] to John Shaw Billings, January 16, 1902

My dear Doctor Billings,

I have had no detailed conversation with President Gilman about the Carnegie Institution, although he has said that he intends to talk with me about the plans. I was greatly pleased to find your name in the list of Trustees to look after the interests of scientific medicine. I am anxious to have a talk with you about the future of the Rockefeller Institute, as I think Mr. Rockefeller is prepared to provide in a large way for the Institute as soon as the Directors are ready to advise him as to the best lines of organization and development. If we could bring together a good working staff, especially could find a man of first-rate scientific and administrative ability as Director, I should favor the establishment of a laboratory in New York, beginning in a rather modest way. This could be done without abandoning our present plan of aiding investigations elsewhere, but I think our general policy will end toward concentration of the work in a separate Institute or laboratory.

In reply to your inquiries I would say 1) as to "the terms on which fellowships and scholarships should be offered," that it seems to me the following classes should be considered in making such appointments: a) those who have demonstrated by published work their capacity to initiate and conduct original scientific investigations; b) those recommended by heads of laboratories who have established reputations as investigators to carry on specified researches under their direction.

I believe that valuable results would come from the appointment of well trained men, capable of independent, original work, to undertake investigations in the best laboratories connected with Universities. At present the staff of these laboratories is occupied very largely with teaching; if to these staffs could be added men who shall give their time mainly to original research it seems to me that our laboratories would be much more productive than at present.

You will have to face the difficulty that the number of those who have any real fitness for original research is small, and that these are likely to come to the front under any circumstances, while there is great danger of encouraging those who have no genuine capacity for such work and who had better be at something else. One can conjure up possibilities of great harm coming from the so-called endowment of research, but I do not believe that the Carnegie Institution will run risk of this under its present management. The immense enlargement of opportunities for scientific work opened up by the Carnegie Institution cannot fail in time to develop good workers and stimulate research. Too much should not be expected in the way of immediate results.

There is one matter in this connection worth your consideration, that is the difficulty of getting good men to take up work on a stipend without any assurance of permanency in their employment, even if they are successful, or of its leading to a definite career.

2. I am not prepared at this moment to recommend anyone for help in continuing research work. I should like to consult with some of our men here, and those elsewhere before making any definite recommendations, but, if you would like to have me then do so, I should be glad to make suggestions in this regard.

3. I would suggest the following names for consideration: Prudden,[2] Councilman,[3] Flexner,[4] Hektoen,[5] Novy.[6] If men devoted more exclusively to bacteriology are desired, Theobald Smith[7] might be considered in place of Councilman, Park[8] or Biggs[9] in place of Prudden (but I should be sorry to see Prudden left out), Abbott[10] in place of Flexner (but Flexner is a more productive worker), and Jordan[11] in place of Hektoen. If Canadians are to be included, Adami[12] should be considered.

If the non-medical side of bacteriology is to be included Sedgwick[13] may be thought of.

If, as you say, I am to be on the list and only four other names are desired I should recommend Councilman (or Theobald Smith), Prudden, Flexner, Novy (or Hektoen).

Walter Reed[14] of the Army is a good man, but I should not be inclined to substitute his name for any of these four; if you think otherwise he might go in in place of Novy.

I have not forgotten about the alcohol paper, and hope to have it ready for you.

Yours Sincerely,
William H. Welch

Billings Papers, New York Public Library

1. WELCH (1850–1934) was a pathologist and first dean of the Johns Hopkins University Medical School, highly influential both within and without the medical world. He had a key role in the early years of the Rockefeller Institute. During the first World War he was president of the National Academy of Sciences.

2. THEOPHIL MITCHELL PRUDDEN (1849–1924) was a pathologist and bacteriologist at the College of Physicians and Surgeons in New York City.

3. WILLIAM THOMAS COUNCILMAN (1854–1933), a pathologist, was at the Harvard Medical School.

4. At this point SIMON FLEXNER (1863–1946) was at the University of Pennsylvania. He became the first head of the Rockefeller Institute. Flexner's investigations were in bacteriology and pathology.

5. LUDWIG HEKTOEN (1863–1951) was a pathologist at the University of Chicago.

6. The bacteriologist FREDERICK GEORGE NOVY (1864–1957) was at the University of Michigan.

7. SMITH (1859–1934) was a physician who worked with the Bureau of Animal Industry of the Department of Agriculture before joining the faculty at Harvard Medical School (1895–1915). He studied parasitic vectors of diseases, bacteria in animals, and the relation between animal and human tuberculosis.

8. WILLIAM HALLOCK PARK (1863–1939), a bacteriologist, directed the New York Health Department Laboratories, 1894–1937.

9. HERMAN MICHAEL BIGGS (1859–1923), public health official, headed the health departments of New York City (1910–1914) and New York State (1914–1923).

10. Perhaps ALEXANDER CREVER ABBOTT (1860–1935) who was a pathologist at the University of Pennsylvania.

11. EDWIN OAKES JORDAN (1866–1936) was a bacteriologist at the University of Chicago.

12. The pathologist JOHN GEORGE ADAMI (1862–1926) who was at McGill University.

13. A Johns Hopkins Ph.D. in biology, WILLIAM THOMPSON SEDGWICK (1855–1921) was at M.I.T. from 1883. He devoted himself to public health problems.

14. REED (1851–1902) was a U.S. Army medical officer best known for work on yellow fever.

Minutes, Board of Trustees, Carnegie Institution of Washington, January 29, 1902

Absent: The President of the United States,[1] the *President of the National Academy of Sciences,*[2] *Mayor Seth Low,*[3] *Hon. J. C. Spooner,*[4] *Hon. Andrew D. White, Justice Edward D. White.*[5]

The *Secretary then read the minutes* of the meeting of the incorporators and *presented the Articles of Incorporation,* after which *Mr. Andrew Carnegie* was introduced by the Chairman, and *spoke* as follows;

Mr. Chairman and Members of the Board of Trustees:

I beg first to thank you for so promptly and so cordially coming forward to aid me in this work by the acceptance of trusteeship. The President of the United States writes me in a note of congratulation "I congratulate you especially upon the character, the extraordinarily high character, of the trustees." Those are his words. I believe that that estimate has been generally approved throughout the wide boundaries of our country.

May I say to you that my first idea while I dwelt upon the subject during the summer in Scotland was that it might be reserved for me to fulfil one of Washington's dearest wishes—to establish a university in Washington. I gave it careful study when I returned and was forced to the conclusion that if he were with us here today his finely balanced judgment would decide that such, under present conditions, would not be the best use of wealth. It has a tempting point suggested to me by the president of the Women's George Washington Memorial Association, that the George Washington Memorial University, founded by Andrew Carnegie, would link my name with Washington. Well, perhaps that might justify such association with Washington, and perhaps it is reserved for some other man in the future to win that unique place; because if we continue to increase in population as we have done it is not an improbability that it may become a wise step to fulfil Washington's wish. But while that may justify the association of any other name with his, which is a matter of doubt, still, I am very certain nothing else would. A suggestion that this gift of mine, which has its own field, which has nothing to do with the University, except as an aid to one if it is established, which has a field of its own, that is entitled to the great name of Washington, is one which I never for a moment could consider. If the coming university under the control of the Nation—as Washington suggested a national institution—is to be established, as it may be in the future, I think the name of Washington should be reserved for that and for that alone. Be it our opportunity in our day and generation to do what we can to extend the boundaries of human knowledge by utilizing existing institutions.

This is intended to cooperate with all existing institutions because one of the objections—the most serious one, which I could not overcome when I was desirous to establish a university here to carry out Washington's idea—was this: That it might tend to weaken existing institutions, while my desire was to cooperate with all kindred institutions, and to establish what would be a source of strength to all of them and not of weakness, and therefore I abandoned the idea of a Washington University or anything of a memorial character.

Gentlemen, a university worthy of Washington, or a memorial worthy of Washington, is not one costing a million dollars, or ten million dollars, or twenty million dollars, but of more. When I contemplated a university in Washington in fulfilment of Washington's great wish I set a larger amount than the largest of these. I take it for granted that no one or no association would think of using the revered name of Washington except for a university of first class rank, something greater and better, if I may be allowed to say so, than we have in our land today—and you all know the sums which are now used for our universities.

Gentlemen, your work now begins, your aims are high, you seek to expand known forces, to discover and utilize unknown forces for the benefit of man. Than this there can scarcely be a greater work. I wish you abundant success, and I venture to prophesy that through your efforts, in cooperation with kindred organizations, our country's contributions through research and the higher science in the domain of which we are now so woefully deficient, will compare in the near future not unfavorably with those of any other land.

Again, gentlemen, from my heart, I thank you, and I will now, with your permission read the deed of trust which has been prepared. I may say that the intended officers of this Institution have a letter from my cashier, stating that the notice of the transfer of the bonds will be sent you early in February. They can not be transferred until the first of the month. They begin to bear interest on the first day of February. Here is the deed of trust . . .

The *purposes of the Trust* are as follows, and the Revenues therefrom are to be devoted thereto:—

It is proposed to found in the city of Washington, an institution which with the co-operation of institutions now or hereafter established, there or elsewhere, shall in the broadest and most liberal manner encourage investigation, research, and discovery—show the application of knowledge to the improvement of mankind, provide such buildings, laboratories, books, and apparatus, as may be needed; and afford instruction of an advanced character to students properly qualified to profit thereby.

Among its aims are these:

1. To promote original research, paying great attention thereto as one of the most important of all departments.

2. To discover the exceptional man in every department of study whenever and wherever found, inside or outside of schools, and enable him to make the work for which he seems specially designed his life work.

3. To increase facilities for higher education.

4. To increase the efficiency of the Universities and other institutions of learning throughout the country, by utilizing and adding to their existing facilities and aiding teachers in the various institutions for experimental and other work, in these institutions as far as advisable.

5. To enable such students as may find Washington the best point for their special studies, to enjoy the advantages of the Museums, Libraries, Laboratories, Observatory, Meteorological, Piscicultural, and Forestry Schools, and kindred institutions of the several departments of the Government.

6. To ensure the prompt publication and distribution of the results of scientific investigation, a field considered highly important.

If in any year the full income of the Trust cannot be usefully expended or devoted to the purposes herein enumerated, the Committee may pay such sums as they think fit into a Reserve Fund, to be ultimately applied to those purposes, or to the construction of such buildings as it may be found necessary to erect in Washington.

The specific objects named are considered most important in our day, but the Trustees shall have full power, by a majority of two-thirds of their number, to modify the conditions and regulations under which the funds may be dispensed, so as to secure that these shall always be applied in the manner best adapted to the changed conditions of the time; provided always that any modifications shall be in accordance with the purposes of the donor, as expressed in the Trust, and that the Revenues be applied to objects kindred to those named, the chief purpose of the Founder being to secure if possible for the United States of America leadership in the domain of discovery and the utilization of new forces for the benefit of man.

IN WITNESS WHEREOF, I have subscribed these presents, consisting of what is printed or typewritten on this and the preceding seven pages, on *(twenty-eighth) day of (January), Nineteen Hundred and Two,* before these witnesses.

ANDREW CARNEGIE
January 28th, 1902

Archives, Carnegie Institution of Washington

1. THEODORE ROOSEVELT (1858–1919).

2. ALEXANDER AGASSIZ (1835–1910), son of Louis Agassiz.

3. LOW (1850–1916) was a merchant reformer. He had recently left the presidency of Columbia University to run successfully for the mayoralty of New York City.

4. A senator from Wisconsin, SPOONER (1843–1919) was a Republican conservative.

5. WHITE (1845–1921) would become Chief Justice of the U.S. Supreme Court in 1910.

Charles D. Walcott to Andrew D. White, May 3, 1902

Dear Sir:

In order that the Trustees may be informed of the steps taken by the Executive Committee of the Carnegie Institution toward formulating a plan to be submitted to the Trustees in November next, the following brief outline of action thus far taken is sent to you:

Various scientific men have been asked to act as Advisers to the Executive Committee. Thus far the Advisers appointed form committees on Astronomy, Physics, Chemistry, Geology, Meteorology, Geographical research, Geo-Physics, Engineering, Zoology, Botany, Physiology, Anthropology (including Archaeology and Ethnology), Economics, and Bibliography. As yet no committees have been appointed on Mathematics, Morphology, Psychology, History, Philology, Art, and Education, action having been deferred until the return of the President in the fall.[1]

Several of the committees have already submitted reports of value, and it is anticipated that preliminary reports will be in from other committees by July 1, 1902.

Arrangements have been made with the United States Trust Company, of New York, by which funds for expenses during the spring and summer are provided in advance of the first payment of interest from the Steel bonds, due August 1, 1902. The Secretary has been authorized to act as the disbursing officer until the November meeting, or until some other officer is appointed.

The more thoroughly the Executive Committee have investigated the subject of working out a scheme for research, the more they have been impressed with the fact that it is desirable to obtain a general view of the whole field before making more than few minor allotments from the $75,000 voted by the Trustees for the use of the Executive Committee prior to January 1, 1903. It is necessary to work out a method of procedure as the Institution is unique as regards its purposes.

The Committee has had its attention called to the desirability of the Carnegie Institution making a grant for the maintenance of the Marine Biological Laboratory at Woods Hole, Mass. A sub-committee was appointed to investigate and report upon the advisability of appropriating $4,000 for the maintenance of the laboratory during the season of 1902; also to inquire into the feasibility of the laboratory being turned over by its Board of Trustees to the Carnegie Institution. The Committee favor doing this if certain conditions are complied with.

A special committee was appointed to investigate the facilities for post-graduate student work in Washington under the direction of the Washington Memorial Institution. This report has been submitted, and the matter will probably be brought to the attention of the Trustees at the November meeting.

Dr. J. McK. Cattell has been authorized to prepare for the use of the Institution a list of the scientific men of the United States, in the form of a card catalogue, containing the name, birthplace, age, education, degrees, present and previous university or other positions, membership in learned societies and principal publications of those who have contributed to the advancement of the natural and exact sciences, whether by research, by important application, or by teaching and writing. He estimates the number to be between three and five thousand. He will also select, with the advice of experts, the thousand who are supposed to be doing the best work, and supply statistics in regard to the distribution of scientific men among the different sciences in different regions of the country. An allotment of $1,000 was made to cover clerical assistance, stationery, printing, postage, etc.[2]

A similar allotment has been made to Dr. Hideyo Noguchi[3] and Prof. Simon Flexner, for the purpose of continuing their studies of the physiological and pathological action of snake venom and allied poisons, to be expended under the direction of Dr. S. Weir Mitchell.[4]

A sub-committee has been appointed on questions relating to publications of the Carnegie Institution.

Mr. Cooper Hewitt[5] has been requested to devote part of his time during his visit in England to the study of the methods providing for original research in that country.

The President of the Institution sailed for Europe on the 17th of April, to be absent until the latter part of September. He will make inquiries in relation to the research work in various countries of Europe.

It is anticipated that the Executive Committee will hold meetings from time to time during the summer for the consideration of matters relating to the formulation of a plan of operations to be submitted to the Trustees in November.

Respectfully yours,
Chas. D. Walcott
Secretary of Executive Committee
Archives, Carnegie Institution of Washington

1. The reports are printed in the first CIW *Yearbook*, 1902, in Appendix A.

2. This is the origin of *American Men of Science* edited for many years by Cattell and later by his son Jacques.

3. NOGUCHI (1876–1928), a Japanese-born microbiologist, was at the Rockefeller Institute from 1904 to his death.

4. MITCHELL (1829–1914), a trustee, was a highly respected physician noted for his research. He was also a novelist.

5. PETER COOPER HEWITT (1861–1921) was A. S. Hewitt's son and a successfull inventor.

Theodore W. Richards to Alexander Agassiz, June 9, 1902

My dear Mr. Agassiz,

I am anxious to hear your opinion of a plan which I have formed to utilize the funds of the Carnegie Institution. Most of the applications which have come to the Chemical Committee[1] are unsatisfactory, and it seems to me that they will continue to be so, for the reasons given below.

A large majority of the more active men in American universities, to whom one would look for important research, are so overburdened with hack and routine work and administrative detail that they have but little time and energy left over for investigation. It is usually impossible for the colleges to avoid this, for their funds have been provided for teaching, and not for research. The average professor in America could not use money for apparatus if he had it. His extra time must be spent on insignificant college tasks and in endeavoring to eke out his too limited income by outside work, not in research. He is, moreover, too much worried by pecuniary embarrassment to give his mind freely to abstract thought.

It seems to me that all these difficulties might be avoided with the help of the Carnegie Institution. A large part of this endowment might be set apart for the establishment of research professorships, to be nominated by disinterested committees of eminent foreign specialists, and appointed by the Trustees of the Carnegie Institution. The salaries should be large enough to enable the holders to live comfortably (which is more than can be said of most Professors' salaries) and the positions should be called "Carnegie Professorships of Research." Because a small amount of teaching is often serviceable to research, they should be allowed to teach one or two hours a week at their chosen universities, and they should be paid for such service by the Universities. They should understand, however, that the positions should be considered as trusts, that their essential duties would consist in the advancement of knowledge: and they should be expected to give occasional obvious proof of serious effort in this direction.

Thus the Carnegie Institution would pay for services rendered to research, and the Universities would pay for such immediate services as were rendered to them—and each should receive the best possible return for the investment.

It seems to me that this plan is one which would not only provide adequately for further poduction from those who have been successful in producing in the past, but would also stimulate production in younger men, who would look forward to the possibility of attaining these highly honorable positions. I hope that you will approve this plan, and support it before the Trustees.

I regret very much that a previous engagement, which cannot be escaped, will prevent my hearing your address next Thursday.

Yours most sincerely,
[*Theodore W. Richards*]

Richards Papers, Harvard University Archives

1. The report of the Chemical Committee was one of the few with a minority report. Although Richards broadly agreed with his colleagues on the committee, his minority report strongly made the same point here expressed to Agassiz.

Ira Remsen[1] to Theodore W. Richards, August 18, 1902

My dear Professor Richards:

Just one word. The case is as I supposed. Your statement "The German Government" etc. rests practically on Van't Hoff.[2] Therefore the rest of that sentence in your report is entirely misleading. I refer to your words "unless America does likewise there is danger of our dropping yet further behind." America has done likewise and to a greater extent than Germany. Practically all the *University professorships* in the Johns Hopkins University were free from routine work. Certainly, so far as Sylvester, Rowland and Gildersleeve[3] were concerned, the conditions were better than in any place I know anything about. I too was told to fix my own duties. Even a younger man will tell you, I think, that they are not over burdened—they are hardly burdened.

Somewhat similar conditions exist in Chicago. Neff[4] could do as he pleased. The same is true of others. My impression is that there have been similar cases at Harvard before yours. So that, I repeat, the phrase "unless America does likewise" is misleading. The fact is America has made immense strides in the last quarter of a century, and those of us who have lived through this period and watched the course of events cannot help being optimistic. The leaven is working. I am gradually getting the impression that the only truly overworked men in America are the Harvard men. It does not seem to me to be wise for the Carnegie Institution to adopt measures for the remedying of difficulties that do not exist in several institutions, however great these difficulties may be in a few other institutions.

Yours sincerely
Ira Remsen

Richards Papers, Harvard University Archives

1. REMSEN (1846–1927), an organic chemist, was Gilman's successor as president of Johns Hopkins University. In this letter, Remsen is arguing that conditions for research in American universities were better than implied by Richards's minority report. Two points are significant in regard to this text: (1) while Hopkins, Chicago, and other schools might have better conditions, if a Harvard lagged, what about far lesser institutions not mentioned here? (2) While Richards favored the university as the site for his research, others could and did cite such complaints as reasons why research should be nurtured outside the university.

2. J. H. VAN'T HOFF (1852–1911) was a physical chemist from the Netherlands. Since 1891 he had been at the University of Berlin with a very light teaching load.

3. J. J. SYLVESTER (1814–1897), the mathematician, H. A. ROWLAND (1848–1901), the physicist, and BASIL GILDERSLEEVE (1831–1924), the classicist, were members of the original faculty at Hopkins.

4. JOHN U. NEF (1862–1915), the chemist, was at the University of Chicago.

Thomas C. Chamberlin to James McKean Cattell, September 10, 1902

Dear Professor Cattell:—

I cordially appreciate your kind invitation to make public in *Science* my views as to how the Carnegie Institution can contribute most effectively to the advancement of science.[1] I am also indebted to you for proof of your interesting paper on the subject. At the moment I am not quite clear as to whether it is altogether appropriate for me to discuss the subject in public, as I am a member of one of the Advisory Committees.[2] The question arises whether I am not in some sense a member of the jury rather than an advocate. I will think over this phase of the subject and take advice before I decide.

If you will permit me to express myself frankly, there are some portions of your statement that strike me unpleasantly, and I am wondering whether a knowledge of my impressions might not be welcome to you in revision, if that is yet possible. The expressions are those which embrace "pauperizing," "lording it over," and less declared implications of sinister influence. It appears to me that any plan that may be adopted will be liable to abuse, and it is not clear that the dangers of one are essentially greater than the dangers of another. It would be easy for one holding views different from yours to point out possible abuses in the plans you suggest. The general effect of giving prominence to these possibilities in a public discussion is to throw cold water upon the entire enterprise. If there is not sufficient presumption that the fund will be administered with reasonable wisdom under any one of the alternative plans that are being considered, to justify discussion on that basis exclusively, slight encouragement is offered to Mr. Carnegie and others to venture their millions in the endeavor to promote science.

If you will permit me to go a step farther, you seem to me singularly unfortunate in opposing the establishment of a geophysical laboratory at Washington and substituting a proposition for laboratories in physics, chemistry and psychology. It seems to be in direct contradiction to the principle you lay down that the Carnegie Institution should not do what other institutions are doing, or can do, for several institutions have admirable laboratories of research in chemistry, physics and psychology, whereas no institution has a geophysical laboratory covering the ground of the proposed one, nor is there any likelihood that any institution will attempt to establish and maintain such a laboratory. The most that can be said is that a very few institutions are incidentally attempting work on a very few of the more available geophysical problems, and that this is being done in a very limited and relatively inefficient way. The problems of geophysics are beyond the reach of the resources of any institution now in existence, as I am sure you would be easily convinced if the full length and breadth of what is urgently desired and immediately needed were fully realized. It does not fall within the proper functions of the government surveys.

I do not suppose it is possible for any one to discuss the Carnegie problem from a strictly impersonal and uninstitutional point of view, much less from one that is wholly impartial as respects the relative value of the respective sciences, but we ought if possible to keep institutional and professional preferences out of public discussions so far as possible, lest they do us harm with the public and with those who are or may become generous patrons of science. *Men of science are now on trial.* If they justify public confidence by largeness of view and administrative competency, the result will be great for science under any special scheme that may be adopted. If not, it may result in as sad a spectacle as the late issue of the National University problem, and be as disastrous to the end sought.

Will you therefore pardon me if I urge that both in your leader and in the supervision of the contributions it calls forth, a wide berth be given to the dangers indicated, for they seem to me serious in their possibilities.

Very truly yours,
T. C. Chamberlin

Cattell Papers, Library of Congress

1. Cattel had sent Chamberlin a proof of his article offering unwanted advice to the Carnegie trustees (*Science* 15 [1902]: 460–469). Cattell was decidedly against the geophysical laboratory, the aid to astronomy, and other aspects of the developing programs of the trustees. The article appeared in the September 19, 1902, issue. The independence of Woods Hole was stressed. Cattell and others feared domination by nonscientists and by scientist administrators.

2. CHAMBERLIN (1843–1928), of the University of Chicago, was on the geophysics committee. He was a geologist then embarking on a grand venture into cosmology. Chamberlin will reappear in the chapter on astronomy.

Excerpts, Minutes, Board of Trustees, Carnegie Institution of Washington, December 8, 1903

MR. CARNEGIE: Mr. Chairman, before you adjourn, as I shall not probably be able to be here until late in the afternoon, I would like just to say to you Trustees that if there be any man here who has received one-half of the satisfaction I have this morning, he is a happy man. You have your troubles and your trials, and I am in a position to look at general results—you at details, all of which are never encouraging or satisfactory. I have listened today and a flood of light has come to me upon this matter. To think that you have already discovered this man, this Japanese man in Philadelphia, this man in Michigan, McLaughlin,[1] and this man in Chicago, and this man in the West, and all this in the short period that you have been carrying on this work! I do not know of anybody that has achieved so much in so short a time, and I wish to congratulate one and all of you upon your transcendent success. I am the happiest man living just now; I think, because I now see that this is a great success. It is all well for you to look at your disappointments and failures and so on, but they are nothing. As the Chairman has said, if you have three or four of the exceptional men, that is a great deal. We do not judge by the number that do not produce, we judge by the number that are successes.

I wish to say in regard to the communication from the President that it was a staggering surprise to me this morning.[2] It was only a few minutes before the meeting that he told me of his intention. I am glad that you can say that it was with unanimous regret that you receive his letter. I thought that the President would perhaps have thought it best to remain his five years, and if he found that the labors were too great for him it could be arranged that a satisfactory assistant—it was the idea from the beginning—a younger man, should relieve him a great deal, yet that he would remain at the head of the Institution. But that is a matter for him and for you to deal with.

I only repeat to you my feelings of elation this morning, and I can not close without telling you another thought that is in my mind: the spectacle that you present yourselves, men of your standing and position, known throughout the world. You are willing to come here and give us your brains and give us your hearts for an object in which you have no personal concern, in which you can receive no reward, except the greatest of all rewards, the consciousness that you are devoting part of your time for a great end, and laboring to leave the world a little better than you found it.

I thank you all from the bottom of my heart. (Applause.)

Archives, Carnegie Institution of Washington

1. An exceptional man but not one of the grantees. ANDREW CUNNINGHAM MCLAUGHLIN (1861–1947), a historian then at the University of Michigan, was on the CIW advisory committee for history.

2. A reference to the resignation of Gilman. For background, see David Madsen, "Daniel Coit Gilman at the Carnegie Institution of Washington," *History of Education Quarterly* 9 (1969): 154–186.

Robert S. Woodward to Thomas C. Chamberlin, July 11, 1904

My dear Professor Chamberlin:

I have been looking ten days for an interval of leisure to write you about Carnegie affairs, but two papers for the press promised for early July have absorbed my attention until to-day.

Soon after returning from Madison I received a letter from Dr. J. S. Billings[1] asking for an interview with me at my convenience. A few days later I called at his office, at the Astor Library, and was surprised to find that his object was to learn whether I would become a candidate for the presidency of the Carnegie Institution.

The interview was a long one, and we discussed many topics with respect to the Institution—its prospects, the attitude of the Executive Committee, the attitude of the Trustees, the merits of several candidates for the presidency, etc., etc.

Dr. Billings spoke very earnestly with respect to these matters and urged me to stand for the position, assuring me that in his opinion any appointee in whom the Trustees could repose confidence would receive their cordial support. He told me that Walcott would be pleased to be elected but that he could not be by the present board. Pritchett,[2] he said, had declined to take the position on the ground that he could not control the Trustees; Billings thought, however, that P. could not be elected for the reason that he has sought the position too ardently.

After our deliberations I told Billings I would stand as a candidate, with the hope that the best man may win.

I have spoken only to Cattell about this matter, my other more intimate colleagues being away from Columbia at present.

Such is the state of the case from my present point of view. How does the matter look to you?

Sincerely yours,
R. S. Woodward

Chamberlin Papers, University of Chicago

1. Before long, the two men would have serious differences. Shortly before Billings's death, Woodward and William H. Welch, the dean of the Johns Hopkins Medical School and a trustee at that time, would correspond about how difficult it was to work with Billings.

2. HENRY SMITH PRITCHETT (1857–1939) was president of the Massachusetts Institute of Technology. An astronomer, he had previously served as superintendent of the Coast and Geodetic Survey. Later he became head of the Carnegie Foundation for the Advancement of Teaching and a trustee of the Carnegie Institution of Washington.

Theodore W. Richards[1] to Henry Lee Higginson, December 1, 1904

Dear Mr. Higginson,

Your question is a very interesting and important one, and I have thought seriously about it since your letter came last night.

There is no doubt in my mind that President Remsen is by far the best of the men whom you name. He combines executive and administrative ability and experience with a real knowledge of scientific research. He has moreover the saving grace of common sense in unusual degree. Besides these advantages, it seems to me that another exists in the subject which has chiefly interested him. I may be prejudiced, but I cannot but believe that chemistry (especially through medicine, which is becoming more and more dependent upon chemistry) has greater possibility of promoting the welfare of mankind in the next twenty years than either physics or astronomy. It is almost inevitable that the president of such an institution should lean a little toward his own subject, and I think that astronomy has its full share of endowment at present.

But entirely apart from such considerations, simply as a man, Remsen seems to me much the most suitable. Smith[2] is a good man, but a narrower one than Remsen. Of Professor Woodward I know little, and on enquiring of some of our instructors in physics this morning I was interested to find that they also knew little of him. This I cannot consider as a good sign. I have an impression, which these gentlemen confirmed, that Professor Woodward may be seeking the position with some energy, and that in the quest he may bring a good deal of influence to bear. This may be quite an erroneous impression; but in any case it seems to me more fitting that the office should seek the man, and not vice-versa. There can be no doubt that there are a number of physicists of higher rank in America as to original work than he is.

I was interested to find that our physicists (I mean the ones whom I consulted) agreed emphatically with me in thinking that Remsen is the best man at present available for the post which you wish to fill. Needless to say I did not give them any clue as to the reason why I led the conversation in this direction, or why I wished to know their opinions.

If I can help you further in this matter, please let me know. It will give me very great pleasure to do all I can for you.

Yours most sincerely,
[*Theodore W. Richards*]

Richards Papers, Harvard University Archives

1. This is in reply to Higginson's letter of November 10, 1904, asking for opinions of possible candidates.

2. Edgar Fahs Smith was the chemist at the University of Pennsylvania. Others mentioned are Hale and W. W. Campbell of the Lick Observatory. HIGGINSON (1834–1919) was a Boston banker and a trustee of the Carnegie Institution.

Theodore W. Richards to Henry Lee Higginson, December 3, 1904

Dear Mr. Higginson,

Since writing to you the other day, I have made further investigations concerning Professor Woodward. It appears that by mathematical physicists dealing with gravitation and triangulation he is ranked high among American students of this subject, on account of some work which he did in connection with one of the Government Departments in Washington. Because this subject is far removed from my work, as well as from that of the two physicists whom I consulted before, it is not surprising that he was not well known to us. His earlier stay in Washington would naturally lead his thoughts again in that direction.

Although I take pleasure in sending you this new intelligence out of justice to Professor Woodward, I still feel that the new President may best be found among scientific men familiar with *experimental* research, rather than among those whose work has been chiefly of an abstract or mathematical nature; for I believe that Mr. Carnegie meant to assist experimentation, and that experimental work is greatly in need of endowment. Moreover, I believe that scientific advance is more likely to occur along inductive than along deductive lines. I cannot imagine that a man who is not himself an experimenter can fully realize the needs of other experimenters.[1]

With kindest regards,

Yours most sincerely
[*Theodore W. Richards*]

Richards Papers, Harvard University Archives

1. Richards has no basis for this statement about Carnegie's intentions, nor is he being quite fair to Woodward who had experimental work to his credit. What is significant here is the bias against mathematical or abstract research.

Theodore W. Richards to Robert S. Woodward, March 28, 1905

My dear Dr. Woodward:—

In the course of your pleasant visit here, you asked my opinion about two points, concerning which I wanted time to consider. One was whether or not it would be advisable for the Carnegie Institution to re-publish in full the works of Dr. Wolcott Gibbs. This is rather a delicate question. In the light of my friendship for Professor Gibbs and in recognition of his great influence upon chemical thought in America years ago, I am tempted to advise this step. I think that it would be an excellent thing to do, if the Carnegie Institution were not in my opinion intended to advance original research. From this latter point of view I can hardly feel that the re-publication of these works would greatly advance chemical research today. Their influence lies mostly in the past. They are accessible in all the prominent libraries both in America and England to the few who may wish to consult them. Therefore, I feel as if their republication would be rather a matter of sentiment than a living contribution to research in the future;—but perhaps my acquired New England conscience warns me too strongly against doing a pleasant thing for an intimate friend.[1]

The other point of which you spoke was concerning the possible cooperation of scientific men in America. It seems to me that such cooperation might in some cases bring very valuable results and in other cases be almost equally destructive. In astronomy particularly I should think that cooperation might be effective and almost necessary because of the enormous difference of opportunity and range of vision afforded by different latitudes and because of the frequent desirability of having simultaneous observations made from very different stations.

Moreover, in pieces of routine work in which the fundamental concepts have been well established cooperation might be useful in other sciences as well. Also, cooperation between a chemist and a physicist for example might be of great value, the one supplying the technical knowledge which the other lacks when both are dealing with subjects on the border line between the two.

On the other hand it seems to me that the making of a great original discovery in science is not unlike the writing of a great poem or the painting of a great picture. The thought and its execution must be hammered out by genius alone and without the multitude of adminstrative duties which cooperation would be likely to bring upon its professor. Such a man would be greatly hampered by the necessity in a moment of white-hot mental activity of signing papers or ordering materials or directing the computations for another man, and his thought, whatever it might be, would be liable to be extinguished. I agree with Mr. Carnegie entirely in his belief that from the individual exceptional man alone is any great addition to be expected to a sum of our original conceptions. Who can imagine Faraday as cooperating!?

In short it seems to me that cooperation may be highly productive of routine work and of a general rounding off of already acquired knowledge; but might be equally destructive of great advance in an entirely new direction. I hope that you will pardon me for writing so fully, but the subject is one concerning which I have strong convictions; and I think that these convictions are justified by the history of science.

Let me express once more the great pleasure which your visit here gave us all, and reiterate our hope that you may come soon again and stay longer.

Yours most sincerely,
[*Theodore W. Richards*]

Richards Papers, Harvard University Archives

1. OLIVER WALCOTT GIBBS (1822–1908). Woodward, perhaps, thought otherwise. He was very conscious of the need to make the scientific tradition more visible. No edition of collected papers was printed.

James McKean Cattell to George Ellery Hale, April 15, 1905

Dear Professor Hale:—

I thank you for sending me a copy of the paper on the Solar Observatory. I should be very glad indeed to print this in THE POPULAR SCIENCE MONTHLY,[1] which would not interfere with its also being published in the proceedings of the Astronomical Society of the Pacific. It would be particularly interesting if it could be accompanied by illustrations, such as were used in *The Astrophysical Journal.* In case there is some reason why you regard this as undesirable, I trust that you will in the near future contribute an article on the observatory to the MONTHLY.

Any distribution of the funds of the Carnegie Institution is likely to cause criticism. There is probably no one in America to whom the majority of scientific men would be more willing to entrust a large appropriation than to you. As a psychologist I naturally feel that we are as much in need of money as the astronomers, and I do not see just how the account can be settled, except on the assumption that as there are about an equal number of workers in astronomy and psychology, they should receive equal consideration. As it happens no appropriation at all has been made for experimental psychology.

Very truly yours,
J. McK. Cattell

Hale Papers, California Institute of Technology

1. Cattell was also editing this journal at this time as well as *Science*.

George Ellery Hale to James McKean Cattell, May 2, 1905

Dear Professor Cattell:—

I am very much obliged for your kind letter, and appreciate what you say about the distribution of the funds of the Carnegie Institution. I do not for a moment pretend to say that it would be better to use a given sum of money for astronomical than for psychological work. As a matter of fact, I have no doubt that very important results could be obtained through the use of a large appropriation for psychology, and as my interests are by no means confined to any single branch of science, it would please me greatly to see such an appropriation made. In arguing that funds could be used to good advantage for a solar observatory, I have not set this up against other departments of science. It has only seemed to me that such an opportunity as the present one has been available on very few occasions in the history of observational astronomy. As my last article makes clear, I trust, the Solar Observatory will in no wise duplicate either the Yerkes Observatory or the Lick Observatory, but will occupy a new field of its own. I can only hope that the results will be such as to bear out the unusual promise that seems to me to exist.[1]

I hardly think that the article[2] I sent you is entirely suitable for the Popular Science Monthly. A little later, however, I can perhaps send you a somewhat similar article which may include some results obtained at the Solar Observatory. Our apparatus is now nearly ready, and if all goes well we should have some good results before many weeks have passed.

Very sincerely yours,
George E. Hale

Cattell Papers, Library of Congress

1. This exchange is on a classic problem in allocating resources. Within a given field it is relatively easy to make judgments between competing demands. Both Hale and Cattell were sincere. On scientific merits alone there was (and is) no way of deciding the issue between different fields of science in any truly objective manner.
2. Apparently never sent.

Theodore W. Richards to William Albert Noyes,[1] November 23, 1905

My dear Dr. Noyes:—

I am glad to reply to your letter of November 18th about the Carnegie Institution of Washington. It seems to me that for chemical purposes, the Carnegie funds can, as a rule, best be used at present in the way in which they have been used in the past; namely, to enable men whose salaries are paid by other institutions, and who have already a large outfit of ordinary apparatus at their disposal, to procure such additional apparatus and efficient assistance as to enable them to carry out in their own laboratories the researches which are of especial interest to them. This method of procedure would probably apply equally well to any other field of investigation which does not need for its experimentation a special climatic or topographical environment.

In those cases where the work might demand the erection of a new building, it seems to me that it would usually be economical for the Institution to build that building at the place where the man in question is already employed and finally to give it to the corporation in that place, because thereby the Institution would not have to bear the constant burden of the investigator's salary and the maintenance of the building, as it would if it called him away from his position to Washington.

Only one difficulty may arise here; namely, the danger that such an investigator has already too many routine duties in the place where he is employed. If this is the case, and the man is clearly the person to carry out the especial research in question, it seems to me that the Carnegie Institution might well pay temporarily a portion of his salary, thus setting free part of his time by enabling the college, or other corporation which employs him, to pay a substitute for part of his work. Often in America assistance of this kind is sorely needed. This

I stated in my original report as one of the first Advisers to the Carnegie Institution.

The economy to the Institution involved by allowing some other corporation to bear a large portion of the expense in both salary and equipment is manifest. With the funds now at the disposal of the Institution it would not be possible to establish and maintain an adequate individual plant covering all the fields of knowledge, even if such a plant were certainly desirable. If the income were so large that economy need not be seriously considered, the problem would of course be different.

Another reason why the money will accomplish more when distributed among already existing institutions of learning is because a great mass of advanced students are already attracted thither, and have grown into intellectual sympathy with their teachers, the investigators. From these students, held partly by their desire for the degree and partly by fellowships and by sentimental considerations, the investigators can obtain adequate assistance at very much less expense to the Carnegie Institution than they could in any other way.[2]

An original idea, like a great poem or a great picture, is not to be produced at the bidding even of the greatest thinker. The association with young and eager minds, and the occasional review in systematic lectures of one's whole subject, afford no small stimulus to new thoughts. The Institution could therefore by no means be certain, if it should call a great investigator from a University to a central "investigating plant," that he would continue to have original ideas. If his ideas should diminish in his new position, the investigator's existence would be a burden both to himself and the Institution. The obvious inference is that the investigator should be assisted by the Institution at the time of his intellectual activity in the position where he happens to be.

A further reason might be urged against transplanting a man of such ability as the Carnegie Institution wishes to foster. A man of great ability will as a rule be found in a position which he likes, because he will already have had plenty of opportunities to go elsewhere. I believe that any man will work most effectively in a position which he likes, therefore, the presumption is against the moving of a man of ability. Moreover, the mere act of transplanting such a person with his accumulated apparatus, not to mention his family and belongings, involves an expenditure of an enormous amount of energy which is a clear loss to the world and might have been put into productive scientific work. The advantages of the transplanting must be great, if this disadvantage is to be counterbalanced.

Finally it seems to me that the method recommended is most nearly in accord with both the letter and spirit of the admirable and far-seeing provisions made in the Trust Deed of Mr. Carnegie.

Hoping that these answers may be clear and satisfactory to you, I am,

Yours very sincerely,
T. W. Richards

I shall send a copy of this letter to the President of the Carnegie Institution.

Richards Papers, Harvard University Archives

1. WILLIAM ALBERT NOYES (1857–1941) was a chemist at the University of Illinois.

2. Later, cynical graduate students would refer to this as the slave-labor theory of research administration.

Robert S. Woodward to Theodore W. Richards, December 1, 1905

My dear Professor Richards:—

I have your letter of November 20th, along with a copy of your letter of November 27th, addressed to Professor W. A. Noyes.

I am very glad to get this full and frank expression of your opinion concerning a question that has required, and promises still to require, much study.

I am coming to think that it may be quite practicable to utilize the services of a considerable number of eminent men connected with educational institutions, especially in case the latter institutions will recognize fully the desirability of permitting the investigator to spend nearly all of his time in work of investigation. It will not be necessary to look very hard for such exceptional men. In most cases they will discover themselves, in the sense in which Dr. Johnson often used the word. It will be much wiser, also, to select such men than to institute experiments with untried men in the hope that some of them may prove exceptional.

On the other hand, I am coming to think that money may be easily wasted by the broadcast system of furnishing funds to men who have laboratories and enthusiasm. The colleges and universities with which such men are connected are almost certain to get the best part of their time and energies. I am satisfied that this plan of distributing aid has not thus far worked well for the Institution. It appears also that this plan is likely to develop friction in some cases; in fact it has already done so; so that in the future it will be necessary for our Institution to have a complete understanding with college and university authorities before awarding a grant to an individual.

While it is possible for the Institution to have a corps of foremost investigators who may be each connected with an institution and an environment of his choice, there remains a large field for work quite outside of any existing institutions. It should be our aim especially, I think, to take up certain large projects which are not likely to be carried on by any other means. We should especially avoid the fields now pretty well worked by men connected with academic institutions, since the use of our income for such a purpose is likely to work grave injury to those institutions. Large as our income is, it is quite insufficient to meet the demands for worthy investigations for which applications are coming all the time from men connected with colleges and universities. You may not be aware of the fact that there is manifest already a tendency to pauperize scientific investigation in some of the best institutions of the United States. Many able investigators are confronted by the dilemma of Omar. The educational institutions say that it is unnecessary for us to furnish funds for investigation if they may be had from the Carnegie Institution. On the other hand, if the latter Institution rejects a proposal of an applicant, the rejection proves that his project ought not to be carried out. The fact is that we are already involved in no inconsiderable difficulties by reason of affiliations which have grown up rapidly during the past four years.

The subject is too large, however, for consideration in a letter, and I shall hope presently to see you and have a long talk with you about some of the difficulties that confront us. I imagine Dr. Noyes has drawn out your letter through some suggestions and questions which I laid before him some months ago.[1]

Sincerely yours,
R. S. Woodward
President

Richards Papers, Harvard University Archives

1. Woodward had embarked on a systematic program of finding facts and gathering opinions, a course he followed very carefully during his tenure.

Excerpts, Minutes, Board of Trustees, Carnegie Institution of Washington, December 12, 1905

[PRESIDENT WOODWARD:] . . . [Burbank] is like a mathematician who never has to refer to his formulas; all information he possesses he can summon in an instant for his use. He has for many years been developing new varieties of fruits, vegetables, grasses, flowers, and so on, and he has achieved greater success in that line than any of his competitors, in fact all of his competitors, the world over. For example, when I was at his orchards last summer he had about three hundred thousand different varieties of plums. It was possible to see on a single tree from twenty to sixty different kinds of plums. Most of these plums he throws away, but he picks out the best ones and puts them on the market. I should say, roughly estimated, if we speak of his plums and peaches alone, that the plums and peaches in his orchards which he will be ready to put on the market in a year or two will be worth as much to the state of California, as the entire endowment of this Institution. . . .

If we can only sustain that man in his work—he does not care for money, he only cares to go on with this work—and if we can enable him to do it he will be of the greatest service to humanity. But in enabling him to do that we desire, if possible to train up two or three young men who can learn Burbank's tricks, so to speak, and we may be able to utilize the experience and extraordinary knowledge which this man possesses in our search for the solution of the greater biological problems which are taken up at Dry Tortugas and Cold Spring Harbor. There is no doubt that Mr. Burbank may be able to throw light on problems of heredity and those abstruse problems which are being worked out now in zoology, more light than—well, it is sufficient to say, than most men. He can throw more light than most zoologists already at work on these problems because he has had a vastly greater experience in them.

A word or two as to his personality. He is not a trained man of science; he lacks knowledge of the terminology of modern science. He often expresses himself in a way quite offensive to many scientific men, if due allowance is not made; but he is a man who unconsciously works by the scientific method to most extraordinary advantage. I think anybody who goes to his orchards and sees what he has produced and who studies Mr. Burbank as I have done will admit at once that he is a most unusual man. But along with his unusual abilities he has unusual peculiarities. Let any ordinary man of science go to his orchards and enter into a discussion with him and they will be at loggerheads in fifteen minutes. It is our duty to make allowance for these things. If we can make such allowance, the work we can get out of Mr. Burbank will be extremely valuable to humanity; if we cannot make that allowance we shall fail. . . .

Coming to the matter of the Solar Observatory, we want to find out how the sun is made, how the source of the heat is maintained and, if possible, we want to predict to those people we expect, as Colonel Higginson has said, to come after us, how the sun will behave in the distant future. I think that it is quite probable that we may be able to make long distance predictions with regard to meteorology as we do in regard to the tides. As you know, we are able to predict the tides fifty years in advance. We can tell today what will be the tide at Sandy Hook at four o'clock in the afternoon in the year 1950 with a high degree of precision. But in addition to that we might call the practical and commercial value of this work, it seems to me we might expect a still higher value coming from a better understanding of our conditions on this planet. Of course there are various ways of looking at these matters. There are at least three different kinds of people on this planet with three different kinds of views as to what we should do. I have not time to explain the nature of these people, but the history of science shows that wherever any progress has been made it has been made to the highest value of humanity. Astronomy, for example, has done more than all the other sciences put together up to date to straighten out the kinks in the minds of human beings, to remove superstition and to enable us to think straight on questions in general.

In this connection perhaps I may be permitted to say a word with regard to the other notable facts to which one of our trustees has referred this morning, that it is much easier in general to get money to carry on astronomical researches than to carry on other researches. That is a startling fact and I have devoted a great deal of time in the past dozen years trying to find out why that is so—and especially during the last ten months. There is a good reason for it. Many people have asked me how it is that, if we speak of shares amongst the different sciences, astronomy is getting the lion's share. I have tried to find out why it is, and it is this: It is the oldest and most highly developed science, and if you will come to our office and look over the applications we have for projects you will find that the certainty, the definiteness of these applications is almost directly proportionate to the age of the sciences represented. Roughly speaking, the order of precedence, which is fixed not by us but by nature, is this: Astronomy, chemistry and physics second, and zoology third. If you will look at the great number of projects placed before the Institution and examine the papers you will find that for definiteness of specification the astronomers come first; they not only know what they want and how they expect to do it, but they will tell you what they expect to get. One of the aims should be to bring up the other sciences to the level of the older sciences. I would not seek to use the word science, because it is offensive to some people, perhaps, but the reason why certain lines of knowledge seem to get preferment in society is, substantially, that they merit it, they are older, they are more definite, and they can give us better assurance of results. I have touched lightly upon this subject in my report. . . .

MR. CARNEGIE: Mr. Chairman, consider me an average man. I think nothing that you have done has attracted such general attention as your aid to Mr. Burbank. That of itself is a gain to the Institution. Now, it strikes me from the President's remarks that you have met a genius, you have met a man who has done something. I think that a man of science, who perhaps has done nothing, should go cap in hand to a genius; and you should not endeavor to harness up that genius to drive in your dog-cart, because he will not drive.
THE PRESIDENT: That is true—
MR. CARNEGIE: What you should do (I probably have no right to speak in this Institution, but you know I do not interfere in that way to any great extent) I cannot but help feeling that Mr. Burbank will have to be used as far as you can use him; but if you allow men of science to interfere with that man, or criticize or dictate to him, you will lose all you can get, and therefore I would perhaps criticize somewhat the remarks of the President, which seem to indicate that you are going to put Mr. Burbank in with others and consolidate the free institutions and that Mr. Burbank was to be one of them. I think when you find the exceptional man you have got to let him work in an exceptional manner, and that you should just see what he can give you and not try to divert him to any regular mode. One of the objections which it seems to me can be made to the work of great institutions with salaried men is that they are apt to depend upon these regular scientific men trained in the terminology of science and you get into a contented state of mind and you go on in a routine manner, and when a genius comes along he may be defective in manner or mode or education or things of that sort, and a body of professors acting in a professorial role, will look askance at the idea of such a man coming here and revolutionizing us and troubling us with his experiments. Therefore it has always seemed to me that this Institution will do more good by taking the exceptional man wherever found and letting him work in his own field, and you will then escape this going along in mere routine. . . .
MR. WHITE: May I say one word? I cannot think that any person here has any doubt as to the value of the investigations and the work which Mr. Burbank is carrying on. There is one feature which I would like to speak of. I shall deprecate as much as Mr. Carnegie does, in the light of my own experience, the turning loose on Mr. Burbank of a lot of college professors or scientific men. They are the best of men, often the most useful, and frequently you will find among them men of genius, but the men we want to put in touch with him in order to get the best results, both now and hereafter, are, it seems to me, those to whom the President has referred—the right sort of young men. I should suppose that the President, conferring with Mr. Eliot,[1] we will say, con-

ferring with President Jordan,[2] with other men about the country whom he knows better than we, would be likely to find just the right sort of young men whom Mr. Burbank would be glad to have about him. They could absorb ideas from him, knowledge from him; they could make suggestions, some perhaps he would not like, and then withdraw them and put new ones there. That, to me, is the part I take most interest in—the selection of some assistants for him whom he will admit, who can come into touch with him, who can take, so to speak, the torch of knowledge from him and carry it on and finally, perhaps decades hence, put it into the hands of others. That is a feature among the suggestions of the President that I hope will not be lost sight of.

I should hope he would put himself in communication with the right sort of men over the country, find just the right sort of young men, put them there and withdraw them if they do not suit and replace them, and in that way the Carnegie Institution will insure the passing on to future generations of what Mr. Burbank has so magnificently begun.

DR. MITCHELL: A word upon this matter. It seems to me it has taken a very practical turn. We did not quite begin that way. Everybody knows that this man, Burbank, must have enormous practical skill in the production of plants and changes in plants, and a mind fertile with suggestions, that he must have worked altogether by rule of thumb. It is difficult to understand that he is anything else more than an exceedingly careful observer. The President I see shakes his head at me. It is a very large rule of thumb I am talking about, but there is something beyond this that we ought to be interested in as scientific people, and that is to know the laws under which these changes have occurred and under what laws he gets his failures. He gets a lot of failures as well as successes. Now, those failures have not been utilized by him scientifically. That is what we want, too; we must have some men there of science, not men who, as my neighbor says, simply pass on this torch of knowledge for future generations but will find out under what laws this man acts, who succeeds where other people fail. We have to determine those laws if we can. That, to my mind, is the largest part of the investigation we are to make—because somebody will learn to do as he has done; but we want to go beyond that and get the scientific reason for what he does and the reasons why he sometimes fails.

THE CHAIRMAN:[3] I have not any doubt that our President, having met Mr. Burbank and being in perfect relation with him, and also knowing something of the peculiarities of scientific men, so-called, does not need any suggestions, and what I am going to say is not at all for advice to him; but it is in answer to some of the general remarks that have been made.

Any man that goes out there with any idea that he will make any suggestions to Mr. Burbank, that he will attempt to explain to Mr. Burbank the laws by which he is acting, and if he gives any scientific explanation that Mr. Burbank will not promptly object to it—I do not care what the explanation is—he is very much mistaken. A man has to go out there like a little child and put aside temporarily what he thinks he knows, and he has got to just learn the business, and he has got to learn through observation if he can. It will take another exceptional man to learn through observation. And I have no doubt you will have to experiment several times before you will find that man. But the idea is to go there to learn from Mr. Burbank and to put down in simple language what is learned from him. The scientific language is rubbish; plain, simple language is what is needed, and it can always be put into that. If you want the scientific terms of course the simple language can be translated. That is a subsequent matter that does not amount to anything. The thing is to get Mr. Burbank's statements of facts. His statements will not amount to so much—that is my impression—as his showing—"This shrub here is going to be a success, that shrub will not be a success." (Indicating) "Why?" You have to get a little further into the "Why" than the answer he gives you. What particular thing is there about that life? I would like to learn. Put yourself in that attitude of trying to learn as far as you can. It will be difficult, because he probably puts together fifty things at once, by reason of his long experience, but that is what the young man has to go there to do first. Mr. Burbank will ask him for explanations. He has heard of Mendel's theory and all the rest of it, and whatever explanation he gives Mr. Burbank is not to be satisfied with it. That is part of the business, he has to accept that.

MR. MORROW:[4] Of course Mr. Burbank is not in accord with any of this; he does not believe in Mendel's theory or De Vries' theory or Darwin's—

THE CHAIRMAN: And he is not likely to, and it is of no importance whether he does or not as far as I know.

PRESIDENT WOODWARD: If I may be permitted to speak again on the subject, I would like to explain what is a misapprehension, I think, in Mr. Carnegie's mind.

I have not sought to force scientists on Mr. Burbank; I have only sought to get the cooperation of our three departments in biological or scientific work, to cooperate with him. The last thing I would think of doing would be to put a man in Mr. Burbank's orchards to discuss heredity with him; that is what I wish to prevent. The only way to do that is to enable the Institution to control the investigation. If we farm it out, as was attempted, to the Department of Agriculture or to some university—there are two universities on the Pacific Coast that would like to do this work for us—I feel sure we shall fail; I feel that unless we can control the situation, unless we can put some young men who are willing to suppress their personalities into Mr. Burbank's orchards at the right time and for the right period, we shall certainly fail. We must make large allowance here for the personal equation; in fact personal equation is the whole thing here. That is what I would seek to do by getting the cooperation of the two gentlemen especially who have charge of the biological station at Cold Spring Harbor and the station at Dry Tortugas and the man whom I would like to see in charge of the Desert Botanical Station. They can give me a great deal of expert advice and expert information, and I feel sure that they will suppress their personalities sufficiently to get en rapport with Mr. Burbank. I am sure that I possess his confidence at present and that I can get him to work with these other men. Now, if we can only get young men whose reputations are yet to be completed, young men occupied with the Desert Botanical Station and with our biological station at Cold Spring Harbor, to come under our jurisdiction so we can place them there when we need them and take them away when we do not need them there, we can control the situation so far as it can be done at all. It is highly essential we should not only be able to see that there is an excessive amount of personal equation involved in this work, but we must appreciate that it is almost the whole thing here. . . .

Archives, Carnegie Institution of Washington

1. ELIOT (1834–1926), the president of Harvard University, opposed the concept of a national university. Before turning to administration, Eliot had taught chemistry at Harvard and M.I.T. See Hugh Hawkins, *Between Harvard and America: The Educational Leadership of Charles William Eliot* (New York, 1972).

2. The naturalist DAVID STARR JORDAN (1851–1931) was the president of Stanford University.

3. Billings.

4. WILLIAM W. MORROW (1843–1929), a federal judge from California, was a trustee.

Robert S. Woodward to James McKean Cattell, February 2, 1906

Dear Professor Cattell:

I have your letter of January 30th.

You are quite right that astronomy is relatively overendowed and oversupported at present in this country; but, on the other hand, it seems to me that you underestimate the difficulties there are in the way of taking up work in most of the newer sciences. It is a curious fact that the obstacles in the way of progress of some of these newer sciences are very seriously presented by many of their devotees, especially in their incapacity to agree on the kinds of work best worth doing. This fact comes out strikingly from a statistical examination of the applications laid before the Institution.

It is not alone sufficient to show the possibility of taking up work in new directions. This is in fact the least of the difficulties. The greatest difficulty is to overcome the prejudice and inertia of the society in which we live. Some lines of work which appear to me to be of the greatest importance to the future of humanity are looked upon with unmitigated contempt by many of our most eminent contemporaries. The Institution is severely criticized for pursuing the obviously advantageous sciences. What would be the stream of criticism and abuse received if we were to take up work in lines less obviously advantageous?

I believe, for example, that we could ascertain, with at least a fair degree of certainty, what are the conditions essentially which produce great men; but do you believe for a moment that our Trustees would regard such an investigation as worthy of considerate attention?

Very truly yours,
R. S. Woodward
President

Cattell Papers, Library of Congress

Edward Bennett Rosa[1] to Arthur Gordon Webster,[2] March 12, 1906

Dear Prof. Webster:

At the last meeting of the Physical Society I moved a resolution appointing a Committee to consider (and report to the Society at the next meeting) the question of securing an endowment fund for research. Prof. Merritt[3] suggests that I write you, as chairman of the committee, my views on the subject. Our present policy (which I think is wise) of giving the members in subscriptions the full value of their dues makes it impossible to devote any portion of the dues to research. The A.A.A.S. does not take the place in this country of the British Association in England in respect to investigation. For half a century or more the latter institution through its committees has done a vast amount of good work, largely because of its grants to these Committees. The French Academy has had an endowment from the beginning, and I think if we put before the public a full statement of what the members of the Physical Society have done and are doing, and what they could do with some additional funds, and especially the advantage of having standing committees on various questions to which committees grants of money could be made for carrying on investigations or preparing reports, that we might get a generous sum in the course of a few years. A hundred thousand dollars would be none too much to ask for at the start; the income of half a million could be wisely used. I think such a fund would be of the greatest benefit to the society, to science, and to the working physicists who would be members of the various committees. It would enable us greatly to increase our output as a society, and give us increased influence in the scientific world.

The Carnegie Institution is not going to do the work in this direction that some of us expected. It is the policy of the institution to undertake a few large projects and to draw out from the smaller enterprises; to give very few grants to individual workers. The President declares that he thinks giving small grants to scientific workers in the Universities tends to pauperize rather than to enrich. I think that rather than attempt to combat this feeling on the part of the administration of the Carnegie Institution, it is better to go to wealthy people who are giving millions every year for educational purposes and seek an endowment for the Physical Society. Indeed, it might be proper to go to Carnegie himself and show how the Washington Institution is not fulfilling the mission that was expected would be filled, and ask him for a generous endowment for research. It would be better, however, if it could be done, to get the money elsewhere, and not ask him at all.

I don't think it is necessary for me to elaborate the idea further. You can see as well as I what could be done, and I felt sure when I made the motion that you would endorse it.

The next meeting is to be in Washington April 13 and 14, a few days before the National Academy meeting.

Yours very truly,
E. B. Rosa

Webster Papers, Archives, University of Illinois at Urbana-Champaign

1. ROSA (1861–1921), a physicist at the National Bureau of Standards. He is reflecting the view of his discipline about CIW policy.
2. WEBSTER (1863–1923) of Clark University directed a small but influential graduate physics program.
3. ERNEST GEORGE MERRITT (1865–1948) was a physicist at Cornell University.

Robert S. Woodward to Theodore W. Richards, March 30, 1906

Dear Professor Richards:

I have your letter of the 28th inst., with a new draft of your proposed letter to Mr. Carnegie.

Please accept my best thanks for your kindly and sympathetic appreciation of the difficulties which beset the development of the Institution. I would be very glad if all my scientific friends could show a like degree of patience and comprehension of the magnitude of the problems before us.

I find only one minor objection to your letter to Mr. Carnegie, and this even need not lead you to make any modification. I think, however, I should state my objection, because it is one which has come up many times already with reference to other projects.

You state in the next to the last paragraph of your letter: "Obviously the building of such a laboratory in Cambridge is outside the province of the Carnegie Institution." This is not quite true, for I see no reason why we should not be as willing to build a laboratory at Cambridge as at Washington, or as at Tucson, in Arizona, if Cambridge were plainly the best place for such a laboratory. What I would object to, however, would be a proposition to build a laboratory at Cambridge which might be used jointly by our Institution and by Harvard University. Ample experience shows that it will not be expedient in the present state of human development for the Carnegie Institution of Washington to cooperate in such a manner with other institutions. This, I think, would go without saying if the proposition were to contemplate a similar alliance between two colleges or universities, although the number of such institutions that would like to have the Carnegie Institution of Washington equip them with laboratories, etc., is very large.

You may be interested to know that, in view of the present disposition of our Trustees, it would probably be easier to authorize the establishment of a laboratory in Cambridge or in Idaho than in the city of Washington.

Your letter suggests a current fallacy to which my attention has been called by many correspondents. Curiously enough, it seems to be tacitly assumed by many eminent men that there is something about a university that will prevent investigators from stagnating more completely than similar influences in institutions organized for the express purpose of research. This argument applied to the Royal Institution of London, for example, would lead us to suppose that Davy, Faraday, Tyndall, and Dewar ought to have undergone deterioration immediately on being given life positions for the purpose of devoting their entire energies to investigations.[1]

An eminent man of science wrote me a few days ago that he viewed "with alarm" the tendency of our Institution to build up departments of work, for, he went on to say, the men at work in these departments will tend to stagnate, and will fall into routine methods that will lead only to commonplace results. Everyone knows, of course, that most men tend to stagnate rather early in life; but I fail to see why picked men should tend to stagnate in an institution whose atmosphere is that of research, any more than men in academic institutions, who are, as a rule, to a much less extent subject to stimulus to original productivity.

I am encountering also, almost daily, another ghastly fallacy, namely, that able men connected with academic institutions who have little time for research, may become truly fruitful if they are enabled to hire one or more assistants. My disposition in every such case is not to depend so much upon hired men who are to be directed by somebody else, as upon those men who are capable of giving directions. It seems to me that it is upon the latter that we should concentrate attention. If an eminent man is worth assisting at all, the assistance should be given in such a way as to secure his entire attention to investigation.

I would like to call your attention also to one more fallacy, namely; that there is in general grave danger in transplanting men of marked abilities from one locality to another. I have taken great pains to look up the history of this subject, and while notable instances illustrating both sides of the argument may be found, my conclusion is that on the whole eminent investigators have generally been made more productive by transplantation.

With the hope that I may presently have a chance to talk with you concerning this nearly endless topic,

I am
Very truly yours,
R. S. Woodward,
President

P.S. The manuscript of your letter, addressed to Mr. Carnegie, is returned herewith.

Richards Papers, Harvard University Archives

1. A more than slight distortion of history. Part of the duties of these investigators was to deliver popular lectures. Although unspoken here, the argument offered implied a lack of involvement with applied concerns, a point made explicitly elsewhere by Woodward and others. This was not true either. For the early years of the institution, see Morris Berman, *Social Change and Scientific Organization: The Royal Institution, 1799–1844* (Ithaca, N.Y., 1978).

Robert S. Woodward to Charles Loring Jackson,[1] June 29, 1906

My dear Professor Jackson:

Your letter of May 28th, 1906, was duly received during my absence on the Pacific Coast, whence I have only recently returned.

Please accept my best thanks for the trouble you have taken to consider the interests of the Institution, and for the helpful way in which you have expressed your views. I esteem very highly your suggestions, although I am quite unable to agree with you in many of your statements and conclusions. This difference of opinion, I take it, shows that we do not understand the questions at issue fully, otherwise men of science ought to agree upon them better than they do. The fact that most college and university men disagree with me concerning matters of administration of the Institution, forces me to proceed with great caution. On the other hand, without undue exaggeration, it seems to me that my personal experience has happened to be of much wider range than that of the average university professor, since about one-half of my life has been spent in contact with affairs outside of academic institutions, and the other half with affairs inside of such institutions.

Quite irrespective of theoretical questions, however, the problem before us is, as you remark, how to get the most and the best returns from our available income. And in finding ways and means for this purpose, we must be quite willing to sacrifice any false ideals we may have heretofore entertained.

My interpretation of the case presented by the Royal Institution of London is very different from yours, and it appears to me that you have not rendered Tyndall due justice. That he deserted abstract research in order to popularize science is in a way quite true; but the same charge may be brought against Huxley and Spencer. And yet, in my opinion, no men did more to advance science during the last half of the nineteenth century than those very men. We Americans, I think, are especially indebted to Tyndall for helping us to get a start in our colleges for the pursuit of postgraduate studies. We should remember also that while Tyndall left the proceeds of his "Lectures in America" to found fellowships at Harvard, at Columbia, and at the University of Pennsylvania, the prince of humanists, Matthew Arnold, refused to go on the lecture platform in some instances before receiving a check in payment for his addresses.[2]

I do not anticipate any of the dangers you seem to foresee to our colleges and universities by offering attractive positions for research. Any institution with a limited income cannot draw more than a few such men from the colleges and universities, and my impression is that as many men might be returned to the educational institutions as are likely to be drawn from them. In any case it seems to me the effect will be advantageous for the universities just as I think it would be advantageous for our Institution if another one for a similar purpose were to be founded independently.

We seem to differ quite markedly in respect to contact with students. I do not undervalue the stimulus coming even from immature minds; but it seems to me that this stimulus counts for nothing in comparison with that which comes from one's colleagues in the same or closely allied fields. My impression is that in America the standard of efficiency amongst academic men is set not so much by academic institutions as by professional societies. In the pursuit of research the influence of one's peers not only ought to be but, so far as my experience goes, is of far greater importance than the influence of immature students however enthusiastic they may be.

This is too large and difficult a subject, however, for adequate discussion in a letter. I shall reserve a section for its consideration in my forthcoming report.

Again thanking you for your generous appreciation of the difficulties of the situation in which I am placed, I beg to remain

Very truly yours,
R. S. Woodward
President

Richards Papers, Harvard University Archives

1. JACKSON (1847–1935) was a chemist at Harvard.

2. Woodward is doing two things at this point. First, he is upholding the desirability of research untrammeled by other duties by defending Tyndall. Second, by noting that the lecture tour proceeds aided scientific education, Woodward is asserting the moral superiority of science over the humanities in the person of Arnold. Woodward had undoubtedly read Arnold's contrary view. Although he wanted to aid the humanities, Woodward was not happy about pressures from that source on him as head of CIW.

Excerpts, George H. Shull's[1] Report on Luther Burbank, July 26, 1906

To the Committee of the Carnegie Institution having in charge the investigation of the work of Luther Burbank:—
Gentlemen:

I have rarely found a man who is more honest to his own experiences than Mr. Burbank. He prefers not to allow the views of others to in any way influence his perceptions as to the relationship of the forms on which he is working. He strives against the effects of his own preconceptions, also, with what he believes is success, but which appears to me to lack much of it.

This question as to the effect a preconception has upon his observations is difficult to determine, but the fact that the views he holds now were stated definitely as long ago as 1893, indicates that these preconceptions do largely dominate his experiences, and they will need to be kept constantly in mind in weighing the value of his statements. He holds that nothing but acquired characters may be inherited and is thus led to attribute every variation to environmental causes in a more definite way than observations would warrant. Again, without knowing just what the various scientists and philosophers have learned or taught, he takes sides strongly,—*against* Weismann, Mendel, and De Vries, and *with* Darwin, and the modern opponents of mutation and Mendelism. He explains everything in heredity on the basis of the mingling of forces, but this is again the result of a purely theoretical consideration, and has little bearing upon the accuracy of his observations, or his statements regarding those observations, except at such time as he is definitely discussing heredity. Another source of bias is seen in his strong faith in teleology, which makes him assign to many characters a use that present knowledge discredits.

Notwithstanding these sources of error in observation, my whole experience with Mr. Burbank leads me to believe that his observations of the things in which he is interested are exceptionally accurate. There are certain kinds of observations, however, which he does not make with his wonted precision. He says he never counts things, and when asked as to the accuracy of the estimated numbers of plants which always figure so largely in the published accounts of his work, he said that those numbers were obtained by counting a small section and multiplying, then scaling down the result to a figure that satisfied him that there was no exaggeration. My own observations upon his cultures have shown me, however, that the numbers given are not at all reliable, and I believe that they would in most cases need to be divided by 2—10 to get the correct number. A single instance will illustrate: He stated that he had planted 4000 varieties of potato out of some 14000 to be planted. It was already late for the planting of potatoes and the ground was occupied to such extent that it became evident that all were planted except some seedlings to be set into the garden at Santa Rosa. After his statement was made to me and was also given to the press by him, I counted all that had been planted out, and found just 390 varieties. So large a discrepancy as this might look to some to be due to dishonesty, but I do not think it is. The numbers simply so exceed his capacity for numerical conceptions that any large number will seem to him to be reasonably moderate.

Honesty is one of Mr. Burbank's consciously strong characteristics, and he says frankly that he does not know, in cases where it might appear to his advantage to know, and where his imagination could quite readily supply the information without the possibility of detection if he so desired. It is a noteworthy fact that he likes to speculate about the things about which his knowledge is least adequate, and in discussing these subjects he deals in a great many generalities that are not sufficiently supported by observation, but the saying of things in an offhand way in such connection, even though he says them with all the vigor that indicates certain knowledge, is not due to dishonesty for he really believes the statement at the time he makes it.

Shull Papers, Library, American Philosophical Society

1. GEORGE HARRISON SHULL (1874–1954) was then with CIW; from 1915 to 1942 he was at Princeton. A pioneer geneticist best known for his work on corn, Shull founded and edited the journal *Genetics*.

Excerpts, Minutes, Board of Trustees, Carnegie Institution of Washington, December 11, 1906

[PRESIDENT WOODWARD:] In addition to the remarks which I have made at great length—too great length I fear—in my report, concerning the larger projects and the minor projects of the Institution, I would like to make a few supplementary remarks. I may say that the larger projects of the Institution, of which there are now twelve in addition to the work of administration and in addition to the work of publication, seem to me to be prospering very favorably. There seems to be little to be desired in regard to these departments. They are all manned by very able and efficient men, and in most cases manned by the very best men available and the work they are doing is progressing very rapidly. So much has been said concerning them in my report and so much has been said by the directors of these departments in the full reports of the Institution, that I need not dwell on them longer. On the other hand, I may say briefly that the work carried on by the Institution under the head of minor projects does not seem to me to have progressed favorably. I regret that I have been obliged in writing my two reports to devote so much space to this matter. The amount of space devoted in my last report on the subject of minor grants or minor projects, seems to me excessive, but I have felt obliged to devote this amount of space by reason of the fact that in my work for the Institution at least three quarters of my time is absorbed by the business appertaining to these small grants and to the business of research-assistantships. My impression is that we shall find it advantageous presently, though I am by no means disposed to proceed hastily in this matter, to change our method of administration of such minor grants. At present, to use a figure which is perhaps more forcible than elegant, we are conducting a species of Havana lottery, with monthly drawings, in which the inexperienced and inexpert man is almost as likely to receive a prize as the expert and the experienced man. This, however, is simply my opinion, but I am not sure that all members of the Trustees would agree with me in this matter. As a consequence, in the recommendations which I felt impelled to make I have had in mind criticisms of this system of awards. In those recommendations I have stated that the remedy which I suggest is proposed with the hope that it may draw out discussion. I have felt bound to call the attention of the Trustees to the actual state of affairs, and, while this actual state is, it seems to me, plainly very unsatisfactory, I feel that we may easily remedy it in such a manner as to relieve the situation of all its embarrassments. The whole subject is complicated to some extent by our unknown relations to colleges and universities. I say unknown relations, because an essentially new institution like the Carnegie Institution of Washington must find out what its relations to colleges and universities should be. I have therefore taken great pains during the past two years to draw out of people who have had opinions to express on this subject, their opinions, with a view to getting all the information possible. It has cost me a deal of labor during the past two years with respect to this subject alone. I have held more than 2,000 interviews, and you can see that that must take a good deal of time from a man who ought to be devoting at least fifteen hours a day to the larger interests of such an institution. As nearly as I can ascertain, about 60 per cent. of our colleagues in academic institutions would like to have us continue a system of small grants which would serve as a supplement to the system of scholarships and fellowships which obtains so largely in the colleges and universities of the United States. About 60 per cent., I say, of the colleges would think it best to continue that system, and about 40 per cent. are opposed to it. Amongst the latter minority some eminent educators of the United States are very strong with me in their opposition to any such system. Amongst those I might mention especially President Eliot of Harvard, whose opinion I drew out a few days ago in the conferences of the Association of American Universities held at Cambridge, Massachusetts. Almost equally strong in confirmation of my own view of this subject were the expressions of opinion from President Remsen, President Schurman,[1] President Wheeler,[2] and several others; so that I think there can probably be comparatively little difference of opinion in regard to this matter when we once thoroughly come to understand the exact status of the case. My own opinions have undergone rather radical changes in the past few years with regard to the administration of any system of minor grants. During the past twelve years I have served as Treasurer of the American Association for the Advancement of Science, and my duties in that connection have been to act as chairman of the committee on grants of the Association. Of course the Association has only a small amount of money to dispense in grants in aid of research; but I learned from my experience with that Association that we were getting, to use a commercial phrase, at least ten per cent. compound interest on our investments, and I thought we could extend the same system in this Institution advantageously. I find, however, that we are not extending the system employed in the case of the American Association for the Advancement of Science in which we deal almost exclusively with experts, with men of proved capacity. My impression is, therefore, that we may learn a lesson from the American Association for the Advancement of Science, and if we apply that lesson we shall relieve the situation of all its difficulties in the way of minor grants. This subject is so large in its ramifications that no one can understand it unless possibly he would take the trouble to read thousands of pages of correspondence which we have in reference to such matters. I think, however, that it may all be boiled down into a compass no greater than that in which I sought to boil it on page 23 of my report. While I would not urge by any means any hasty action concerning this matter, I would like to ask the careful attention of the Trustees to the propositions laid down under the heads (a), (b), (c), and (d) on this page 23 of the report. The difficulties in the way of administration I think lie chiefly in two matters: In the first place, we need carefully to choose our men who are going to receive minor grants; and, in the second place, we need to treat them equally carefully in order that they may esteem it an object to work under and for the Institution. In the case of the system as it has been administered hitherto, grantees have been mostly young men and women without experience. . . .

MR. MITCHELL: I would like to say a word or two on one point upon which the President has dwelt a great deal and upon which there is a mild difference of opinion between him and some members of the Executive Committee. I sent a paper representing my views and those of Dr. Billings to the Chairman of the Executive Committee,[3] whose report will doubtless give you some information in regard to this matter, but whether it will be as full as that which I desire to make I do not know. I have not had time to read the whole of it, but I wish to say that in Mr. Carnegie's original gift to us he especially dwelt on the desirability of experimenting on the finding of competent men. He did not expect to do what the President desires to have done, to limit our gifts only to those who had through years proved their capacity. What he wanted was to expend a certain amount of money in finding people. Now I want to say that the President has expressed his difficulties on this subject very well, and has pointed out how much of his time is taken up by examining the subjects presented and the individuals who wished to deal with these subjects, and I can only say that Dr. Billings and I once spent a great deal of time in doing this thing, and the President has my sympathy; but, nevertheless, I regret to say that that is part of his business, and is "what he is here for," to carry out Mr. Carnegie's views as well as our modifications of them, in the best possible manner. I will say also that his request for research associates and none others, to have these grants, would be cutting off a certain amount of valuable future human material. I feel strongly about this subject and Dr. Billings and I have gone over the statistics of success and failure in these minor grants and we believe that we have had a sufficient amount of success in these gifts to entitle us to continue for some years longer this search for competent men. I feel very strongly in regard to it, because there have been times in my young life when certainly, under the President's rule, I would not have been a proper person to give a grant to from his view of the matter, and when I should have longed to have one, and applied to institution after institution without getting it. I would not have gotten it from you if the President's views had been carried out.

He wishes the affiliation of the associates to be such that they should be advisers of the Institution. Of course that limits our gifts to men who are capable of giving advice and only to men of middle age and assured competence by success in many investigations. I want to say that I do not want the President to have his way about this entirely. Also I desire to say that last night I took home and looked over the list of fifty applications which had come to him for minor grants within the last four months. A great many of them would be thrown out by a glance; it was not worth while to consider them at all. A certain proportion beyond that were worthy of more or less consideration. I should not think there would have been much trouble in finding out who they were. Among those which he rejected were only three about which I had any doubt at all. One of those, at least, I think ought to be given a grant, perhaps two of them, and we went over all the minor grants to which he desired to give money yesterday and we did not differ as to one of them, so I hope there is not quite so much difference between the President and the Executive Committee as to lead to any serious warfare. . . .

Archives, Carnegie Institution of Washington

1. JACOB GOULD SCHURMAN (1854–1942), originally from Canada, was president of Cornell 1892–1920.

2. BENJAMIN IDE WHEELER (1854–1927), a classical philologist, was president of the University of California 1899–1918.

3. CARROLL D. WRIGHT (1840–1909) headed the Bureau of Labor, the predecessor of the Department of Labor, from 1885 to 1905.

Excerpts, Minutes, Board of Trustees, Carnegie Institution of Washington, December 14, 1909

[PRESIDENT WOODWARD:] Passing now to quite a different branch of the work of the Institution, I would like to explain to you, confidentially, the status of the Department of Economics and Sociology, which was in charge of our late colleague, Dr. Carroll D. Wright, from its inception up to the time of his death in February last. Soon after his death, I called a meeting of his collaborators. He had associated with him eleven collaborators. Nine of these collaborators were present at a meeting held in our office, I think on the 20th of March last. The object of this meeting was to learn from them the status of the work of the department, and to consult their advices as to what would be best to do in continuing or in closing up the work of the department. I regret very much to state that we found what I would consider an exceedingly unfortunate state of affairs. It will be recalled that there had been appropriated for that department, directly, $150,000, and, indirectly, for the purpose of the preparation of an index to state documents, $17,500. Of course, the figures were all in our hands, so that we knew essentially what the financial status of the department was before the meeting was held. Up to that time, nearly $100,000 had been spent, and the Institution had received nothing in the way of publications, except those special publications prepared under the auspices of the appropriation of $17,500 just referred to. It appeared, moreover, that most of the work which had been done under the auspices of this department had been carried on or executed by young men or women who were, at the time, candidates for the higher degrees in colleges and universities. It appeared, on further examination, that many of these candidates for degrees had been supported in going through colleges and universities, and had actually had their dissertations and other papers printed at the expense of the Institution.

To make a rather long story short, I may say that my estimate at present is that the Institution as a research institution will lose about a round hundred thousand dollars in this investment. The reason for that is this. It appeared from the confessions of these collaborators of Colonel Wright, all very able and eminent men, that nearly all of the work which had been collected by their assistants was regarded as untrustworthy, defective, much of it defective to such an extent that it could not be used at all. All of it will have to be done over, they tell me. The department at present, therefore, altho it has yet about $50,000 to its credit, is suggesting that if they finish the work in accordance with the original scheme, more money will be necessary. It is perhaps not the time now to give special advice concerning this matter, because the report of Professor Farnam,[1] who has been appointed Chairman of the department, is just now in print and I think it should be carefully examined by the Board of Trustees before any action is taken; but I may go so far as to suggest that we do nothing during the coming year, either in the way of seeking to give this department additional financial aid or seeking to withdraw the money already to their credit. It seems to me that we should give them perhaps further opportunities to materialize. Whether their work will turn out to be advantageous to the Institution is somewhat doubtful to me. Perhaps, however, the experience is worth all it cost. You will recall that in the early days of the institution there were many people, especially colleagues in academic institutions, who thought that great work could be accomplished if only a sufficient number of students, and especially graduate students, could be set to work. This experience, conducted on a rather large scale, because there have been at times no less than two hundred people at work under the auspices of this department, is very unfavorable to such a view.[2] . . .

The advices from Dr. Shull, who has spent nearly the whole year at Santa Rosa, are that that work is substantially complete and it will not be necessary for us to cooperate with Mr. Burbank any longer on that score.

Further than that, it has appeared to be impossible to avoid entangling alliances with Mr. Burbank and his numerous exploiters. I think they, rather than he, have brought the institution somewhat into disgrace. And in view of the extensive difficulties of getting on with Mr. Burbank, I have heartily made the recommendation which is contained in section 9 of my report, under the head of the budget.

THE CHAIRMAN: Are there any further remarks on the motion to approve the budget?

MR. LOW: May I ask the President a question suggested by his verbal report a moment ago. He tells us that through the Department of Economics and Sociology we have been making year after year large expenditures for which we are likely to get small returns.[3] It would be very interesting to me, and perhaps to the other Trustees, to hear what steps are taken or can be taken to prevent a similar result in other departments.

Take the Department of History,[4] for example. How is it known that our money is being spent in that direction any more responsibly than in any other?

PRESIDENT WOODWARD: That is very easy, gentlemen. The Department of Economics and Sociology was the only department organized after the fashion explained. All of our other departments have each its head and each its staff of paid officers who work on salaries and who are directly responsible to the institution. In the case of the Department of Economics and Sociology the collaborators have been paid no salaries; they have volunteered their services and have received nothing except for some minor expenses for traveling and so forth. But they thought they would get a large amount of work done by having money which they could expend in helping graduate students.

MR. LOW: It was literally sociology and economics?

PRESIDENT WOODWARD: It was literally sociology and the negative of economics.

MR. MORROW: I would like to ask one further word about Mr. Burbank. Do I understand correctly that you have secured from Mr. Burbank thru Dr. Shull a substantial statement of his methods?

PRESIDENT WOODWARD: Yes.

MR. MORROW: And the work accomplished since his connection with the Institution?

PRESIDENT WOODWARD: Yes. . . .

PRESIDENT WOODWARD: I think under the head of new business, I may offer to the Board of Trustees a memorial received a day or two ago, thru Professor Andrew F. West,[5] a memorial emanating from something like ten national societies. It is from the following societies: American Historical Association, American Institute of Architects, American Philological Association, American Philosophical Association, Archaeological Institute of America, Modern Language Association of America, National Academy of Design, National Institute of Arts and Letters, Society of Biblical Literature and Exegesis and the American Dialect Society. This memorial is also signed by the presidents and ex-presidents of nine different universities. I do not know that I can do anything further than submit it to the Board of Trustees, unless they may desire to have the two pages which contain the substance of the memorial read.

The two pages containing the substance of the memorial read as follows:

"With most hearty appreciation of the vast good already accomplished by the Carnegie Institution and with high hopes for its even greater usefulness in the future, we ask the privilege of expressing the strong desire of these learned societies to cooperate with efforts of the Carnegie Institution for the advancement of knowledge in the fields of literature and art in the broadest sense, and also in the humanistic sciences. To this end we respectfully ask that properly approved projects of historical, archaeological, philosophical, linguistic, literary and artistic investigation and publication be admitted in the apportioning of grants of the Carnegie Institution to a recognition similar to that given approved projects of research in the physical and natural sciences.

We are encouraged to express this desire by the very liberal provisions of the Articles of Incorporation of the Carnegie Institution, especially as stated in the following words of the second section of the Articles of Incorporation:

"That the objects of the corporation shall be to encourage, in the broadest and most liberal manner, investigation, research and discovery, and the application of knowledge to the improvement of mankind; and in particular

"(a) To conduct, endow, and assist investigation in any department of science, literature or art, and to this end to cooperate with governments, universities, colleges, technical schools, learned societies, and individuals."

We desire, moreover, to urge our respectful request without the slightest criticism of the generous allotments made to projects in the physical and natural sciences, which embrace, as the Year Book for 1907 (page 21) shows, about five-sixths of all the moneys appropriated, while one-sixth has gone to all other projects. We fully realize that very liberal grants have been needed to carry out these great projects in the domains of the physical and natural sciences. The sole object of our memorial is to plead earnestly that similar recognition may be extended to other sciences and especially to all the departments of literature and art."[6]

THE CHAIRMAN: I think it will be plain to the Board from what we have heard in regard to our funds, and the action we have taken in regard to our budget, that it would not be very profitable to discuss this proposition presented by this memorial at the present time. It will be remembered by the Board that most of these propositions for work such as solar observatory, terrestrial magnetism, astrometry, and so forth, were made with a definite understanding that the work should continue for about ten years. Astrometry was the only one that will probably close in less than ten years, and therefore it would seem doubtful whether we want to change that. We have engaged these men and we have set them at work, and the results have been very satisfactory; but it is a matter that would seem would require more investigation, more information, than we are able to present to the Board today, and perhaps the proper thing to do would be to refer this to the Executive Committee for consideration and for report hereafter as to what can be done.

Archives, Carnegie Institution of Washington

1. HENRY WALCOTT FARNAM (1853–1933) was a Yale economist.

2. This was proof positive, to Woodward, of the desirability of funding proven investigators and of separating research and education.

3. Woodward continued funding the ongoing projects which eventually produced substantial works. In 1916 he tried to revive the department, but without success.

4. Headed by JOHN FRANKLIN JAMESON (1859–1937) and phased out after his retirement from CIW in 1928. Largely because of Jameson's position in the historical profession, the department was fairly influential.

5. ANDREW D. WEST (1853–1943) was a classicist, dean of the Princeton University Graduate School 1901–1928.

6. Woodward was sympathetic but could do very little because (1) funds were limited, and (2) his beliefs sanctioned the priority of the sciences. To the end of his tenure, he tried to allocate funds for specific worthwhile projects in the humanities.

2
Astronomy: The Lure of the Sun

Woodward became aware of the logical and chronological primacy of astronomy through the history of science. By 1900 it was one of the fields in which Americans were established in the international world of science. In positional astronomy the elderly Simon Newcomb continued his search for the structure of the solar system. In astrophysics American productivity had surpassed the Germans. Productive astronomers at observatories and in universities constituted a significant element in the U.S. scientific community. Given the views of the Carnegie trustees and President Woodward, major funding of the field was inevitable, particularly as two very exceptional men applied successfully for support. Although markedly dissimilar, Thomas C. Chamberlin and George Ellery Hale had no clashing encounters; their relations were amicable. Both were interested in the sun but in complimentary ways. As an obvious feature of the heavens, the sun had attracted scientists for ages. In the last century in the United States, a few members of the emerging scientific community began to specialize in its study: Stephen Alexander and C. A. Young at Princeton,[1] the theoretician Jonathan Homer Lane,[2] the astrophysicist S. P. Langley[3] at the Allegheny Observatory and the Smithsonian Institution. Hale came to his obsession with the sun from the bustling field of astrophysics. Chamberlin turned to astronomical questions because of problems in his original field of geology. Their differing views on scientific policy never intersected in practice.

Questions about the glacial ages led to Chamberlin's concern over the history of this planet. That produced a spectacular foray into cosmology, a great influence for several decades before giving way to criticism from Young's successor, Henry Norris Russell.[4] At the time Chamberlin started, the nebular hypothesis of Laplace (and Kant) was encountering difficulties. Chamberlin boldly attacked the widespread view that the earth had originated in a molten state and then cooled down. A quite untypical geologist in many respects, Chamberlin had no qualms in asserting the validity of geological evidence in the face of confident assertions from physics and astronomy. He had previously challenged Lord Kelvin's estimate of the time span of earth history, adroitly carrying the attack to weaknesses in the physical argument. That is, he did not accept the widespread belief in the primacy of the abstract fields of physics and chemistry. He did not feel any compulsion, apparently, to reduce geology (or other sciences) to the laws of physics and chemistry. It was an older view which assumed that physics, chemistry, and mathematics were tools for the solution of problems of geophysics and astronomy.[5]

But if he was to challenge the astronomers on their own ground, Chamberlin needed a mathematical astronomer to develop and to test his ideas. Forest Ray Moulton[6] performed that function. The two worked out a planetismal hypothesis for the origins of the solar system in which a passing star caused an eruption from the sun (or proto-sun), producing aggregations of matter (planetismals) from which the planets and their moons developed. It was not the first dualistic cosmogony; Buffon had a related but different theory in the eighteenth century. Markedly different from the monistic view of Laplace, it took into account both geologic and astronomic data. Perhaps more important, according to a recent study, it tacitly repudiated the unilinear evolution so widely accepted after Darwin and Spencer. In a real sense, Chamberlin was a precursor of the complex, random, chance-laden world views which became so prominent in this century. Although the specific hypothesis was overturned before World War II, when "planetary sciences" arose in the wake of the space program, a revival occurred of some of Chamberlin's attitudes and a number of specific views.

In reaching his conclusions, Chamberlin responded to his deeply held beliefs about proper scientific method. He authored an article about the need to methodically work with alternative and competing hypothesis, an influential piece perhaps more honored than followed in practice. In the work on the planetismal hypothesis, Chamberlin specifically favored a "naturalistic" method against what he viewed as capricious mathematical model building. He favored a form of qualitative reasoning sticking close to the physical data. Hale, not a mathematical theorist, in practice hewed fairly close to Chamberlin's views.

But the two men were very far apart on what we now call science policy. Chamberlin's views, in the letter to George Darwin[7] of June 30, 1906, reflected a successful career in geology. In the United States, on national and state levels, the leaders of that discipline had succeeded in convincing many politicians and members of the lay public that theory and practice went hand in hand. While the path was not always smooth, the pattern was well established in the United States—a broad definition of mission stretching along a spectrum from the abstruse to the narrowly applied. Hale, Woodward, and many others were opting for enclaves of research shielded from the concerns of society.

Hale always viewed his field from the perspective of physics and thought of his observatory at Mt. Wilson as being close in spirit to an experimental laboratory. Like Chamberlin, but for different reasons, Hale was alert to developments in physics. In a technical sense, his attitudes make him very modern, perhaps almost radical. But in most other respects he was of the nineteenth century, a confirmed follower of Herbert Spencer in a world about to undergo intellectual, political, and social cataclysms. A cultured conservative gentleman, Hale saw evolution everywhere, from the heavens to the minute aspects of life. He wanted to make Alexander von Humboldt's *Kosmos* the model for a new integrative education, a great burden for a four-volume work already obsolete on publication in the middle of the last century.

He hoped that the evolved complexity of science would serve as a model of social action. For his own field, Hale promoted cooperation between investigators on a worldwide scale—with considerable success. His long-term goal, in the United States and throughout the world, was to extend these cooperative actions to all the sciences and then to all forms of learning. He was a great visionary and organizer of visions. No doubt it was even Spencerian evolution, from the simplicity of individual observers to the complexity of networks of research institutions. Hale's vision was also very American, quite parallel to the attitudes of the men from whom he sought funds. Clearly, Hale viewed himself and his allies inside and outside the scientific community as being trustees for the evolutionary process.

Hale wanted to organize the entire scientific community through the agency of its leaders. That meant the National Academy of Sciences. It was a high-minded view but very paternalistic. Although the emphasis was on great or exceptional men, cooperation implied large scale with a tacit hierarchy of scientists, most being destined for humble tasks. Hale was not promoting a pure democracy. Although Hale succeeded at Mount Wilson and Mount Palomar, his larger plans for science in the United States and the rest of the world never quite came to fruition. Events were contrary; so too were long-term trends in the United States.

1. ALEXANDER (1806–1883) was the brother-in-law of Joseph Henry. YOUNG (1834–1908) was the son of a Dartmouth professor, Ira Young, who was also interested in astronomy.

2. See the entry in the *Dictionary of Scientific Biography* for Lane's curious career.

3. LANGLEY (1834–1906) was then the head of the Smithsonian.

4. HENRY NORRIS RUSSELL (1877–1957) was a theoretical astronomer at Princeton, best known for his independent discovery of what is now called the Hertzsprung-Russell Diagram, which shows stellar evolution from giant red to dwarf stars.

5. "A Geologist among the Astronomers: The Rise and Fall of the Chamberlin-Moulton Cosmogony," *Journal of the History of Astronomy* 9 (1978): 1–41, 77–104; and "The Overthrow of the Nebular Hypothesis," Technical Note BN-836, Institute of Physical Sciences and Technology, University of Maryland, College Park, June 1976, both by S. Brush.

6. MOULTON (1872–1952) was a mathematical astronomer whose career was spent at the University of Chicago.

7. DARWIN (1845–1912), the son of Charles Darwin, was a geophysicist at Cambridge.

George Ellery Hale to Hugh Frank Newall,[1] January 28, 1903

Dear Mr. Newall:

The Carnegie Institution has appointed a Commission, consisting of Professor Boss, Professor Campbell[2] and myself, to report on the practicability of establishing a large astrophysical observatory at some particularly favorable site, preferably within the United States. In general, the choice of a site, and the provision of a suitable instrumental equipment would be made with the purpose of pursuing investigations along the following lines:

(1) Solar radiation problems, including the measurement of the solar constant at frequent intervals throughout the Sun-spot period. This investigation would also involve the measurement of the absorption of sunlight in its transmission through the atmosphere of the earth and that of the Sun, and also the radiation of different portions of the Sun's disk, such as spots, faculae, and prominences. For this work it would be necessary to have two stations, one at the summit of a high mountain (Mt. Whitney—14,887 feet—is being considered for this purpose), the other near its base, equipped for simultaneous observations.

(2) Solar investigations, principally of a spectroscopic nature, which require atmospheric conditions and instrumental facilities superior to those hitherto employed. These might include photographic studies of the structure of the photosphere, spots, and prominences; investigations of the spectrum of the chromosphere, spots, etc.

(3) Various stellar and nebular problems, such as could be undertaken to the best advantage with the aid of a large reflecting telescope. For example, determinations of motion in the line of sight of the fainter stars, investigations of the ultra-violet region of stellar spectra, investigations of the spectra of the brighter stars with very high dispersion, photography of nebulae, measurement of the heat radiation of the brighter stars, etc. For this work a site must be selected which offers the best possible advantages for night observations.

Provision might possibly be made for laboratory investigations on such radiation problems as are intimately connected with the solar and stellar work.

In the selection of the site, it is not yet certain whether two or three stations may be required. It is, of course, desirable to concentrate the work as much as possible, but it is doubtful whether the conditions near the base of a high mountain would be such as to favor good definition both by night and day. For this reason it is not unlikely that it would be necessary to establish the principal observatory, containing the large reflecting telescope and the equipment for solar spectroscopic work,* at some favorable point at an altitude of from 2,000 to 4,000 feet, perhaps in Southern California. For special solar radiation work a station at the summit and one at the base of Mt. Whitney, mainly for use during the summer months, would probably be desirable. Of course no definite decision can be reached on these points until a careful investigation of various sites has been made.

The Commission is also requested to report on the practicability of establishing an observatory in the southern hemisphere. It is probable that an equipment of a general character, not specially designed for astrophysical work, will be recommended for this station, though a reflector of large aperture would eventually be needed there. It is deemed important to carry on as much as possible of the astrophysical work within the United States, partly because the development of new instruments and methods requires convenient access to instrument-makers and other sources of supply.

It is not now certain that the Carnegie Institution will be able to provide for the establishment of this observatory. But, after considering a brief report on this subject which was presented to it last year, the Institution has thought well enough of the idea to appoint this Commission to prepare a careful statement of the specific advantages to be anticipated from a well equipped observatory of this kind, and has also authorized this Commission to undertake a preliminary study of possible sites. We believe that we should meet this evidence of interest on the part of the Institution with a carefully studied examination of all the leading questions involved; and we should greatly appreciate advice from those of our colleagues who are versed in the matters under consideration. We should feel especially obliged for your suggestions as to work to be undertaken, and as to the equipment best adapted for the purpose.

Thanking you in advance for any suggestions you may be able to offer, I am,

Yours very sincerely,
George E. Hale
Secretary of the Commission

* Including a complete bolographic outfit.

Hale Papers, California Institute of Technology

1. NEWALL (1857–1944) was the Astronomer Royal of England.

2. WILLIAM WALLACE CAMPBELL (1862–1938) was the director of the Lick Observatory in California. From 1923 to 1930 he was president of the University of California; he then had an unhappy term as president of the National Academy of Sciences.

Thomas C. Chamberlin to George F. Becker,[1] April 11, 1904

My dear Dr. Becker:

I appreciate your kind letter of the 5th relative to the decrement in the moment of momentum of the solar system. I had it in mind to write you on this point as soon as I could find a spare moment. The Laplacian hypothesis really has for its corner-stone the doctrine of the essential constancy of the moment of momentum during the evolutionary stages, for the acceleration of rotation which is the prerequisite of the separation of the rings is but a concrete expression of the doctrine, and it seems to me perfectly legitimate to criticize the hypothesis on the basis of its own fundamental assumption, whether these are explicitly declared or merely implicitly assumed. A quantitative study of the known agencies that are at present operative in the system gives no ground for supposing that they are of sufficient value to modify in any appreciable way the application of fundamental doctrines. This has perhaps been best brought out by Darwin's study of the possibilities of modification of the orbits of the planets by tidal influence.

I had it also in mind to write you relative to the statement that Buffon[2] had anticipated my hypothesis regarding the origin of spiral nebulae. On critical reconsideration, I am sure you will see that the idea advanced by Buffon and that advanced by me are rather radically different. Buffon's view that the parent nebula of the solar system arose from a comet striking the sun and expelling a stream of matter from it which reunited into globes and became planets, seems to me fundamentally different from the close approach of the ancestral sun to some other sun or massive body, and the consequent discharge of protuberances in opposite directions by virtue of internal elasticity, aided by relief of pressure due to differential attraction. The two ideas are so distinct in my own mind that I regard the one as untenable and the other as plausible, if not probable. I have regarded them as so distinct that I abandoned a hypothesis of my own analogous to Buffon's and adopted the one I have recently presented. At one stage of my studies, I tried to develop a hypothesis analogous to that of Buffon, in which the essential feature was an eccentric collision between a smaller and a larger spheroidal nebula, the postulated result being an irregular stream of scattered material, chiefly derived from the smaller nebula and partially from the larger which ultimately evolved into planets. But after developing the hypothesis so as to bring out some of its leading dynamical qualities, I abandoned it as unworkable, partly for reasons closely similar to those urged by Laplace against Buffon's hypothesis, and partly for others. No such nebulae, that is, no nebulae with *a single spiral arm,* are known, and I do not see how the bi-lateral characteristics of spiral nebulae, which are sharply developed in many cases and more or less obscurely indicated in others, can arise from a collision of the kind postulated.

It of course may be true that the view I have advanced is not wholly unanticipated, but it is at least wholly original with myself, and I have not as yet found anything in the literature that is, in any discriminative sense, to be regarded as identical with it, nor have any of my astronomical friends so far as I know. My theory is mentioned favorably, and without any indication of previous views of the same sort, in a recent notable work on "Problems in Astrophysics" by the author of the "History of Astronomy during the Nineteenth Century," Miss Clerke,[3] who is regarded as a specialist in astronomical history, and is esteemed by astronomers generally. One thing seems certain at least, that no one has heretofore presented the view in such a definite and concrete form as to give it any appreciable place in the literature of the subject or in the history of thought upon it, and this is the true historical criterion by which the inauguration of a hypothesis is to be judged. There is a wide difference between a vague, undeveloped suggestion of what may be, and a definitely organized hypothesis. So, too, there is a radical difference between a successful and an abortive hypothesis. I do not here refer to the truthfulness of the hypothesis, but merely to its successful organization and propagation. It is perhaps as much a labor of knowledge and skill to develop a definite well organized hypothesis with its logical basis and coherence of parts, as it is to develop other works or art. The marvelous success of Laplace's hypothesis, as a hypothesis, is a brilliant example of successful organization and propagation. The hypotheses that preceded his are perhaps to be regarded as typical examples of abortive hypotheses. But I must not launch out into this field which belongs rather to the historian than to the constructive scientist.

In reflecting upon the whole discussion, I have become more and more impressed with the justness of my first feeling that it was unfelicitous to undertake the serious consideration of a hypothesis that was so complicated, previous to the publication of a careful statement which could be studied at leisure. The mere opportunity to run over manuscript that had to be passed on to a colleague was an inadequate basis for a discussion that could be regarded as more than an expression of first impressions. I was at first inclined to advise the postponement of the consideration of the subject until after publication, but as interest has been awakened, I rather reluctantly decided to accede to the request to present the matter to the Academy. The prominence which my good friend Fairchild[4] has given my hypothesis has not been altogether to my wish, much as I appreciate his enthusiastic adoption of it. For myself, I have been anxious that it should develop slowly and cautiously, because I am fully mindful of its intricacies and difficulties.

Cordially appreciating your frankness and the openness of your criticisms, I remain,

Sincerely yours,
T. C. Chamberlin

Chamberlin Papers, University of Chicago

1. GEORGE FERDINAND BECKER (1847–1919) was an unusual geologist for his day because of his interest in applying mathematics and physics to the field. Like C. S. Peirce, Becker was a schoolboy friend of Henry Adams. Becker's example had much to do with the founding of CIW's Geophysical Laboratory.

2. GEORGE LOUIS LECLERC, COMTE DE BUFFON (1707–1788), had another encounter theory of the creation of the solar system, but it was not the same as Chamberlin's.

3. Agnes Clerke's *History of Astronomy during the Nineteenth Century* (1885) was a widely influential work of popularization. *Problems in Astrophysics* (London, 1903) was not the right kind of authority to cite among the leading scientists.

4. Perhaps the geologist HERMAN LEROY FAIRCHILD (1850–1943), a prolific writer in geology and of pedestrian institutional histories.

George Ellery Hale to William Wallace Campbell, May 19, 1904

My dear Campbell:—

Your kind letter of May 17 reaches me just as I was about to write to you. Mrs. Hale and I hope to be able to accept your welcome invitation very soon. The unexpected outcome of my trip to Washington, which consists of a grant of $10,000. from the Carnegie Institution for immediate use on Mt. Wilson, delayed my return, and will prevent me from going to Mt. Hamilton as soon as I expected. However, I am pushing along preparations for the work as fast as possible, with the assistance of Ritchey, Ellerman and Adams.[1] Ellerman and Adams are to be with me on the mountain, and Ritchey will spend some time in Pasadena in charge of an instrument shop which we expect to arrange for very soon. I have a strong intimation from the Carnegie Institution that considerable more money will be forthcoming in December, so I am planning to go ahead with my solar work. Nothing can be done about the 5 foot reflector unless the large scheme goes through, but there is a rather general opinion that Mr. Carnegie will make the additional gift before very long.

At the meeting of the National Academy a Committee on Solar Research was appointed, consisting of yourself, Langley, Michelson, Young and myself. It seemed impracticable to hold a meeting of this committee at once, but I have seen the other three members, and fully intended to have seen you before this. From your letter, however, I know that you are in favor of securing the appointment of other committees by societies here and abroad, and I am accordingly taking the chance of sending to Agassiz the enclosed letter, with the request that he send it, as President of the National Academy, to the Royal Society and Royal Astronomical Societies of England, the Astronomische Gesellschaft and the Berlin Academy of Germany, the French Academy and the Astronomical Society of France, the Spectroscopic Society of

Italy, the Astronomical Society of Russia and the St. Petersburg Academy, and the Academies of Vienna, Stockholm, Amsterdam, as well as the Astronomical and Astrophysical Society of America. A slightly modified letter, of the same general import, is sent him for the physical societies of America, Germany, and France. In the discussions among various members of the committee, it seemed absolutely necessary to invite all of these societies to appoint committees, in order that there may be a sufficiently general representation of the important men. The undoubted necessity of considering anew in the near future question of standards of wave-length of course greatly increases the range of the work, and multiplies the number of societies concerned.

If you have any objections of any kind to the letter, which is supposed to be signed by Agassiz as President and Newcomb as Foreign Secretary of the Academy, please telegraph me at once at 678 St. John Ave., Pasadena, Cal., which will be my address for sometime to come. I can then wire Agassiz to wait until I hear from you. The reason for haste lies in the fact that several of the societies adjourn about the middle of June for the summer, so we must act quickly if we wish to secure an organization during the present period of solar activity.

I enclose copies of most of the replies to my circular letter, and will send others very soon.

Very sincerely yours,
[*George Ellery Hale*]

Hale Papers, California Institute of Technology

1. Hale's colleagues at Yerkes Observatory and later at Mt. Wilson: FERDINAND ELLERMAN (1869–1940); GEORGE W. RITCHEY (1864–1945) in charge of the construction of the reflecting telescopes at Mt. Wilson; and WALTER S. ADAMS (1823–1946), Hale's principal lieutenant and his successor at the Mount Wilson Observatory, serving from 1923 to 1946.

Thomas C. Chamberlin to George Darwin, June 30, 1906

Dear Professor Darwin:

It gives me great pleasure to respond to your letter of the 17th inst. relative to the extension of geodetic work for the sake of its geologic applications.

As to the political wisdom of urging geological serviceability as a reason for larger appropriations for geodetic work—the essence of the practical side of the question—I should perhaps admit at the outset that what is wise with one legislative body may perhaps be folly with another, and that I can give you only the results of my personal experience and observation subject to this qualification. I have stood in a parental relation to about a score of legislative measures involving appropriations for the promotion of scientific work. The legislative bodies involved however were few and in new and peculiar communities and they may not be representative. Such experience as I have had, however, quite strongly supports the view that a scientific motive, *added* to a commercial one, greatly strengthens the case, and in not a few cases is apparently decisive. A certain percentage of legislators cannot be convinced that the commercial returns from certain classes of work (e.g., such refined measurements as geodesists attempt) are commensurate with the expense involved, and it is hard to convince them that they are not right in thinking that a less close approximation, involving much less expense, would not serve practical purposes equally well. When, however, it is made to appear that work of a high degree of precision, while only slightly, if at all, more serviceable for commercial purposes, subserves scientific ends in correlated fields as well as that immediately involved, they are ready to favor the necessary appropriations. Some legislators are even ready to vote money to advance science when they are hesitant to take similar action for commercial reasons, on the ground, not unjustly taken, that science is universally beneficient, while commercial work is usually helpful merely to individuals or classes, at least in its obvious and immediate applications. I do not think that these individuals are by any means numerous, and certainly they are not a majority of legislators, but I believe their number is underestimated, and that they often hold the balance of power or at least of influence. At any rate I know of nothing that is more genuinely flattering to a legislator than the delicate assumption that he and his constituents are interested in the advancement of science, nor any safer reliance in urging legislation.

I may sum up my views on the practical politics of the case by saying that if the cause of geodetic appropriations were committed to me as an advocate before such legislative bodies as I have been familiar with, I would rely upon the scientific appeal as well as, and perhaps quite as much as, on the commercial appeal, but I should feel the need of both suitably conjoined to make the case a strong one.

I do not need to urge that, in the larger view, the economic, the educational and the scientific, usually blend in a very inextricable and often unexpected way, but perhaps I may cite a fresh illustration growing out of the San Francisco earthquake, which is peculiarly applicable to the present question. The earthquake was caused by a *longitudinal* slip of about ten feet on the average (20 to 0) with an associated drag of undetermined amount. This affected a line some two hundred miles in length. Not only along this line, but in susceptible localities on either side within a tract fifteen miles or so wide the surface was more or less disturbed, and the boundary lines of farms and lots distorted. Some of the land so affected is held at a very high valuation and the reestablishment of just and lawful boundaries is a matter of no small economic importance. There seems to be no way of determining just what the nature and measure of the distortions have been except by geodetic work based on stations in the surrounding region that were not affected. Such geodetic work has been proposed. I do not know whether it will be carried into effect or not, but its possible serviceability at once to land-holders, to the government (by restraining expensive litigation), to seismology, to geology, and, by reflection, to the geodetic work as the source of the benefits, is obvious. Nor does the series end here. Seismology and geology in their turn, when by due aid their evolution shall have reached the productive state, should be able to make invaluable contributions to the material interests, the safety and the peace of the people of this and similar regions.

If, for instance, it shall later be shown, as I think not improbable, that the earth is now in a general way receding from a period of special deformation into one of relative quiescence, and that catastrophic action is on the decline, it will be a contribution of no small value to the comfort of mankind. The public is now very generally depressed by needless apprehension of great impending disasters, if not a universal and final catastrophe, apprehensions derived from the narrow and pessimistic views of the past. From my point of view, which is doubtless a partial one, a contribution of supreme value to the happiness and well being of mankind is likely to grow out of rectified views of the prospective history of the earth to be derived from the prosecution of the earth-sciences. In reaching this consummation I think there is a great function for geodesy, whose contributions, correlated with those of the other earth-sciences, are scarcely less than indispensable. Of course, I do not need to enter into details as to their contributions, but I may venture to express the belief that the ways in which the solution of the fundamental problems of the earth-sciences are to be affected by geodetic contributions are more varied and profound than perhaps any of us now anticipate.

It is, therefore, my quite firm conviction that both from the viewpoint of practical politics and from that of inherent value, material and intellectual, the geological applications of geology may be urged upon legislators with advantage, and that enlarged appropriations for requisite extension of the work may be wisely and hopefully urged.

I thank you for the privilege of giving my impressions in this important matter. I most heartily reciprocate your kind expression of pleasure at meeting in Philadelphia.

Very cordially yours,
[*Thomas C. Chamberlin*]
Chamberlin Papers, University of Chicago

George Ellery Hale to William Wallace Campbell, November 1, 1913

My dear Campbell:

I wish I could supply you with what you wish, but I have never written a general account of the work of the Observatory, and the best I can do is to refer you to our published papers and annual reports, all of which you have, I think. I enclose a MS copy of my last annual report, which kindly return when you are through with it, as it is the only one I have.

The policy of the Observatory remains as at the outset: to devote its main efforts to the study of the physical problems of stellar evolution, giving special attention to the sun, as a typical star, and to laboratory experiments for the interpretation of solar and stellar phenomena. The 150-foot tower telescope is the outcome of our attempt to place the solar work on a laboratory basis, through the use of a long-focus fixed spectrograph, of great stability and constancy of temperature, and the adoption of the means required to give increased sharpness and steadiness to the sun's image. The 100-inch telescope will also be provided with a long-focus fixed spectrograph, to continue the study of the spectra of bright stars with high dispersion, successfully begun with the 60-inch reflector.

The developments of Kapteyn's[1] investigations on the structure of the universe, and your fundamentally important proof of the relationship between stellar motion and spectral type, have made it advisable to defer our main attack on some of the stellar physical problems until a large number of stellar spectra have been photographed for the determination of type and velocity. We are also giving considerable attention to the absorption of light in space, and have found strong evidence in favor of it by two different methods since my last report was written.

The work on the general magnetic field of the sun is advancing very satisfactorily (we now have 18 lines that show the effect), and the working hypothesis of sun-spots briefly outlined in my last published report has been decidedly strengthened by recent work, which favors the horseshoe vortex explanation of bipolar spots.

Professor Einstein of Zürich has written to inquire whether stars can be observed near enough the sun without an eclipse to detect the deflection (0″.84 at the sun's limb) possibly due to solar gravitation.[2] I see no hope of this, but wish you would let him know what you think might be done at eclipses, as Dr. Epstein[3] said you had this under consideration.

With reference to other California observatories, I think of none to add to the list you give in your letter.

With best regards,
Yours very sincerely,
[*George Ellery Hale*]

Hale Papers, California Institute of Technology

1. The Dutch astronomer JACOB CORNELIUS KAPTEYN (1851–1922).
2. This would yield a spectacular confirmation of Einstein's work shortly after World War I.
3. PAUL S. EPSTEIN (1883–1966) was a physicist from Germany who visited the Mt. Wilson Observatory during the summer of 1913. In 1921 he joined the faculty of the California Institute of Technology.

William Wallace Campbell to George Ellery Hale, November 4, 1913

My dear Hale:

My best thanks for yours of November 1st, containing typewritten copy of your annual report. This I shall read within a day or two. No doubt it is of the highest interest. I should like to keep it two or three weeks before returning, but if you should need it earlier, let me know, and I will make the necessary notes at once.

I shall write to Professor Einstein concerning the eclipse problem of recording stars whose rays pass close to the sun, to test the gravitational deflection effect. We have undertaken to secure photographs of this kind for Dr. Freundlich[1] of the Berlin Observatory, using some of the lenses which we employed at past eclipses in search for intramercurial planets. I think the chances for success in this are good. Granted good weather conditions, the principal difficulty will probably relate to an accurate determination of the scale value of the photographs, and probably we shall have to count on the *differential* deflection effect, by comparing the deflection of stars close to the sun with the deflection of stars at a greater distance from the sun.

Yours sincerely,
W. W. Campbell

Hale Papers, California Institute of Technology

1. ERWIN FINLAYSON FREUNDLICH (1885–1964), whose work supported Albert Einstein's theories to the detriment of his own career. Of particular interest are his findings on the orbit of Mercury.

George Ellery Hale to Forest Ray Moulton, May 26, 1914

Dear Professor Moulton:

Absence on the mountain and Ellerman's illness have delayed me in complying with your request. He has now recovered, and will have some prints ready to send you in a few days. If you need other, photographs of the Yerkes Observatory and the 40-inch telescope might be appropriate.

I don't like to appear in the guise of a celebrity, as I am no genius, and what I may have accomplished is merely the result of a very deep interest in science and a lot of hard work. The popular notion, which I have encountered abroad, that ample funds have always been directed my way, would not impress any one familiar with the details of the early history of the Yerkes or Mount Wilson Observatories. The whole project of a solar observatory was dropped by the Carnegie Institution trustees in the autumn of 1903, and all of their funds appropriated for other departments. The point was that they had expected an additional gift from Mr. Carnegie, which was not forthcoming until many years later. So the beginnings on Mount Wilson were of the humblest, and a fair share of the carpenter and other construction work was done with my own hands. This is nothing to brag about, as I had a lot of fun doing it, but I do not like to have the idea prevail that I have had nothing to do all my life but spend money for research. The Yerkes Observatory was a parallel case, as the University was hard pressed for funds and I had to raise money here and there for salaries, instruments, and running expenses, and even to lift the mortgage on the property given for the site.

As you know only one side of me, it is perhaps only fair to say that I am not wholly bound up in astrophysics. Noyes[1] or Osborn or Donaldson[2] could tell you that some other departments of science interest me, and Breasted[3] could add that I have found many attractions in his field. My general library, the collection of which is one of my chief amusements, is strongest in biography (the life and work of men who have done things in my line appeal to me particularly): art and architecture; poetry; history, especially the history of thought and civilization; European politics; Greek art and life; and Egyptology, ancient or modern; truly a motley throng! As president of a local art association I am just completing the organization of an art school, and working on plans for an inexpensive exhibit representing the evolution of art, especially in its earliest development. I have also had a hand in the reorganization of Throop Polytechnic College, which has gone over from a mixed school of low standards and more than five hundred students to a good technological school, with only about fifty students at present.[4]

Outside of such matters, which I have found it possible to do when this infernal neurasthenia has limited scientific work, I enjoy a game of golf or tennis, and especially fly-fishing for trout—the acme of sport in the open, to my way of thinking. The literature of angling is also peculiarly attractive to me.

Not one of these points is worthy of mention, as they do not differ from the ordinary interests and occupations of most men.

As for the Observatory work, I will send you in a few days an elementary account I have been writing lately, to meet a considerable demand for information regarding our object and methods. This is not addressed to scientific men, but may be of some service in illustrating how our policy has been developed.

With kind regards,

Yours very sincerely,
[*George Ellery Hale*]

Hale Papers, California Institute of Technology

1. ARTHUR AMOS NOYES (1866–1936) was a physical chemist. He was to join Hale in California and become one of the builders of the California Institute of Technology. Henry Fairfield Osborn was the paleontologist.
2. The neurologist HENRY HERBERT DONALDSON's (1857–1938) stay at the University of Chicago coincided with Hale's.
3. JAMES HENRY BREASTED (1865–1935) was the great Egyptologist, another colleague at the University of Chicago.
4. Now California Institute of Technology.

Thomas C. Chamberlin to Ernest Willliam Brown,[1] December 13, 1915

My dear Professor Brown:

I want to thank you very cordially for your full letter relative to the irregularities in the rotation of the earth and for the citations you have made, and for your comments thereon. For the last two or three years,

I have been trying to work out a reasonable view of the evolution of the earth's rotation and of its consequences on the growth and segmentation of the juvenile earth under the assumptions of the planetesimal hypothesis. From such considerations as seem to grow out of the cosmogonic conditions postulated and from genetic interpretations of the present configuration of the earth, there seem to be reasons for believing that the rotation of the earth has been quite appreciably irregular and has probably suffered alternations of acceleration and retardation throughout its whole history. The changes of rotational rate were probably more pronounced in its juvenile than in its more mature stages. It is therefore a matter of very special interest to find that even at this late, relatively steady state of the earth's evolution, there are at least intimations of the persistence of such irregularities. It is a great comfort to entertain a definite hope that, with the refinements of observation and computation now attained, the precise nature of the irregularities may be determined, and thru these perhaps their sources.

Before great progress can be made with tidal phenomena, it will be necessary to re-work them from the new point of view now made imperative. The demonstration—if not already determinate, at least foreshadowed—that the earth is highly rigid and elastic, and responds with great promptness to tidal stresses, forces the consideration of water-tides of two types; (1) tides produced by the direct attraction of the tide-producing bodies on the water bodies—the familiar concept of the tides—and (2) tides produced by the tilting of the basins that hold the water bodies, as an incident of the body tides, that is, tilt tides or inertia tides. As these latter are actuated by a body which responds promptly to tidal stress, whereas the attractional tides suffer the lag inevitable in viscous bodies, the two tides are not wholly coincident. The actual tide is composite. The disentangling of the two contributary tides will no doubt add grave difficulties to a problem which, even in its supposed simpler form, has heretofore taxed somewhat too severely the resources of physico-mathematical inquiry. (See "The Tidal and Other Problems," Carnegie Publication No. 107, p. 23.)

It is my judgment, as a merely naturalistic student of the tides, that the tilt tides are the more important. This grows out of a study of certain peculiarities of the tides, among which some of the more declared are the higher tides on the *east* side of the Atlantic than on its west side, and similarly on the *east* side of the Pacific than on the west side, the low tides in the Gulf of Mexico and the Caribbean Sea notwithstanding the converging shores between Cape Race and Cape St. Roque, contrasted with the very much higher tides in the shallow landlocked Hudson Bay. (See "The Tidal and Other Problems," Carnegie Publication No. 107, pp. 31–33.)

When experimental determinations of the Michelson type shall have fixed the values of the lithospheric tides in a sufficient number of well selected localities to give an adequate basis for computation, the tilt-tides can probably be deduced, with a fair measure of approximation, and their effects on rotation derived from these. We are trying to cultivate influential opinion favorable to further determinations of the body tides and would appreciate any good words you may drop as occasion offers.

Pending such determinations, I am making tentative endeavor, by deductive and naturalistic methods, to estimate the possible influences of deformative processes on rotation, and the reactive effect of rotational inertia on such deformative processes. Results so far are encouraging, but they are not yet mature enough to justify much attention from busy men.

Very truly yours,
T. C. Chamberlin

Chamberlin Papers, University of Chicago

1. BROWN (1866–1938), a mathematical astronomer from Britain, served at Yale from 1907 to 1932.

3
Chemistry: Between Theory and Practice

By the turn of the century chemistry in the United States had become a large, amorphous activity. It lacked the kind of prestige which enabled astronomy, for example, to loom large in the national scene and enjoy a measure of renown in the world scientific community. Astronomy's high intellectual status, thanks to Newton, had produced a harvest of observatories. Many were unconnected with universities; even the Harvard College Observatory under E. C. Pickering largely evaded teaching for research. Astronomy depended largely on its own intellectual laurels for support, not on funding from those who felt obliged to support the missions of higher education, industry, or government.

For more than a century chemistry had been recognized as the best example of a useful science in Western Europe and the United States. Industry, agriculture, and medicine all provided occasions for applications. By 1900 there were a large number of individuals who described themselves as chemists to the census takers.[1] Very few were engaged in research, most probably being concerned with routine tasks now assigned to technicians. Yet among the applied chemists were well-trained individuals, some with German doctorates.

The ubiquitous presence of chemical processes in so many areas gave its practitioners entry to the growing body of colleges and universities. Here the chemists encountered the dilemma present in T. W. Richards's exchanges with the Carnegie Institution of Washington. Although providing employment, the great pressure to train chemists created obstacles to the advancement of research, particularly as abstruse research might appear unrelated to practical needs. The leaders of academic chemistry wanted to practice a discipline with a clearly theoretical research component while retaining fruitful ties with present and potential users of their knowledge.

Richards would get a Nobel Prize in chemistry for his work in 1914. Although best known for the experimental precision of his investigations, his unpublished correspondence discloses a great concern with theory and a strong drive for recognition as a theoretician. Richards had studied physical chemistry in Germany with WALTHER NERNST (1853–1932); his work was in the tradition of that specialty. Particularly proud of his theory of compressible atoms, Richards became increasingly hostile to Nernst who gained recognition for the third law of thermodynamics without acknowledging Richards's contribution. At least, Richards thought he deserved more credit.

The correspondence about the offer of a Göttingen chair to Richards also is a continuation of themes in the story of the Carnegie Institution of Washington. Here and in the letter about the offer from the National Bureau of Standards, the drive to replicate one's own kind emerges as a principal motive in the policies pursued by the scientific community. It was not enough simply to increase and to diffuse knowledge. Notable chemists did come from Richards's laboratory, perhaps the most significant being Gilbert N. Lewis.[2]

Another student of Nernst was quite different both in the range of his research and in the locus of his career. IRVING LANGMUIR (1881–1959)[3] also received the Nobel Prize (1932). Almost all of his professional life (1909–1950) was at the General Electric Research Laboratory in Schenectady, New York. The quantity and range of his research output is remarkable. Although a meticulous experimentalist, there is no doubt about Langmuir's theoretical frame of mind. The Nobel Prize was for his work in surface chemistry, but he was notably active in atomic theory and other topics. Toward the end of his life Langmuir became interested in physical limnology and in cloud seeding for weather modification.

The Irving Langmuir who wrote home from Germany protesting his inability to deal with practical matters involving patent rights becomes, in this selection of letters, the General Electric employee who can shrewdly assess the strategic implications to his company of the work of others. Langmuir had notable success in getting patents for his company. Perhaps his most remunerative achievement was the development of the electric light filled with nitrogen, a clear outgrowth of his theoretical studies. Langmuir was the first industrial scientist to get a Nobel Prize. Foreign observers were astonished and impressed by the willingness of some U.S. companies to permit its employees to engage in basic research. Richards at Cambridge, Massachusetts, was in a familiar pattern. Langmuir at Schenectady represented something novel, something peculiarly American in the mixture of the practical and the abstruse.

1. The large number of chemists is briefly considered in the statistical appendix to Nathan Reingold, "Definitions and Speculations: The Professionalization of Science in America in the Nineteenth Century," pp. 55–64 of *The Pursuit of Knowledge in the Early American Republic,* ed. A. Oleson and S. C. Brown (Baltimore, 1976). Also see Edward H. Beardsley, *The Rise of the American Chemistry Profession, 1850–1900* (Gainsville, Fla., 1964).

2. LEWIS (1875–1946) was a lively theoretically inclined chemist who developed a notable Department of Chemistry at the University of California, Berkeley.

3. The authorized biography, Albert Rosenfeld, *The Quintessence of Irving Langmuir* (Oxford, 1969), is informative and chatty but simply fails to grasp what is important in Langmuir's life.

Charles W. Eliot to Theodore W. Richards, July 15, 1901

Dear Doctor Richards:

I have your *Zeitschrift* safe at my house in Cambridge. Where does it belong; at Boylston Hall or at your house in Follen Street? May I return it to your house when next I am in Cambridge, or is that closed?

I thought Dr. van't Hoff a decided acquisition at Commencement. He distinctly contributed to the German quality of the celebration. I was glad to get from him during that warm dinner at the Colonial Club, the details of the facilities given him by the German Government, and of his public duties.

Very truly yours,
Charles W. Eliot

P.S. Professor Jackson has told me of the Göttingen proposal, which strikes me as unique and desirable—especially to decline. I should like to talk it over with you

C. W. E.

Richards Papers, Harvard University Archives

Theodore W. Richards to Charles Robert Sanger,[1] July 17, 1901

Dear Sanger,

I am much moved by your kind letters. It is more pleasant to have one's friends wish one to stay, than to have outsiders wish one to go to them, even; and I thank you for your expression of feeling.

The question is, as you say, a wide one. If I went, it would not be with the intention of spending my life in Germany. I am not sure, too, that I might not have more effect on the chemistry of America by going than by staying. Undoubtedly there would be American students over there; indeed, just those to whom I could be most useful would probably come to me. The unhampered opportunity to form a school of sound precise chemical study is something unique, and unless President Eliot can offer me something similar here, I am much afraid that I must go, at least for a time.

Since writing the above I have had half an hour's talk with the President. The talk had no definite outcome, but the general tendency of it made me think that Eliot would be glad to do his best toward keeping me in America. I think myself that we shall probably strike an acceptable compromise, and that I shall stay in Cambridge. I hope you all will not think me too selfish if I try to get as much freedom as possible.

With renewed kind regards,

Ever yours,
Theo. W. Richards
Richards Papers, Harvard University Archives

1. SANGER (1860–1902) was a chemist at Harvard.

Theodore W. Richards to Charles W. Eliot, July 27, 1901

Dear President Eliot,

Mrs. Eliot has been so good as to let me know of the generous inclination of the Corporation. You already know that my decision has been made; indeed, the letter to Dr. Wallach[1] was sent on the evening of our last conversation.

If I remember aright, the measures which you so kindly recommended to the Corporation were as follows;—

In the first place, more time is to be granted for research. This time is to be gained by reducing the number of my lectures to perhaps three a week, by providing $200 or $300 more for suitable assistance, and by relieving me as much as possible from Committee work—especially from the Admission Committee. In the next place, I am to be promoted to a full professorship, and to receive a salary of $4000. While nothing was said about future salary, I venture to believe that this will be subject to the normal increase with time.

As I said before, these measures seem to me to be very generous. Taken together, they constitute a position so attractive that I have every reason to be satisfied and grateful. Of course, I should feel safer about the future, and freer to buy books and apparatus, if the salary were yet higher; but in comparison with the salaries of other teachers this too is generous. I hope that the considerate action of the Corporation may bear fruit in better work than has been possible in the past.

You spoke of the probable endowment of a new Laboratory. One point in this connection I did not mention, but undoubtedly you have thought of it. The running expenses of a plant as large as that which we desire (unquestionably the largest laboratory in the world) would be very large also. Not only must a number of thousand dollars a year be spent on chemicals and apparatus, but competent service, including a mechanician & glass-blower, should be provided. There should be also a fund for the purchase of chemical books. Our present expenses are largely borne, as you know, by laboratory fees. These fees are a handicap which makes study of chemistry an extra drain on the often limited funds of the students. It is possible that occasionally a man is deterred from studying chemistry by the extra expense. Hence several hundred thousand dollars might well be put into provision for the running expenses of the laboratory, besides the half million for its erection, if ideal conditions are desired.

With kindest regards, and renewed thanks for your interest, goodwill, and generous action,

Yours very sincerely,
[*Theodore W. Richards*]
Richards Papers, Harvard University Archives

1. The Göttingen chemist OTTO WALLACH (1847–1931).

Charles W. Eliot to Theodore W. Richards, July 29, 1901

Dear Doctor Richards:

The measures which the Corporation are prepared to adopt on your behalf are correctly stated in your letter to me of July 27th. They are as follows: a full professorship, and a salary from September 1st, 1901, of $4,000.; reduction in the weekly number of your lectures, the new number being not definitely stated, but physical chemistry to remain in your charge, and you to be brought into contact in the lecture-room with a considerable number of students as well as with the few students who work in your laboratory; better pay for the assistant, say four or five hundred dollars; and as much relief as possible from administrative work. As to future increase of salary, it is supposed that you will have the usual increase of $500. every five years, or thereabouts, till the salary reaches $5,000. I am obliged to say "or thereabouts" because the Corporation has no definite rule on the subject, but only a tolerably well-established practice.

I entirely agree with you that a large endowment would be almost indispensable for the satisfactory carrying on of a new and adequate chemical laboratory for the University. For a building costing $500,000., the endowment should not be less than $250,000. I think benefactors and boards of trustees are beginning to understand that a building by itself, if not endowed, may be an embarrassing gift. This was the reason

I gave some account to the assembled alumni of Mr. and Mrs. Nelson Robinson's thoughtfulness in managing their benefaction.

Sincerely yours,
Charles W. Eliot

Richards Papers, Harvard University Archives

Theodore W. Richards to James McKean Cattell, August 13, 1901

My dear Professor Cattell,

According to your recent request for intelligence about University affairs, it may interest you to know that I have recently been invited to fill a new professorship of inorganic chemistry in the University of Göttingen. The position was to be entirely free from routine teaching, and was to be devoted entirely to research, with the assistance of such advanced students as might prove acceptable. If you should care to make a note of this fact, I think that it might be worth while to point out the broad-mindedness and freedom from prejudice which the Germans exhibit by this action, as well as by the calling of van't Hoff, another foreigner. The nature of these two positions is an index of the growing importance assigned to research in Germany. This phase of the question also is worthy of discussion.

I had almost forgotten to say that I have decided to remain at Harvard, on account of the generous attitude of the President and Fellows of Harvard College.

Yours faithfully,
Theodore Wm. Richards

In haste.

Cattell Papers, Library of Congress

Charles W. Eliot to Theodore W. Richards, August 26, 1901

Dear Doctor Richards:

I propose to relieve you immediately from the Board of Advisers and the Committee on Admission.

The enclosed letter is a very pleasant one, and the paragraph on the fourth page particularly so. To the best of my knowledge and belief, yours is the first invitation given an American to a full professorship in what he calls the Old World. The case of your brother-in-law Gregory is entirely different; and the professorial career abroad of Fitz-Edward

Hall, Harvard A.B. 1846, began in India. I have not been able to recall any other instances than these two.[1]

Very truly yours,
Charles W. Eliot

Richards Papers, Harvard University Archives

1. HENRY DARWIN ROGERS (1808–1866) had a chair in natural history at Glasgow in the last decade of his life. B. A. Gould, the astronomer, received an offer from Göttingen. Fribourg made an offer to E. W. Hilgard, the agricultural chemist. CASPER RENÉ GREGORY (1846–1917), a Harvard graduate, became a New Testament scholar. He was on the faculty at Leipzig for 33 years and became a German citizen; in 1914 he volunteered for the German army as an enlisted man. He became a lieutenant and subsequently died in combat. In 1862 HALL (1829–1910) became a Sanskrit scholar and expert on English usage at King's College, London.

Charles W. Eliot to Theodore W. Richards, January 10, 1902

Dear Dr. Richards,

I was so occupied with meetings of one kind and another yesterday that I was unable to answer your letter of January 8th. It seems to me that you might reasonably accept the appointment in the National Bureau of Standards, provided that you are not under contract to stay until June, 1904. You ought to be free to drop the job, if for any reason it does not agree with you. As to the healthiness of Washington, for Yankees, I cannot at all agree with Dr. Stevens. Possibly your Philadelphia experience may have prepared you for the hotter climate of Washington. I hope you will keep firmly in mind the general principle that a scientific investigator should keep in contact all his life with students, in order to found a school and have worthy disciples to carry on his work.

Sincerely yours,
Charles W. Eliot

Richards Papers, Harvard University Archives

Frederick Soddy[1] to Theodore W. Richards, September 29, 1902

Dear Sir,

In the recent paper by yourself and Mr. Merigold on the Atomic Weight of Uranium[2] I was interested especially in your examination of the Radioactivity of your pure preparations. It is of interest to know whether your observations can be accounted for by the subsequent developments that have recently been made in the subject,—viz. the reproduction of the active constituent after separation, and the inactivity of the nonseparable radiation to the photographic plate—or whether your refined methods had succeeded where previous attempts had failed, in actually removing a specific radioactive element from the uranium, analogous to radium in pitchblende.

I trust you will therefore pardon my bringing the matter to your notice in this letter, in the hope that you and Mr. Merigold would have no objection to our submitting a sample to a quantitative investigation. Professor Rutherford[3] has all the facilities here for such an examination and if you can send us a sample, the point could easily be settled. The oxide U_3O_8 would perhaps be the most suitable compound, and a very small quantity would suffice for a rough examination.

A complete examination of both types of rays would require from 2 to 5 grams. Of course the sample will be returned to you intact, and I will undertake that it suffers no contamination in the investigation. The approximate date of preparation should be known.

I enclose copies of papers[4] by Professor Rutherford and myself bearing on the subject, and have the honour to remain,

Your obedient servant,
Fred[k] *Soddy*

Richards Papers, Harvard University Archives

1. SODDY (1877–1956) was then at McGill with Rutherford, at Oxford from 1919 to 1936. He won the Nobel Prize in 1920.

2. Both men at McGill and B. B. Boltwood at Yale were pursuing topics in radiation chemistry. See L. Badash, ed., *Rutherford and Boltwood: Letters on Radioactivity* (New Haven, 1969). BENJAMIN SHORES MERIGOLD (1873–1962), a Harvard Ph.D. (1901), taught chemistry at Clark University 1908–1946 after a brief stay at Worcester Poly.

3. ERNEST RUTHERFORD (1871–1937).

4. Cited below signature as "Trans. Chem. Soc. *81* (1902), 837 and 860."

Irving Langmuir[1] to Sadie Langmuir, November 1, 1903

Dear Mama,

At last everything connected with the University is in running order. I have been working every afternoon in the laboratory and in the mornings I heard lectures. At first I had a lot of trouble deciding what lectures to hear. There are so many that I want to attend and so few that I have time to attend, that the task was not easy.

The first few days I went to a good many different lectures to see about what the courses were like. A good many I found much too advanced for me. One course in particular, on the general theory of electricity, started out the second day with so much mathematics that I was completely at sea.

So I have finally come down to more elementary courses:—[2]

1 Nernst. Theoretical Chemistry with special reference to the Molecular Theory 3 hours per week
2 Nernst. Theory & Practice of Accumulators or Storage Batteries 1 hour per week
3 Prof. Klein a general course on Diff & Int. Calculus. 4 hours per week.
4 Prof Minkowski
Mechanics & theoretical Physics 4 hrs p[er] week

I wanted to take some courses on Organic Chemistry but this semester there are none except two very advanced courses. Prof. Wallach is giving a good course on General Chemistry Inorganic part, but it comes at the same hour as Nernst's lectures so I decided not to take that.

This semester then I am going to study Organic Chemistry at home & review Inorganic Chemistry, so as to pass the Vërbands Exam. which most of the students say is very hard. Then the rest of the time I am going to study calculus so as to be able to understand some of the many fine lectures on special branches of science.

I am going to write Arthur soon & send him a list of the lectures given here & describe the work I am doing here more in detail.

The life here is so totally different from that in America that it is always interesting. Instead of working all day in college as I did at home and only studying during the evenings, here I have an hour every morning for study and all day Wed. and Saturday for there are no lectures on these days that I want to hear.

Then at Columbia I had to study certain definite subjects as preparation for recitations, but here there are no recitations so you are left entirely to yourself to decide what to study. I have bought Kiepert's Calculus a book in 2 vol. with 1400 pages and then Voigt's Theoretical Physics which latter is so chock full of the most difficult mathematics that I certainly cannot read much over a page an hour, and this book has 1410 pages requiring 1410 hours to read so I feel as though I have quite some work ahead of me.

I take my dinners at Herr Schlote's, a pension. When I started there, there were no Americans but Clement;[3] but now there are two others, but we all speak German.

Yesterday, Sunday, Tufts, Hutchinson & myself were to take a walk to the Haustein & Teufelskanzel, but the weather was just as it always has been here, that is hazy cloudy drizzly weather with a humidity of at least 100%. So we decided to wait. In the afternoon a few little patches of blue sky appeared so I took a very muddy walk of about ten miles mostly through fine woods.

I was surprised to find how many pretty paths there are. The woods were perfectly beautiful with the dark reddish-brown beech leaves and other bright yellow leaves contrasted with the evergreens.

I have sent to Munich for a pair of Norwegian snowshoes (skis) and when we have snow here I will try to learn. From what I heard this summer it is one of the most glorious sports imaginable.—travelling over deep snow about 6–7 miles per hour on the level and 10–20 miles per hour down hill. But it is very difficult to learn. It has become a very popular sport in Germany within the last few years. They say the Hartz is full of "Schneeschulaufer" every Winter. Tufts and Hutchinson are also going to get "skis" so we can continue our Sunday walks or slides.

The skating here is not very good I think; they flood a good sized field nearby & have skating most of the Winter on this, but I have been spoilt for that sort of skating by the grand stretches of the Hackensack & Rockland lake.

Herbert's letter is received & will be answered in a few days.

Yours affectionately
Irving

Langmuir Papers, Library of Congress

1. This letter to Sadie Langmuir, like the ones that follow, were written from Göttingen.
2. Nernst, Felix Klein, and Hermann Minkowski—an impressive trio indeed for an ambitious graduate student. KLEIN (1849–1925), a mathematician at Göttingen, opposed the increasingly abstract nature of mathematics. MINKOWSKI (1864–1909), a mathematician also at Göttingen, was known for his work in geometry.
3. JOHN CLAY CLEMENT (1880–1942[?]), a physicist, received a Ph.D. from Göttingen in 1904. During World War I, Clement became an ordnance officer in the U.S. Army, retiring as a full colonel shortly before his death.

Irving Langmuir to Sadie Langmuir, June 13, 1904

Dear Mama,

This week I have been working on my arbeit night & day practically. I had no idea I would find it so exceedingly interesting. I have really been very lucky in getting such an arbeit & particularly in having Nernst so much interested in it himself. As I wrote last week, he set the apparatus up in his own lab. & for the first three days I worked with him most of the time. Week before last I had not seen Nernst at all, but was spending all my time reading all the literature he had given me

& in working over the part that I had had the argument about, so that I could know what I was talking about when I saw him again. Monday when I saw him he said he had started laboratory work on my arbeit on his own account as he wanted a few general results for a paper he is going to publish soon. He had set up the apparatus & had made only one preliminary experiment. So I made the rest of the experiments of that set. They consisted of leading a current of air over the filament of a Nernst lamp & then determining the quantity of Nitric Oxide found. I made four such experiments keeping all other conditions constant except the velocity of the air current & from these found that the slower I passed the air over, the more nitric oxide I got & almost in proportion. Now what I want to find is what the maximum possible quantity is. If I led it over a thousand times as slow or ten thousand as slow it would not make any difference in the amount formed, to speak of, for in such a long time as that would take, the reaction would have had time to go as far as it ever would. From these four results I calculated about what this maximum value would be & found that it would be a very slow process to get an accurate result by this method, so Nernst proposed using sealed bottles & leaving a filament in them for a certain length of time & then analysing the gas. Nernst then went to work & passed air through his iridium furnace heated to 2000° (or even 2100° in one experiment) & cooled the gases down from this temperature very rapidly by passing them into a capillary quartz tube. This gave him .8% of the air changed to nitric oxide whereas my results had so far only shown .3%, although the filament had a temperature of 2200°C.

Last Saturday I started to set up apparatus for experiments by myself, & have everything ready now to begin to-morrow morning. Nernst comes around to see me every little while & sees that I get all the apparatus I need, while most of the other students have to wait their chance to see him once a week.

These experiments on air form only the first part of my arbeit, & the simplest part. The work that Nernst is doing with his electric furnace will give me a basis on which to work. Otherwise I would have had to start with steam instead of air & the results would be much less satisfactory. The arbeit is one very sure to give good results & it may result in the development of one of the best methods of finding equilibrium of gases at very high temperature. As yet no one has worked on the subject at all as far as I can see, except from an entirely false point of view.

My arbeit or at least the part of it that I am now working on, has been suggested by Clement's arbeit, who found that no ozone is formed in air by heat alone no matter how intense it is & that what many great chemists & physicists have claimed to be ozone is really nitric peroxide.

The arbeit ought also to have a technical importance for it should lead to a much better understanding of the production of nitric acid from the air as is being done at Niagara Falls now.

I don't suppose you have understood very much of what I have said, but can, perhaps, see that I am mighty lucky & that I am mighty well satisfied all around with the arbeit. When I get along a little further I will write Arthur & tell him just what I am doing. I don't believe he gets the faintest idea of the real object of the research from what I have written. I am sure I wouldn't.

Nernst will probably not be so much interested in my work when he finishes his along the same line. He says when he publishes his paper he will write that I am continuing the researches along that line, so that, as he says, I may have a free field. This is the great advantage of being at Göttingen rather than at Leipzig, for there I would never see Ostwald, to speak of, & with so many students no one would take any interest in any one arbeit.

Before I left America people told me that Nernst was spending all his time automobiling & working on making the lamp a commercial success. Well, he is doing both, but does not spend one tenth of his time at them. He does very little experimental work himself. In fact I do not think he is a good experimenter, but he has a lot of assistants some of whom work on improvement of the lamp, but most of whom seem to be carrying certain investigations under his direction. He goes around from one to the other & does a little work himself when it particularly interests him. Nernst's ability lies principally in drawing good conclusions & having an enormous number of brilliant ideas. When talking with the different men about their arbeits he will make a lot of suggestions, many of which are not good, but some of which are exceedingly good. In his own work he gets a tremendous number of ideas & tries them all; a few prove of value.

Yesterday I took another trip to the Harz, with Gillespie:[1] didn't go very far. The weather was good & the trip very pleasant, but I am getting a little tired of going to the same region every time. I only went there this time because Gillespie had never been there. There are so many other pretty places around outside of the Harz that it is rather a shame to go there so often. I think from now on I will leave the Harz till next Winter unless someone wants me to go with them. The trouble is that I have a sort of reputation as a guide to the Harz, I think.

Well, this is the last letter I will write you to America. I hope you have a fine trip over. I wonder if you are going to stop at Gibraltar. If it is not already too late I wish you would bring over the leather bag of Papa's, which I ought to have brought with me last year. And please bring an eye-cup.

Bon Voyage!
Irving Langmuir

Langmuir Papers, Library of Congress

1. DAVID C. GILLESPIE (1877–1935) was a mathematician, later at Cornell.

Irving Langmuir to Sadie Langmuir, October 31, 1904

Dear Mama,

Your letter asking for information about the Nernst lamp for Mr. Young[1] came a few days ago. I have inquired about a few points that I was ignorant of, but have not had an opportunity to make any careful investigations.

The lamp has been manufactured & for sale for a long time in America. The Westinghouse Electric Co have the American rights I believe. So I very much doubt whether there would be any chance to get any useful concessions from Nernst.

As I understand the matter Nernst sold out the patents to the Allgemeine Electrische Gesellschaft for about one million dollars & at present has nothing whatever to do either with the manufacture or sale of the lamps.

He is however still trying to improve the lamp by finding some mixture for the filament which will enable the lamp to light itself quickly without the use of the present mechanical contrivance.

So far I do not think he has met with any success in this line.

The Allgemeine Electrische Gesellschaft has been advertising the lamp very extensively in the electrotechnical journals. They offer it principally as a substitute for the arc light for interior lighting. It is particularly adapted to lighting stores & public buildings, in general for all large rooms which are in constant use. In its present form I doubt very much whether it will ever become popular for household purposes.

The lamp itself consists of a cylindrical filament, about 3/4 inch long, for a 220 volt circuit or double that length for 440 volts, made of a mixture of oxides of several rare elements. The diameter of the filament varies with the candle power of the lamp, being a little less than 1/32 inch for a 50 candle power lamp with 220 volts.

This filament, being non-combustible, does not need to be enclosed in an exhausted glass bulb as in the case of the carbon filament of the

Edison lamp. It has the great advantage that it can endure a very much higher temperature than the carbon filament without injury, but it has the very great disadvantage that when it is cold it is a non-conductor of electricity and it must be warmed to a dull red heat by some external means before enough current can go through it to cause it to light itself. This makes it necessary to have a coil, made of porcelain, wrapped around with fine platinum wire. When the switch is first turned on, to light the lamp the current passes through the heating coil, which by radiation gradually heats up the filament until the lamp lights. This takes from 30 seconds to two minutes. As soon as the lamp is lit an electromagnetic device turns the current off from the coil, thus introducing a part which although simple, may get out of order. In this respect the lamp then is better than the arc lamp but worse than the Edison.

The two great advantages of the Nernst lamp are

1st Very pleasant light, being probably more nearly like daylight than any other practical artificial light.

2nd Low consumption of current. About 1.8 watts electrical energy are needed per candle power as against 3.5 watts for the Edison lamp.

The lamp itself, because of the expensive heating coil & electromagnets, costs a good deal, but does not wear out. The only part needing replacement is the filament which costs about the same as an Edison lamp & has about the same life.

Tell Mr. Young that if there are any other points he would like to know about I will be glad to find out from Nernst, or others here, all I can about them.

But as for obtaining the right of sale in America or making any similar arrangement with Nernst or the Allgemeine Electrische Gesellschaft, I feel that I am entirely unsuited for that. It would not be scientific work in any sense, but a pure business proposition & in matters of that nature I am absolutely inexperienced. I know of no improvements which I could make in the lamp. That would probably be possible only after years of investigation and practical experience in problems of electric lighting.

This last week I have gotten well started in my arbeit & have not got much better apparatus than last year so that an experiment takes a much shorter time than before.

As yet I have only developed about two dozen of this summer's pictures. They all came out satisfactorily & a few of the 5 × 7s are really most excellent pictures.

Yesterday (Sunday) I went to the Harz again leaving Göttingen 7:41 A.M. & getting to Sonnenberg for a meeting of the Oberharzer Ski-klub at about 5:00 P.M. The weather was glorious although in Göttingen there was fog & low clouds all day. We had an interesting & lively business meeting. I got to bed about 12 & three o'clock in the morning got up again, walked 6 miles to Andreasberg & took the train to Göttingen getting there in time to have breakfast & attend a 9 o'clock lecture.

Tomorrow night I am invited to Nernst's to supper. Am sorry that it conflicts with a fine symphony concert which I wanted badly to hear.

Since you left Hanover there has been in Göttingen one bright day, six days which were mostly cloudy but the sun appeared hazily for a few minutes, and eleven days on which the sun or blue sky were not visible once. Most of this time the weather has been fine in the Harz.

Yours affectionately,
Irving

Langmuir Papers, Library of Congress

1. Perhaps this is Owen D. Young, then in private practice, and later counsel for General Electric (1913–1922) and chairman of its board (1922–1939, 1942–1944).

Willis R. Whitney[1] to Irving Langmuir, February 24, 1909

Dear Dr. Langmuir:

Yours of February 22nd has just been received and we are glad to know that our suggestion met with your approval. I can readily see that you could interest us a great deal in the matter of the formation of ozone, and I would be glad if you could talk to us on that subject on the date you suggest—March 13th.

Our talks are entirely informal, and I should not advise you to prepare more than simple notes for reference, as it is much more interesting if the talk can be in a conversational style. There are usually from fifteen to thirty men at the Colloquium, and our average knowledge of such a subject would be about that to be expected from a graduating chemist of one of our colleges.

Yours very truly,
W. R. Whitney

Langmuir Papers, Library of Congress

1. When WHITNEY (1868–1958), the head of the General Electric Laboratory at Schenectady, wrote this, Langmuir was teaching at Stevens Institute in Hoboken, N.J. A physical chemist who also had been trained in Germany, Whitney believed ardently in the value of industrial research. He left M.I.T. for G.E. to found what became one of the great industrial research laboratories. In World War I he tried to improve education and

research in physics and chemistry by promoting the unsuccessful movement to establish engineering research stations at the state universities using the model of agricultural research stations already functioning successfully.

Irving Langmuir to Willis R. Whitney, March 15, 1909

Dear Dr. Whitney:—

After seeing the Laboratory and some of the work being done there I am more desirous than before to put in a couple of months work with you this summer.

Work on colloids would be very attractive to me altho I have not done any work in that line as yet. If you could let me know in advance along what line you would wish me to work, I would gladly study up the literature in that field and be better prepared to take up active experimental work on arriving in Schenectady.

An investigation into the properties of metallic uranium or of some other of the rare elements would, I believe, be a field in which I could obtain useful results.

My expenses on the trip to Schenectady were just ten dollars.

Thanking you for your kindness and hospitality at Schenectady, I remain

Yours sincerely,
[*Irving Langmuir*]

Langmuir Papers, Library of Congress

Charles W. Eliot to Theodore W. Richards, October 22, 1910

Dear Mr. Richards:

I enclose a copy of the letter you wrote me about Henry Adams's book entitled "A Letter to American Teachers of History". Henry Adams's brother Brooks, who joined his brother in Europe this summer, undertook to ascertain whether Henry would like to have the book reviewed by a competent chemical physicist; but I have not yet heard from him on that subject.

In any explanatory letter which you write about it, will you please mention the fact that the book was privately printed, and therefore is not a subject for public criticism. It is not for sale, and if it had not been for Brooks Adams I should never have seen it.

Sincerely yours,
Charles W. Eliot

[October 22, 1910, enclosure to Richards of his June 10, 1910, letter to Eliot]

To me it seems that the apparent contradiction between the second law of Thermodynamics and Evolution is not really a contradiction. The second law says only that "heat cannot *of itself* go from a lower temperature to a higher temperature." But with the help of some other manifestation of energy a large quantity of heat by falling through a small temperature range may raise a small quantity to a very high temperature—for example, a steam engine by driving a dynamo can raise a Tungsten lamp to 1800°. The case is somewhat (although not exactly) parallel to the working of a hydraulic ram, where gravitation apparently works against itself with the help of inertia.

It seems to me perfectly plausible, therefore, that while the continual and inevitable degradation of energy proceeds on its course, small quantities of energy may be raised to a very high potential in other forms. Now life is a manifestation of chemical potential, for the living creature is a chemical machine; and just as the efficiency of the hydraulic ram depends upon its construction, so the efficiency of the brain must depend upon its chemical structure (not upon its weight or outward form). This chemical structure might well increase in potential in succeeding generations, at the expense of large amounts of solar energy, without contradicting the second law. The act of thinking is the running down of this potential; the loss must be built up by chemical energy taken in from outside.

Adams confounds the *total* degradation of one kind of energy with the individual *degradation* of another.

I should like very much to talk it over with you. When can I see you?

Yours most sincerely,
T. W. Richards

Richards Papers, Harvard University Archives

Irving Langmuir to Willis R. Whitney, February 6, 1911

Dear Dr. Whitney:

Experiments made during the last few months have led us gradually to the conclusion that the life of a tungsten lamp at any given efficiency is limited by the rate of evaporation of the tungsten from the surface of the filament, and that the present tungsten lamps have already nearly reached this limit. It appears that no improvement in the vacuum is capable of prolonging the life more than a small amount.

Future improvement in the life (or efficiency) of the tungsten lamp must therefore depend upon the prevention of evaporation of the metal or upon the removal of the bad results of such evaporation. Since the filament must be heated in a vacuum, it is hard to believe that it is possible actually to prevent evaporation. The following suggestion of Mr. E. Q. Adams,[1] however, would seem to offer a way by which the bad effects of the evaporation could be entirely eliminated, thus opening up new possibilities for the tungsten lamp.

The suggestion is, that there be introduced into the lamp a compound of tungsten, volatile but stable at the temperature of the bulb and unstable at the temperature of the filament. It is also necessary that the compound be readily formed from its elements at the temperature of the bulb. There seem to be several compounds that would answer this description; perhaps the most promising is tungsten flouride WF_6, which is a gas condensing to a liquid at 19°C (at atmosphere pressure). Tungsten undoubtedly reacts vigorously with gaseous fluorine at room temperature. Further, since this compound would form with a very high heat of reaction, it must become much less stable at very high temperatures and would probably be dissociated into its elements at the temperature of the filament (like the other tungsten halides).

In operation, the tungsten fluoride would act as follows: The filament, running at a very high efficiency (probably about 0.4 W.P.C.), would react with the small amount of WF_6 present and decompose it into tungsten metal which would deposit on the filament and fluorine which would stay in the space or be driven onto the glass. Dry fluorine gas almost certainly will not react with dry glass, for it is known that dry HF will not do so and fluorine should be still less active in this regard. The tungsten of the filament would gradually evaporate and deposit on the bulb where it would immediately combine with the fluorine and form WF_6, which would again be decomposed by the filament, etc. In this way, although the evaporation of the tungsten from the filament would not be prevented, the tungsten would be returned from the bulb to the filament; so the lamp, as far as can be seen, would have an indefinite life.

A great deal of experimentation will probably be necessary before a successful lamp of this type will be made. There are many different compounds that can be tried and, in case of failure with tungsten, success may be had with tantalum or molybdenum, so the chances of ultimately producing a lamp involving this principle seem very encouraging. I would therefore recommend that the matter be referred to the Patent Department.

Yours very truly,
[*Irving Langmuir*]

Langmuir Papers, Library of Congress

1. ELLIOT Q. ADAMS (1889–[?]), a Ph.D. (1914) from the University of California, Berkeley, who spent most of his career in industrial research.

Emil Fischer[1] to Theodore W. Richards, December 23, 1911

Dear Friend,

Today the Christmas vacations started here and thus I have an hour of leisure to talk with out-of-town friends and to take the opportunity to send them my best wishes for the New Year.

The past year has not been a happy one for Europe since we were troubled by heavy political fears and an extremely hot summer which caused heavy damage to the crops.

I do not know what will happen to the German-English relations. Here, day by day, one is more afraid of a warlike development, which in my opinion would be a real disaster for Europe.

In contrast, America is in a lucky position since political problems are apt to arise only from within, while Europe, at the present time, resembles a powder keg, where only one small spark would be necessary to cause an explosion.

My sincere congratulation to the great success you have had with the Faraday-Lecture.[2] I read your paper with great interest and I am sure that you will find many more interesting things during the practical experiments of the problems that you discussed.

Unfortunately, I will be unable to attend the International Congress in New York next year since both the trip and the many social activities are too strenuous for me. In my work I am still productive, but I am lost as a representative. I explained this also to Dr. Nichols,[3] who visited with me and invited me very graciously.

Your move into your new laboratory next year should make you very happy since you undoubtedly provided it with all thinkable resources.

Our own Kaiser Wilhelm Institute for Chemistry, which also will be dedicated solely for research, is expected to be opened next October. Beckmann from Leipzig, Willstaetter from Zurich and Otto Hahn[4] from Berlin will be appointed for this Institute.

Furthermore a second Kaiser Wilhelm Institute for physical chemistry is under construction and F. Haber[5] from Karlsruhe is supposed to work there.

As you can see Germany is providing some money for scientific purposes in spite of the enormous expenses for war-preparations, since both these institutions are being financed totally with private contributions.

I, myself, have been working diligently this winter and hope to be able to publish some interesting observations soon. It involves some new reactions in the carbohydrate chain, furthermore about tannic acid and medical questions.

Also, I continued with the stereochemical experiments and am now convinced that van't Hoff's descriptions are useful only in the field of stereochemical statics, while they are a total failure in dynamics.

My 3 sons are presently here together and we do hope to have a very merry Christmas, although it will be just the family.

The oldest will soon take his examination as a doctor of chemistry. The second one is studying medicine and the third one will be graduating from school next September and will most likely study physics.

Kind regards from all of us, including Miss Barth,

Yours truly and faithfully
Emil Fischer

Richards Papers, Harvard University Archives

1. Translated from the German. FISCHER (1852–1919) was perhaps the greatest chemist of Germany's great age of chemistry, a Nobel Prize winner with a great range of interests.

2. "The Fundamental Properties of the Elements" which appeared in 1911 in the *Journal of the Chemical Society of London, Science,* and *Annual Report* of the Smithsonian Institution.

3. Perhaps WILLIAM HENRY NICHOLS (1852–1930), the chemical industrialist.

4. ERNST OTTO BECKMANN (1853–1923), RICHARD WILLSTAETTER (1872–1942), and HAHN (1879–1968) were a very impressive trio. Willstaetter and Hahn would later receive Nobel Prizes.

5. FRITZ HABER (1869–1934), who received the Nobel Prize in chemistry in 1919, played a key role in the German war effort in World War I and in the politics of German science until the advent of Hitler.

Irving Langmuir to Owen Richardson,[1] June 18, 1914

My dear Richardson:

I have delayed answering your letter of April 21st until I knew something more definite about the thermionic emission from carbon at high temperatures.

During the last few months we have been continually at work on electron emission and have obtained a great many interesting results. We are more convinced than ever of the general correctness of the theory that I put forward in the Physical Review paper, and more explicitly, perhaps, in the two papers in the Physikalische Zeitschrift.[2] It needs, however, a few extensions. I have never yet found a case, except with the Wehnelt cathode, where positive ion bombardment has increased the electron emission from a hot cathode to a value higher than that characteristic of the pure metal. In case, however, the electron emission from the metal has been reduced by presence of a film, it is then fairly often observed that the presence of gas will increase the electron emission slightly—the more so, the higher the voltage. We have been making quite a detailed study of some of the lag effects observable in the presence of gases, and in this find the best possible confirmation of the formation of films on the surface of the wire. For example, in nitrogen at very low pressures and a cathode at rather low temperatures (say, 1800°K.), we find that increasing the voltage on the anode first of all increases the current slightly, and then, after 10 or 15 seconds, starts to decrease it until it may become a small fraction of its former value. When the filament is brought into any particular condition in this way, this condition is not altered if the filament is allowed to cool and the vacuum is changed and then it is heated again to the same temperature as before. With oxygen, when the bulb is cooled in liquid air, there are some very remarkable effects produced in this way. I hope later to be able to publish these results in more detail.

We obtained interesting results with CO, and found that with this gas, electron emission from tungsten is cut down generally to about 1/3 of its value in vacuum. The interesting fact developed, however, that a filament which had been treated in this way was no longer sensitive to the effect of nitrogen or to small amounts of water vapor or oxygen. We suspected that this was due to the presence of carbon on the surface of the filament, and have thoroughly verified this by heating the filament in low pressure in methane. Hydrogen is evolved and the carbon is taken up by the filament, as is readily shown by the increase in resistance of the wire. After carbon has thus been introduced, we find that small amounts of oxygen slightly *increase* the electron emission, until finally a point is reached where a further addition of oxygen causes an enormous decrease. This occurs when all the carbon on the surface of the filament has been oxidized. The behavior with nitrogen is similar. We thus conclude that carbon on the surface of the filament decreases the electron emission, but only slightly, and that as long as carbon is present, oxygen or nitrogen react only with the carbon and do not, therefore, decrease the electron emission from the tungsten.

In these experiments, as well as in many others which I might describe to you, we have the best of evidence that chemical reactions in the case of a tungsten filament at least, do not cause more than a negligible electron emission, in comparison with that due to thermionic emission.

Personally, after having read over Fredenhagen's papers on the thermionic emission from the alkaline metals,[3] I feel that there is no evidence that your originally high values are due to chemical reactions. It seems to me that Fredenhagen's results are abnormally low because of secondary effects.

In the case of tungsten, we find that certain impurities present in the metal may cause an enormous increase in the true electron emission from the metal, and it may be that in many other cases the presence of other impurities which normally have a high electron emission may cause irregular results, but I have never yet been able to find any evidence that the electron emission due to chemical reactions is anything more than a very small effect which can cause an electron emission corresponding to much less than the electrochemical equivalent of the reacting substance.

In regard to the electron emission from pure carbon filaments (that is, metallized carbon filaments which are of high purity), our most recent values have given a current of .0028 ampere per square centimeter at 2000°K., and value of B of 48,700. I think that our former value of 32,000 for carbon is undoubtedly incorrect.

I am enclosing some reprints of recent papers published in the Physikalische Zeitschrift.

Yours very truly,
[*Irving Langmuir*]

Langmuir Papers, Library of Congress

1. RICHARDSON (1879–1959) won the Nobel Prize in physics for his work on thermionic emission of electrons. From 1906 to 1913 he taught at Princeton, then returned to his native England. One of his sisters married Oswald Veblen; the other married C. J. Davisson who later won the Nobel Prize in physics.

2. "The Effect of Space Charge and Residual Gases on Thermionic Currents in High Vacuum," *Physical Review*, 2d ser. 2 (1913): 450–486.

3. K. H. H. P. FREDENHAGEN (1877–1949) was a student of Nernst. He held a professorship at Griefswald from 1923 to 1949.

Irving Langmuir to Willis R. Whitney, August 9, 1915

Dear Dr. Whitney:

Last Friday, Mr. Armstrong,[1] Professor Pupin's[2] assistant, came to the laboratory to try to obtain some pliotrons for experimental purposes, but more particularly to impress upon us the desirability of not concluding any negotiations by which Pupin would be permanently prevented from obtaining pliotrons.

He stated that Pupin had definitely solved the problem of eliminating static and that this had been proved both theoretically and experimentally. He outlined to me the method by which this has been accomplished. It consists essentially of the use of an untuned antenna containing a resistance of about 500,000 ohms in series with a negative resistance. The negative resistance has a maximum negative value of about 500,000 ohms for some definite frequency (that of the signal to be received), but for other frequencies its negative value is much smaller. Another feature of the negative resistance is that it takes a certain time for it to develop its full negative value. With this combination, when a static pulse, or other disturbance of a frequency different from the signal, falls upon the antenna, its energy is consumed in the 500,000 ohms resistance. On the other hand, when a signal of sufficient duration and proper frequency is received, it meets with practically no resistance in the antenna, for the 500,000 ohm resistance is balanced by the negative resistance.

Armstrong did not disclose in detail the nature of the negative resistance, but told me the following about it. He says they have two methods of obtaining a suitable negative resistance, one by mechanical means (rotating machines) and the other by means of vacuum tubes. The mechanical means has a theoretical advantage, in that there is no limit to the degree to which static may be reduced in intensity. The practical disadvantages of the mechanical method are the cost inflexibility and delicacy of the machine used. In the second method they have been using experimentally six De Forest audions with very good results, but the fluctuations in the characteristics of the tubes are so great that two men are needed to keep them in adjustment. He says that from the curves we have published on the characteristics of the pliotrons, two pliotrons would be equivalent to six audions and he hopes that the necessity for constant adjustment would disappear.

From what Mr. Armstrong has told me, I feel confident that they have really succeeded in reducing the strength of static relatively to the signals. The introduction of a high resistance in the antenna has appeared to me for a couple of years to be the first requirement of any method of eliminating static. The use of the "tuned" negative resistance seems to make this successful. During the last year Dr. Hull[3] has devised and experimented with a negative resistance of an entirely novel type which we have used in connection with pliotrons for receiving signals. Our negative resistance, however, has a value independent of frequency, so that it cannot be directly used in the way described by Armstrong.

If it is true that Pupin has been successful in even partially eliminating static, the importance of the result cannot be over-emphasized. This problem has been worked on for many years by many wireless engineers, but so far entirely without success. The methods of amplification which have recently been developed (audion and pliotron) have not been of very great commercial importance for this reason: static has always been amplified at least as much as the signal. The only means of extending the range of radio communication has been the use of more and more powerful stations. Thus the use of 130 KW at Tuckerton. Sayville is now being equipped with 400 KW generating apparatus (120 KW in the antenna). To radiate such amounts of energy, huge antennas are necessary. Thus, Sayville, which now has one tower about 600 ft. high, is being equipped with four additional masts 500 ft. high.

If it is possible to reduce the intensity of static to 10%, present results may be obtained with a 10 KW sending station, instead of a 100 KW.

If Pupin is able to get fundamental claims on methods of eliminating static, his system should supercede all others. It would therefore seem very important that we should not give *exclusive* right to the pliotron to anyone at present.

We have considered that the pliotron is already covered by some fundamental patents of De Forest[4] and that we therefore could not sell pliotrons to Pupin. I believe, however, that De Forest has publicly stated that he has licensed the Western Electric Company to his patents only for use with *wired* circuits, reserving for his own company the wireless rights. If so, Pupin could deal directly with the De Forest Company.

Finally, I should like to point out that by the use of Dr. Hull's negative resistance we apparently have another method of eliminating static, quite different from that described by Armstrong (altho probably covered by Pupin's general theory).

Yours truly,
Irving Langmuir

Langmuir Papers, Library of Congress

1. EDWIN HOWARD ARMSTRONG (1890–1951) was an electrical engineer at Columbia University who was noted for many contributions to radio technology. See Lawrence Lessing, *Man of High Fidelity* (Philadelphia, 1956).

2. MICHAEL PUPIN (1858–1935) was a physicist at Columbia University who also worked on problems of electrical engineering and electronics.

3. ALBERT WALLACE HULL (1880–1966) was a physicist at the General Electric Research Laboratory.

4. LEE DEFOREST (1873–1961) was a Yale Ph.D. For an account of the events referred to here, see H. G. J. Aitken, *Syntony and Spark: The Origins of Radio* (New York, 1976).

4
The Germination of Physics

No one could have argued in the United States of 1900 that physics was a strong science. The contrast with Germany and with Great Britain was too obvious. In retrospect, the development of the study of physics in those countries is seen as the beginning of a great intellectual revolution—the rise of modern physics upon the foundation of Newton's legacy. Many observers—scientists, philosophers, and others—regarded the field as the model for all sciences, perhaps even all knowledge. Astronomy, which had enjoyed such prestige, was now partially eclipsed.

Not that the United States was bereft of physicists. There were even some of note: the first American citizen to win a Nobel Prize (1907) was the physicist Albert Michelson. Yet a feeling of professional inferiority seemed to linger among the American physical scientists during the years 1900–1940. Certainly a downgrading of the professional past propelled many physicists to strenuous efforts to close the gap. As classical phyics gave way to modern physics, the disparity must have appeared even greater to many.

Recent research modifies greatly the group memory of American efforts in physics. By 1900 the United States was engaged in a substantial effort to expand the field of physics.[1] Although not yet a serious rival to the hegemony of the Germans, the physics community in the United States was distinctly in the same league as countries like France and Italy. Given the scope of the social and physical investment, there was every likelihood of easing into a stronger position internationally within the next few decades. What was noticeable about this effort initially was the relative provinciality of its practitioners. Since the leading edge of research was overseas, the U.S. physicists continued to seek not only advanced training abroad but also to maintain contacts with European intellectual leaders. The principal American universities invited prominent physicists to deliver lectures, to spend brief periods in residence, and even to accept permanent posts. In what follows are instances of various kinds of contacts, starting with examples of correspondence between R. W. Wood[2] of Johns Hopkins and Lord Rayleigh.[3]

In addition to provinciality, physics in the United States in 1900 had another important characteristic, one which greatly influenced the attitudes of the scientific community. Although the gross output of scientific papers was respectable, in relative terms the United States lagged. One could argue, with some justification, that this was a consequence of the familiar national strategy to erect a broadly diffused base rather than a small number of concentrations of excellence. Most of this 1900 activity took place in the bustling American system of colleges and universities. Although research had an assured, even honored role in the leading institutions, by and large teaching still had pride of place—even among many of the strongest universities. The pointed and curious confirming piece of evidence is the relatively low salaries of American academic physicists in comparison to their European colleagues. Society in the United States tacitly regarded the staffs of colleges and universities as chronological extensions of the teachers in primary and secondary education. Salary scales reflected this viewpoint. In Europe the professoriate had a quite different, elevated position in the social hierarchy.

American physicists facing their society in the next half century were not simply engaged in efforts to increase and improve their research output. That they could achieve by introducing elements of the research institute into the universities. They also strived to increase the socioeconomic status of the leading investigators and then the entire body of researchers. The high prestige of their field was none too subtly reinforced by citing heroic figures; first Faraday, then a few decades later, Einstein. Initially, physics had a modest place in the industrial research laboratories. Whitney and Langmuir were both physical chemists. As the century progressed, physicists became highly visible in all manner of useful roles. The inevitability of future tangible benefits became an increasingly common argument for the support of physics and of a better life-style for physicists. In all these respects, this field resembled other scientific disciplines. But the high prestige of physics and the success of its practitioners, intellectually and otherwise, reinforced the strivings of other scientists.

1. Paul Forman, John L. Heilbron, and Spencer Weart, *Physics circa 1900* (Princeton, N.J., 1975) (vol. 5 of *Historical Studies in the Physical Sciences*). A massive, baroque compilation of statistics about the discipline in the leading countries around the turn of the century. Many of its findings support the discussion here. A key point to remember is that the large investment by the United States was relatively recent, an effort to overcome the leads of other nations, *and* other nations were also increasing support of physics. Complaints about physics in the years covered by this volume should always be appraised against the status of other fields. Only medical and astronomical research can compare with physics in support; all other fields were relatively poorly supported.

2. WOOD (1868–1955) was an experimental physicist best known for work in optics. After a stay at Wisconsin, he succeeded Rowland at Johns Hopkins in 1901, remaining until 1938.

3. JOHN WILLIAM STRUTT, 3d BARON RAYLEIGH (1842–1919), was a British physicist. He headed the Cavendish Laboratory at Cambridge from 1879 to 1884 and was at the Royal Institution in London from 1887 to 1905. Rayleigh received the Nobel Prize in physics in 1904 for his role in the discovery of argon.

Robert Williams Wood to Lord Rayleigh, March 31, 1900

My dear Lord Rayleigh:

I have been unable to find any reprint of your R. Inst. lecture on "Transparency & Opacity" in Nature,[1] and we do not have the Institution's publications in our library. If you have a reprint to spare, or one which you could lend me, I should very much like to see it. I wrote you from Paris how much I regretted not being able to accept your most kind invitation to lunch, to meet Lord and Lady Kelvin.[2]

Lipmann[3] was kind enough to give me one of his remarkable pictures. I was surprised at the progress he has made. The colors are as vivid as aniline dyes, and really quite true to nature.

Just now I am not a little interested in the so called diffraction shadow bands which appear on the ground for the minute preceding and following totality, during solar eclipses. Their distance apart has been recorded at from 3 to 20 inches, and the speed of drift about 12 feet per second. Before totality they move from west to east, after totality from east to west. How they are formed I cannot imagine. Have you any theory? I am planning to view the eclipse of May 28 through a rotating disc with perforations. When the perforations pass the eye *nearly* in step with the passage of the bands, fluctuations in the intensity of the source of light producing them ought to be seen. Possibly in this way it may be possible to determine whether they are produced by light coming from the whole exposed crescent or the diffraction ring surrounding the moon.

Is it possible that the moon acts as a Loyd's[4] single mirror on a large scale, (This is a crazy idea that has just entered my head) the comparatively slow motion being the widening of the system as the similar sources approach.

This is of course absurd, and I am sorry that I let such a notion run out of the tip of my pen. Please consider it as "Something One would rather have left unsaid."

I should like your opinion of Hasting's[5] diffraction theory of the corona. It seems to me to be wholly inadequate. Inside of the geometrical shadow does not the diffracted light appear as if coming from the *immediate edge* of the obstacle? I cannot see how the space adjoining the obstacle can appear luminous.

Very truly yours,
R. W. Wood

Rayleigh Papers, Air Force Geophysics Laboratory

1. "Polish," *Nature* 64 (1901): 385–386.

2. WILLIAM THOMSON, LORD KELVIN (1824–1907), was one of the great Victorian physicists.

3. GABRIEL JONAS LIPPMANN (1847–1921) of the Sorbonne received the Nobel Prize in 1908 for a method of color photography.

4. This is HUMPHREY LLOYD (1800–1881) of Trinity College, Dublin, who devised a method of producing interference fringes with a single mirror.

5. CHARLES SHELDON HASTINGS (1848–1932) of Yale.

Robert Williams Wood to Theodore W. Richards, December 8, 1901

My dear Richards:

I have been intending to write to you ever since I heard of the very gratifying offer which our German friends made you. Please accept my most hearty, though somewhat tardy, congratulations. I am glad that you decided to stand by your country. We have had two or three interesting lectures by Sir Robert Ball.[1] He is very amusing is he not! As a popular lecturer he certainly understands his business.

You will be interested to hear that I feel pretty certain that I have found the long sought electromagnetic resonators for *light waves.* I form deposits of metallic sodium, corpuscular in structure, as shown by the microscope (1/12 inch oil imm.), the diameter of the metallic corpuscles being only about 0.0002 mm, which is about the right order of magnitude for a spherical metallic resonator for waves of the length of the light waves. The films show selective reflection and transmission, the colors being as brilliant as films colored with aniline. Warming the film causes partial evaporation, reduces the size of the corpuscles and alters the color of the transmitted light. There are many other curious things about them. They are of course in high vacua, and the admission of the *least trace* of air causes the instant disappearance of color. I am trying in every possible way to refer the color to some one of the many known causes, but every test that I have applied thus far, points to resonance as the only possible explanation. I am going to look into the optical properties of colloidal gold & silver, for it seems to me that they should show anomalous dispersion and there may be some connection between them and my sodium corpuscles. Please give my best regards to your wife, and do not forget that we expect you to make us a visit whenever you are in the vicinity.

Very sincerely yours,
R. W. Wood

Richards Papers, Harvard University Archives

1. SIR ROBERT STAWELL BALL (1840–1913) was an astronomer and mathematician, professor of astronomy at Cambridge, and author of popular works on science.

Robert Williams Wood to Lord Rayleigh, June 12, 1904

My dear Lord Rayleigh:

You may be interested in seeing a sample of the new and very remarkable alloy of manganese aluminum and copper, which is as magnetic as cast iron. I secured a small amount in Berlin a few days ago and enclose a sample. Try the filings with a ∩ magnet. I suppose the al. and cu. in some way loosen up the Manganese molecules so that they can turn around.

I have just finished up the work on the dispersion of sodium vapor, and find that the results fit the formula

$$n^2 = 1 + \frac{m\lambda^2}{\lambda^2 - \lambda^2 m}$$

most beautifully. Values of n ranging from 1.38 (near D, on red side) to .619 on green side of D_2 have been found. The work has been done for the most part with the interferometer. The interesting fact was brought out that the D_3 fringes, which become invisible when the path difference is about 2 cms, can be achromatized by introducing a column of sodium vapor into the path of one of the interfering pencils. In this way I have been able to see the fringes with a path dif. of 6 or 8 cms. I do not understand the physical properties of the vapor, which behaves more like a viscous liquid. A dense mass of it in a high vacuum does not immediately spread out and fill the vessel, but only diffuses very slowly. In a tube like this, with a spiral of iron wire suspended by a glass fibre, the whole highly exhausted, when the metallic sodium is heated, the spiral is carried to one side by the vapor mass which however only distills very slowly to the colder parts of the tube. I have obtained many other curious results which I hope to study quantitatively next Fall. The diffraction color-photographs I have also brought to a high degree of perfection. They are now superior to all except the Kromograms of Ives,[1] and could I think be made to equal them, if someone like Ives cared to spend a year or two in finding out a little more about printing gratings.

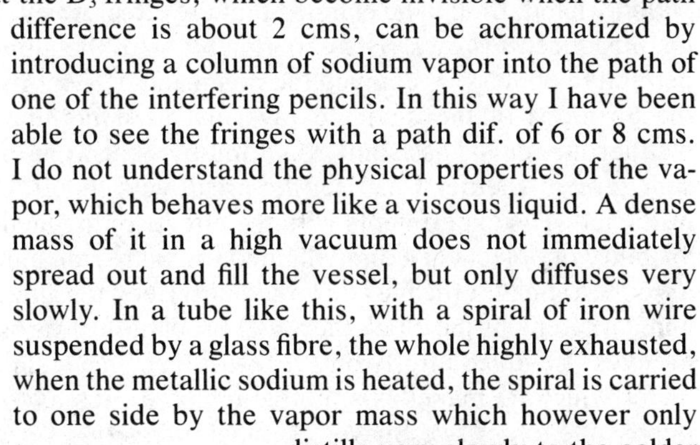

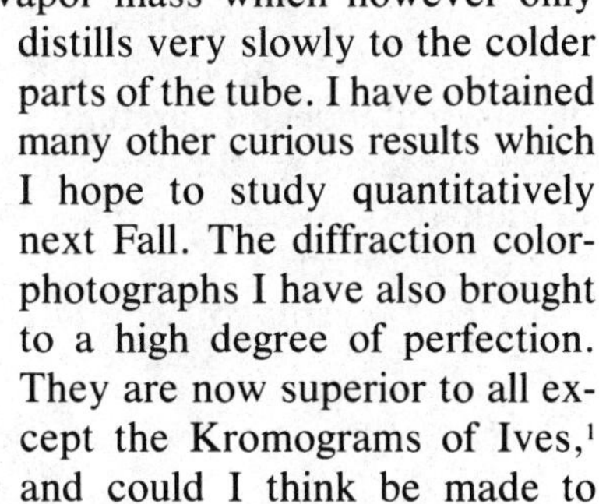

By the way I have brought over with a lot of other curiosities the grating which gives the dark and bright bands in the spectrum. We are to spend the summer on the Brittany coast at Beg Meil, from which point I shall probably make short side trips.

Very truly yours,
R. W. Wood

Rayleigh Papers, Air Force Geophysics Laboratory

1. The inventor FREDERICK EUGENE IVES (1856–1937) worked on color photography.

Robert Williams Wood to Lord Rayleigh, September 24, 1904

My dear Lord Rayleigh:

I have been to Nancy and fear that Blondlot[1] is very badly deluded. Not only did I fail to see any of the effects of the n rays, but I soon discovered that all tests made of what Blondlot *said* he saw failed. He was *absolutely unable* to tell by his spark when I intercepted the rays with my hand, though he claimed that the effect was very marked when he introduced his own. His experiments with the aluminum prism are perfectly hopeless. I removed the prism and he went right on observing the positions of the deviated rays just as if nothing had happened. Neither he nor his assistant could tell by observing their phosphorescent screen whether I had set the prism with its ref. edge to the right or the left. His experiments with the effects of the file in increasing the visibility of objects in a dimly lighted room are ridiculous. I substituted a wooden stick for the file and he still described the alternate brightening and darkening of the clock face, as I brought it towards and away from his eyes. He failed utterly on *every* test that I made, and the conditions under which he makes his photographs are such as to make it almost impossible to get two images of equal intensity.

I have written up my visit for Nature[2] where you will find further details, and though I carefully avoided mentioning names, I think that everyone familiar with the question will perceive that I have been to headquarters. We are on our way home and I am going to try to drop this off at Dover.

My best regards to Lady Rayleigh and your sons.

Very truly yours,
R. W. Wood

Rayleigh Papers, Air Force Geophysics Laboratory

1. RENE-PROSPER BLONDLOT (1849–1930) of the University of Nancy had previously announced the discovery of a new kind of radiation, the *n*-ray.

2. "The n-Rays," *Nature* 70 (1904): 530–531. This article recounts Wood's visit to Blondlot in which Wood convincingly reported that no such radiation existed. Blondlot, we now assume, was self-deluded. Since the initial reports created a great sensation, Wood's refutation was a devastating blow to the prestige of French science.

Ernest Rutherford[1] to Jacques Loeb, February 3, 1907

Dear Dr. Loeb

Just a line to remind you that I am still alive & kicking as I have no doubt you are also. You may have heard that notwithstanding all your good endeavours, I have accepted the directorship of the Physical Laboratory in Manchester. I shall have a fine laboratory, a good salary & be in the center of things physical. I hope to get round me a few lively spirits to keep me from rusting. I leave here in May. There is much lamentation here but I think I am doing the right thing. Between ourselves, I was informally approached re Leland Stanford—verb sap.

I read an underlined annotated document from California recently re a course of lectures by Slate[2] *assisted by Professor Lewis* on Electrons & Radioactivity. I enjoyed it as much as the sender, whom I gather was Taylor. I feel that my journey to California was not in vain "Cast your bread etc."

I am feeling very fit & hard at work examining the recent evidence of the parentage of radium—latest report still uncertain though no doubt there is a productive parent. I did not hear whether you made your trip to Germany but hope you did. Please convey my best wishes to Mrs. Loeb. I will never forget your many kindnesses to me and have still an eloquent remembrance of that most excellent salmon chowder which Mrs. Loeb thought I did not eat enough of. I hope your boys are well & flourishing.

Kindly remember me to Taylor[3] & Robertson[4] & tell them that I am sorry I won't be able to look them up again shortly. I still regret Robertsons tea & have a lively remembrance of my feelings in eating the Club dinners—the former statement of regret is meant well & not to be taken in conjunction with the remark about the club. I meant it as an inverted contrast. Let me hear how you are getting along & how the origin of life is looking today.

Yours very sincerely
E. Rutherford

Loeb Papers, Library of Congress

1. Loeb had tried to get Rutherford to come to the university at Berkeley. RUTHERFORD (1871–1937), a New Zealander, was a pioneer in the study of radioactivity and in the development of nuclear physics. At this time he was at McGill University in Montreal

where he had been since 1898. In 1908 he received the Nobel Prize. He went to the Cavendish Laboratory at Cambridge in 1919. The physiologist Jacques Loeb will appear in a significant role later.

2. The physicist FREDERICK SLATE (1852–1930) was on the faculty at Berkeley.

3. ALONZO E. TAYLOR (1871–1949) was then a professor of pathology at the University of California, Berkeley, later at Pennsylvania, and during World War I he was Hoover's principal scientific advisor in the war food program. For 15 years after the war he was head of Stanford's Food Research Institute.

4. THORBURN BRAILSFORD ROBERTSON (1884–1930) was an Australian. He had been Loeb's student at Berkeley and his successor there. After a brief stay at Toronto, Robertson, a physiologist and biochemist, went to the University of Adelaide in 1919.

Ernest Rutherford to Jacques Loeb, January 24, 1908

My Dear Dr. Loeb,

It was very good of you to write and congratulate me on the Nobel award. I thank you & Mrs. Loeb for your good wishes. My wife went with me to Stockholm and we had an excellent time abroad—Arrhenius was in great form & we saw a good deal of Frau Arrhenius who is a charming woman. I was very glad to renew my acquaintance with the Taylors who looked after us & helped a great deal to give us a pleasant time.

When your time comes, mind you & Mrs. Loeb come over to receive the award; you will have the time of your lives.[1]

We spent a quiet Xmas after our festivities but work is in full swing again. The University is giving a dinner in my honour in a week or so. I don't much care for these formal functions. It is hard for me to repress my natural volatility.

I find the Lab here suits me very well while the climate, although far from ideal, agrees with the health of all of us & especially with that of my wife who is looking much better for the change: I have been laid up for a week with "housemaid's knee" acquired in some unknown unorthodox fashion. It is fortunately nearly alright again. I hear from Robertson and Lewis occasionally. I hope the former is growing fat under maternal care.

Arrhenius doesn't alter much with time and seems as brief[2] as ever. He generally saw to my welfare in Stockholm and piloted me on the social round. I called at Berlin on the way home & saw most of the Physicists and Chemists there.

They gave me a good time. When your boy[3] has reached years of discretion, send him along to my paternal care. I hope all is well with you & your family. With best remembrances to Mrs. Loeb and the boys.

Yours very sincerely
E. Rutherford

Loeb Papers, Library of Congress

1. The prize never came to Loeb.
2. A conjectural reading.
3. Leonard Loeb did go to study with Rutherford at the Cavendish Laboratory in Cambridge before starting a successful career in physics.

Robert Williams Wood to Lord Rayleigh, May 13, 1909

My dear Lord Rayleigh:

I am very sorry to hear that you went through such a severe sickness, and trust that you are quite recovered by this time.

Last summer I sent you a companion volume to my first book on Nature study,[1] which I hope you will run across when you look over your second class mail matter.

A lot of new things have come to the surface this year. I have at last discovered a phenomenon which I have long looked for, the selective reflection of monochromatic light from an absorbing vapor. Mercury vapor at 20 Atmospheres in a quartz bulb shows the thing nicely. At lower pressures the absorbed light of λ 2536 is re-emitted as diffuse resonance radiation. At high pressure it is reflected *specularly,* just as if the inner surface of the bulb was coated with silver. If we reflect the iron-arc from the vapor, and photograph the spectrum, one line is greatly enhanced in brilliancy! It is a question of the suddenness of the absorption I think rather than proximity of the resonators to enable Huygen's principle to come into play.

I had quite an argument with Planck[2] who thinks one may have reflection in a medium in which the ref. index changes gradually. I say "no" and I believe that you agree with me, (Your paper on subject makes me think).

The mathematical people divide the medium into an ∞ number of thin layers, each one of which reflects a minute amount. Even If this were so the wave trains would gradually change phase thus  and we should get nothing. Or we can take any two planes $\frac{\lambda}{4}$ apart; disturbances from these destroy each other, and the same can be done for all the planes. Is not this reasoning sound: Please let me have your opinion, as I am revising my optics.

I am also interested in the question of the flow of energy in a system of interference fringes. It seems to me that the energy flow is along the hyperboloids, but there are some interesting points, as to the type of wave which travels along a bright fringe. Have you any ideas on the subject.

With best regards to Lady Rayleigh.

Very truly yours
R. W. Wood

Rayleigh Papers, Air Force Geophysics Laboratory

1. *Animal Analogues* (San Francisco, 1908) had Wood's poesy. The year before Wood had sent Rayleigh its predecessor, *How to Tell the Birds from the Bees* (San Francisco, 1907), with more examples of his poesy.
2. The theoretical physicist MAX K. E. L. PLANCK (1859–1945), Nobel Laureate in 1918.

George Ellery Hale to Ernest Rutherford, June 1, 1914

My dear Sir Ernest:

Your kind letter of May 11, telling me of your return to England, was very welcome. I assure you that we most heartily appreciate the great favor you have done the National Academy, and American science generally, by inaugurating[1] this series of lectures. I never before saw so much interest or enthusiasm at any scientific lectures. Professor Loeb, in a great state of delight, said that Faraday never did better. The splendid thing was that you not only brought out points of the greatest general interest and the most fundamental importance, but you stirred the imagination of every investigator worthy of the name who heard you, and stimulated them to try and do likewise. I never was so stirred in my life, and I have wondered a hundred times since how I could learn to think and to do things in such a big and effective way. The rush of your advance is overpowering, and I do not wonder that Nature has retreated from trench to trench, and from height to height, until she is now capitulating in her inmost citadel. Much as I enjoy astrophysical research, it is after all only an applied science, and its results seem poor and meagre in comparison with the superbly fundamental conclusions which you derive.

I wish I could adopt your kind suggestion regarding Professor Eliot Smith,[2] but the last two lecturers have already been chosen: Professor Henry Fairfield Osborn[3] and Professor James H. Breasted, the latter to deal with "The Evolution of Civilization in the Nile Valley."[4] Can you tell me of a great and inspiring lecturer on Geology, to give the third course in April, 1915? I am finding considerable difficulty in selecting a suitable man for this occasion.[5]

Campbell is anxious to read your lectures as soon as possible, as he wishes to prepare his course on evolution in astronomy for the autumn meeting of the Academy. So I hope it will be convenient for you to revise and return the stenographer's notes before long.

With repeated thanks and kindest regards.

Yours very sincerely,
[*George Ellery Hale*]

Hale Papers, California Institute of Technology

1. "The Constitution of Matter and the Evolution of the Elements," in the Smithsonian Institution *Annual Report, 1915* (Washington, D.C., 1916), pp. 167–202.

2. Smith was a professor of anatomy at Manchester and an authority on ancient Egypt.

3. Osborn's lecture was published in an expanded version as *The Origin and Evolution of Life, on the Theory of Action, Reaction and Interaction of Energy* (New York, 1917).

4. Apparently not published in this form.

5. The National Academy of Sciences *Annual Report, 1914* (p. 24) notes that a European geologist will talk on the evolution of the planet at the 1915 meeting. With the intervention of the war, Osborn's talk expanded to cover other projected topics in his 1917 book.

Owen Richardson to Robert A. Millikan, March 25, 1915

My dear Millikan

Your letter has reached me at last. I am glad to find that you have taken my little note in such a delightful spirit. Abstracts are always dangerous and require very careful wording. What I felt after reading your abstract was that the uninformed would come to the conclusion that previous work on this subject was of practically no value and in particular that the points you mention as essential to success had been overlooked altogether by previous workers. At the same time I felt sure that you had no intention of creating such an impression and your reception of my note has established that conclusion firmly.

After Compton[1] and I had finished our experiments and after weighing all the evidence very carefully I felt that there was no escape from the conclusion that the real law of photoelectric action was the one you have arrived at and that the apparent deviations were due to experimental difficulties and to conditions which were not fully under control. There are, of course, two methods of establishing a law of this character. One is to investigate a large number of cases somewhat roughly and the other is to concentrate on a particular case and follow it with extreme precision. I think both are necessary to put the law on a firm basis and the former method of attack usually, often indeed necessarily, precedes the latter.

One of the points you mention explicitly we did verify with very considerable accuracy, namely that the threshold frequency ν_0 agrees with the intercept of the energy line on the ν axis.

As regards the point you raise in your letter about the contact potential of sodium, I am not certain whether the value we finally took was from a table or was a value given by the curves. As a matter of fact our attempts to determine the maximum point with sodium were not satisfactory, as we explained in the paper, but they would have enabled us to tell if the value taken was far wrong. In any event you will be quite safe in describing our value of the contact potential for sodium as only roughly determined.

In making these experiments I have always felt that there was another difficulty in getting the true energies, in addition to the changing contact potential. I have felt that there must be a reduction in the energy of the emitted ions due to films of gas or something on the surface of the metal. You do not appear to have experienced this difficulty. You must have managed to get better vacuum conditions than we did.

If you have time I should welcome a note expressing your opinion as to the existence or non-existence of contact potential difference between clean metal surfaces in a good vacuum. Some time back I felt sure of its existence but the recent experimental evidence, which is in many respects more reliable to me than [the] old, seems very conflicting. The question is a very important one in connection with the theory of the emission of electrons from hot metals as well as from the photoelectric effect. As it is a matter to which I know you have devoted a good deal of attention I should greatly welcome an expression of your opinion.

I am looking forward with great interest to your full paper on the photoelectric effect. Judging from the results given in the abstract it is a splendid piece of work.

Yours very sincerely,
O. W. Richardson

Millikan Papers, California Institute of Technology

1. KARL TAYLOR COMPTON (1887–1954), while a student at Princeton, had assisted Richardson in his experiments on thermal emission of electrons. The work on the photoelectric effect supported Einstein's theory of that phenomenon. Compton was on the Princeton faculty from 1918 to 1930, reaching the post of chairman. In 1930 he became president of the Massachusetts Institute of Technology, resigning in 1948 to become chairman of the Research and Development Board of the Defense Department, where he succeeded Vannevar Bush. He was part of the short-lived Science Advisory Board in 1933. During World War II, he headed one of the principal divisions of OSRD (14, Electronic Devices). With Bush, James B. Conant, and several others, Compton was one of the leaders in developing the new relationship between science and the federal government that came out of World War II.

Ernest Rutherford to Gilbert N. Lewis, December 13, 1915

Dear Lewis

I have received your letter of November 19th. asking me my opinion about Debye[1] and Bohr,[2] from the point of view of lecturers in Mathematical Physics.

I do not know Debye personally, but I believe he is regarded as one of the best mathematicians of the day, and was recently appointed Professor in Göttingen in succession to Professor Voigt.[3] As you know he is a Dutchman by birth, and is undoubtedly a man of great ability both on the Mathematical and Physical side. I am unable to express any opinion about his lecturing qualifications etc., but my experience is that most Dutchmen speak English quite well.

Bohr, as you know, holds the position of Reader in Mathematical Physics in my Laboratory. I regard Bohr as one of the coming men in Mathematical Physics, and I think he has a better grip of Physics than any of the Mathematical people I have come across. He is a man of great originality, and, as you know, his work has already attracted wide attention, and I am confident will do so even more in the future. As a matter of fact, I think Bohr is just the type of man that would fill your bill. He is thoroughly au courant of all the modern physical problems, and has an extraordinarily wide knowledge of experimental as well as of theoretical Physics. He is a pleasant fellow, speaks English quite well, and is quite a clear and interesting lecturer.

I do not, of course, know how he would regard a visit to California, especially in war time. As you probably know, he is a Dane, the son of the late Professor of Pathology[4] in the University of Copenhagen, and I believe he is likely soon to be given a Professorship in that University. He has a very charming wife.

I am at present very much occupied with Admiralty work, and have practically no time for my own investigations. My Research School has vanished to Flanders or to the Dardanelles, but we still have a good number of women students, and men physically unfit and under age, so we have to keep all our classes going.

Yours sincerely,
E. Rutherford

Lewis Papers, University of California, Berkeley

1. PETER JOSEPH WILLIAM DEBYE (1884–1966) was a Nobel Prize–winning chemist. He came to Cornell University in 1940 from the University of Berlin.
2. NIELS BOHR (1885–1962), the Danish theoretical physicist, was then at Manchester with Rutherford.
3. WOLDEMAR VOIGHT (1850–1919) of Göttingen was best known for work on crystals.
4. Christian Bohr.

5
Mathematics: The Devotees of the Abstract

Like many of their colleagues overseas, American mathematicians of the nineteenth century concentrated largely on applying their knowledge and skills to such fields as astronomy and geodesy. Benjamin Peirce of Harvard and his son, Charles Saunders Peirce, are examples. Both men, however, also participated in varying degrees in newer developments in mathematics; C. S. Peirce also contributed significantly to logic.

The newer developments often were accompanied by higher standards of rigor. Throughout the course of the nineteenth century there was an increasing concern for the development of mathematics as a self-contained intellectual endeavor. It was a world of abstractions governed by its own rules and operations.

The older relationship with other fields was not replaced by simply characterizing mathematics as a pure field whose products could serve as powerful tools for other disciplines, a position taken by some individuals. Another traditional viewpoint persisted, one claiming for mathematics a role as the unifier of knowledge. From this position, mathematization of scientific fields was neither merely providing a common language nor a pragmatic use of a convenient tool; it was a development toward a powerful, all-encompassing insight into the natural world. Views like this were strong in Germany but also appeared in other national settings. Among the Germans especially, the concept of mathematical sciences tending to some future unity countered the strong trend to proliferation of scientific specialties and subspecialties, each tending to yield a separate intellectual domain with associated supporting institutions. American mathematicians of similar views were less vociferous on these matters than their German colleagues.

Almost all the American mathematicians, even those stressing the intellectual independence of the field, believed, in varying senses, in the possibility of applying mathematics. By this, some meant the use of mathematics in the conceptual growth of other fields (i.e., theoretical physics and theoretical astronomy were applications of mathematics). Nor were utilitarian applications as such derided. What counted was the elegance and sophistication of the mathematics. Applications, whether for theory or for utility, merely using uninteresting techniques did not excite the leading mathematicians. In the twentieth century both relativity and quantum mechanics attracted some eminent individuals in large part because of the challenges posed to their own intellectual ambitions as mathematicians.

Here we encounter three notable American mathematicians. The first, ELIAKIM H. MOORE[1] (1862–1932), was a Yale Ph.D., a student of Hubert Anson Newton, the close friend of J. Willard Gibbs. After a brief period of study abroad, Moore eventually obtained a post at the University of Chicago in 1892 when that school opened. There he built a very strong department. Moore is best known for his work in general analysis. Among his honors was an honorary doctorate from Göttingen in 1899. Another honor, indicative of the changing climate, was his selection to the National Academy of Sciences—the first pure mathematician so honored, according to C. S. Peirce. Despite the abstruseness of his intellectual concerns, Moore conceived his responsibilities as encompassing both pure research and the education of teachers below the collegiate level.

The second, a student of Moore's, OSWALD VEBLEN (1880–1960), was a nephew of Thorstein Veblen. The following letters give indications of the University of Chicago scene and the problems of making a career in mathematics. Oswald Veblen went from Chicago to the far-different environment of Princeton where he helped build a strong program in mathematics. In these letters Moore and Veblen, men of different intellectual generations, handled basic problems of definition verging not only on logical but also on philosophical issues. Veblen will appear later in other contexts. His career is important in a number of respects.[2]

Another mathematician who had a lively relation to philosophic issues is Norbert Wiener, who completed a Harvard doctoral dissertation on the relations of mathematics and philosophy in 1913 before his nineteenth birthday. He was a child prodigy, a *Wunderkind,* who was taught at home until age nine, when he entered high school from where he graduated two years later. (His father, Leo Wiener, was a polyhistor at Harvard.) After a brief, unhappy fling at philosophy in Cornell, Norbert Wiener entered Harvard Graduate School in zoology. The letters that follow were sent home by Wiener while studying abroad on a Harvard fellowship. They not only introduce a prodigious contributor to mathematics but also describe the Cambridge scene and, briefly, the Göttingen mathematical scene. Being a *Wunderkind* clearly formed Wiener's personality and career. He clearly stood alone, typical of no one else in style and range of interests.[3]

In historical perspective, what is most interesting about mathematics in the United States from 1900 to the outbreak of World War II is its growth and its decided orientation to pure mathematics. The growth of the field obviously benefited from the expansion of higher education in the period; very few mathematicians found careers in activities analogous to those so visible after World War II. In the universities and colleges, mathematics existed as a cultural subject—by which mathematicians usually meant pure research—and as a practical field. This usually denoted very modest uses usually of older, staid computations. As we shall see in a later chapter, the formidably abstruse Norbert Wiener was a notable exception. The growth also depended on the devotion of a small number of individuals, great teachers like Moore, whose department at the University of Chicago dominated the production of doctorates in the four decades, and men like Oswald Veblen, an eminent topologist with great entrepreneurial skills. In a very real, significant sense, the theoretically committed pure mathematicians proved a most successful, if not relatively the most successful, scientific group in practical terms in the United States in the twentieth century. None of this is visible, of course, in these documents from the early years of this century.

1. For Moore, see vol. 17 of the National Academy of Sciences' *Biographical Memoirs*.

2. Much of what is given in this volume is implicitly or explicitly based on Veblen's papers in the Manuscript Division of the Library of Congress. Veblen is one of the key figures in the history of the American scientific community in the first half of this century.

3. See Wiener's two autobiographical volumes: *Ex-Prodigy: My Childhood and Youth* (New York, 1953), and *I Am a Mathematician: The Later Life of a Prodigy* (Garden City, N.Y., 1956).

Eliakim H. Moore to Oswald Veblen, January 21, 1903

My dear Mr. Veblen,

I am writing to Dr. Carus,[1] and hope that something may come of it. I should have foreseen the possibility that they might print off without a determination of the question of reprints. I suppose they may consider the reception of the article in some measure a recognition, of the nature of return for its preparation. We shall see whether anything comes of my letter.

Next year—If no more desirable post opens, if you are willing, I think a post at Chicago, with rank of associate in mathematics, with partial work in the Junior College, with additional work in the Secondary School or in the University College may open for you. You understand the laboratory method plans of the department in general outline. In proper relation with the Physics department, I wish to establish a unity of spirit in the staff in mathematics from Secondary School through Graduate School. Indeed, in this initial stage, there is relation of problems of the Junior College and of the Secondary School. So I should wish to depend upon your cooperation in the working out of these problems all along the line, and also in some executive ways, in particular in connection with the library. And I should presuppose that you intend to go forward into research in mathematics, and I would be pleased to follow out some plan of direct cooperation in work on the foundation of mathematics, for we both ought, I think, to proceed in this direction at least for a term of years, and perhaps by joint work we can do considerably better than otherwise. As to the future, in Chicago or elsewhere, much depends on circumstances, more, I think, depends on you. Surely I shall support your interests so far as I am able in consideration of all things.

Yours very sincerely,
E. H. Moore

Veblen Papers, Library of Congress

1. PAUL CARUS (1852–1919) was a German-born philosopher, publisher of the journal *Open Court* and of works designed to further the development of a scientific basis for religion and ethics. The reference here may be to Veblen's "Euclid's Parallel Postulates" (*Open Court* 19 [1905]: 725–755).

Bertrand Russell[1] to Oswald Veblen, May 30, 1905

Dear Sir

Your post card has just reached me, & I am very glad there is a likelihood of your being able to come to Oxford on your way home. In order to see Whitehead,[2] you would have to come to Cambridge (where I am staying for a few days at present); in both places there are probably other people whom you would like to meet, & both places are well worth seeing on their own account.

I should be glad to know some little time beforehand when you are thinking of coming to Oxford, as I am fairly often away from home, & I should be very sorry to miss you through not knowing when to expect you.

Owing to the fact that I have not lately been working at Geometry, I have not yet read your work attentively; but Whitehead has praised your work very highly to me, & has, in consequence of it, entirely recast the treatment of Geometry which we had before agreed upon. I am very glad indeed that Peano's[3] methods are finding adherents in America, for I am more & more persuaded that the principles of mathematics can be far better dealt with by those methods than by any others.

Yours sincerely
Bertrand Russell

P.S. Whitehead asks me to convey the pleasure it would be to him to see you. It would be well to let him know beforehand, as he lives two miles from Cambridge.

Veblen Papers, Library of Congress

1. RUSSELL's life (1872–1970) is curiously tied in with the United States. He was, of course, a mathematician, logician, and philosopher of the first rank.

2. ALFRED NORTH WHITEHEAD (1861–1947) was Russell's collaborator at this time. From 1924 to 1937 Whitehead was at Harvard.

3. The Italian mathematician GUISEPPE PEANO (1859–1932).

Eliakim H. Moore to Oswald Veblen, May 22, 1907

My dear Mr. Veblen,

I am delighted at the contents of your two letters, and congratulate you and us all who are interested in you. Tho I am delighted and surprised pleasantly, I am not deeply surprised, because we all are believing very fully in you and your future.

Now you are likely to have a very important decision to make, and all I can do will be to indicate considerations which with others will have occurred to you. I was talking this afternoon, on receipt of your second letter, with Mr. Bolza.[1] We were rather inclined to suggest that your Princeton environment might be fully as valuable for you as the Yale environment would be, and that at Yale your real function interests would perhaps be blanketed by James Pierpont's[2] specializing in that direction. Besides you have the definite assurances given by Fine[3] of appreciation of your scientific purposes and of steady advancement. At Yale you perhaps would not have so good a chance, altho of course Brown's[4] advent at Yale will mean a new life there and so a new chance for everyone. Now I don't know just what the future of the graduate work at Princeton is likely to be. At Yale you would have a considerable body of students (Do you know how many? Very probably more than we have here) at work in advanced courses. You would enjoy Brown and Pierpont and Mason[5] and several others, and you would be working in the tradition of Newton and Gibbs, and Gibbs' scientific position is steadily rising. If you feel quite full of enthusiasm for the opportunities of teaching advanced students at Yale and with the thought of making Yale stand on a par with Harvard and Chicago—full of enthusiasm, with a sense of power to achieve your desires,—then I think I should advise you to take the Yale post, with the expectation of proving yourself there as at Princeton worthy of steady advancement. However, it seems to me you ought to get $2000. To change on $300 advance seems hardly worth while. You must, of course, have care to confer frankly with Fine before you commit yourself in any way to Yale; that understood, if Fine really meets the $1800 offer from Yale, so that you are assured of the $1800, I think I should write the Yale people that you felt it to be inadvisable to accept the offer at $1800, but that in case they made the salary $2000 or gave definite assurances of advancement above the $1800, you would accept their offer. Did you ever talk with Pierpont intimately? Since you would be working much in conjunction with him, I should advise you to make a trip to New Haven to confer with him and others. Don't decide abruptly or under constraint of presence of others. I wish you all success, whether at Princeton or at Yale. You are certainly a cause of much joy to lovers of Chicago.

Yours very truly,
E. H. Moore

I shall be greatly interested to hear of progress of events.

Veblen Papers, Library of Congress

1. OSKAR BOLZA (1857–1942) was a German mathematician who taught at the University of Chicago from 1893 to 1910. He then returned to his native land.
2. PIERPONT ([?]–1938) was a mathematician on the faculty at Yale.
3. HENRY BURCHARD FINE (1858–1928) was dean of the faculty at Princeton from 1903 to 1912.
4. E. W. Brown, the mathematical astronomer.
5. MAX MASON (1877–1961) was a mathematician with a Göttingen Ph.D. (1903). From Yale Mason went to the University of Wisconsin (1908–1925). From 1925 to 1929 he was president of the University of Chicago. In 1929 he joined the Rockefeller Foundation. He was its president from 1929 to 1936.

Eliakim H. Moore to Oswald Veblen, August 19, 1910

My dear Mr. Veblen,

It wasn't until my return for the second term's lectures this summer that I heard of your advancement to the full professorship, and I write to express, what you must know just as well without the expression, my extreme pleasure and affectionate congratulations. And I hope that as the years go by the department will move steadily forward, as it has these last years.

Will the Projective Geometry be out this autumn?

I am hoping you will find time to try out some of the mathematical-logical notations I use in the "Introduction," for I desire to know how you find they actually work. Of course, for me they seem to work very well. From the standpoint of pure logic they are not as adequate to express the niceties of thought as Peano's symbols, but for the mathematician I fancy that is not the prime necessity; I mean that more is expressed by connotation and not explicitly. However, you let me know please how you find them when you have the opportunity to consider them.

With cordial greetings to Mrs. Veblen, and to Miss Veblen also, Yours faithfully,

E. H. Moore

Have you noticed the articles in *Crelle:* 135 by Hessenberg (Kettentheorie and Wohlordnung); 137 by Steinitz (Algebraische Theorie der Körper)? They are based in part on Zermelo's axiomatic Mengenlehre, and proceed with the general conviction that the Zermelo's auswahls-princip deserves to be accepted as working hypothesis.

Veblen Papers, Library of Congress

George David Birkhoff[1] to Oswald Veblen, November 28, 1912

My dear Veblen:

I had meant to have written to you some time ago but recently Marjorie and the children took colds, and for a few days I had no time but to look after them. At present we are all in good health and spirits again; it is Thanksgiving Day and we are quite prepared to give thanks. Our first snow came today but the weather still remains mild.

When Borel[2] was here he gave us a talk and the department gave him a dinner at the Union Club, Boston, to which Coolidge[3] belongs. I enjoyed his talk and on the whole was considerably impressed with the work he has done with "fonctions monogènes mais non analytiques." Of course it is obvious enough after he writes down the formula

$$f(x) = \sum \frac{Ai}{x - \frac{pi}{qi}}$$

where [pi/qi] are all the irreducible proper fractions. To express the idea in a somewhat uncouth way, one sprinkles in a set of *very weak* poles every where densely, but each pole limited in its effect to a region so small that the sum of the measures of these regions is very small. I take it that your impressions of the importance of Borel's work is not very different from my own. Osgood[4] is not greatly impressed although it was apparent that Borel was anxious to convince him. It does not impress Osgood as wholly new, and he had some references to show that such functions had at least occurred before; in particular I seem to remember that Osgood remarked that Borel considered almost identical questions some years ago.

Borel has the intention to develop these ideas systematically, but I doubt if his programme will be of interest, beyond the few simple facts which he presented in the lecture.

It was a great pleasure for us to have Volterra[5] here and I wish that I had heard his Princeton Lectures of which you write. I also think that the fact that many of his ideas were familiar is in a measure due to his earlier work; but there is much of the formal side of the Calculus of Functions which had been recognized before his time by various men. Almost every mathematician who did great things must have noticed the remarkable formal analogies between sums and integrals, and have been guided by them. Before Volterra's time none of them would have thought of attempting to consider these analogies on their own account. But is not this because there were more concrete and simple things to be done first?

I hope that you are going to publish your lecture on Poincaré which you will give before the American Philosophical Society.[6] In case of publication (say in the *Bulletin*) the trouble would be, I suppose, that it would involve too much labor on your part. But we ought to have a review of Poincaré's work in the Bulletin, and I should like to see it done by an American. The power of treating deformation which strikes you is one that Poincaré possessed to a remarkable degree. I should also add as equally important his talent for generalization from a simple instance. In this day and age so many have the latter talent that it is only the first that takes our eye. Poincaré lacked somewhat in the type of analytical power possessed by men like Euler and Jacobi,[7] that is, great formal power. Perhaps he had little occasion to use and develop it. Won't you let me know the general course of your lecture when you write to me again?

We are looking forward to having you and Mrs. Veblen with us during the holidays, for a week or as much of your time as you can give us. I hope that you will find it convenient to come at that time. Our hearts are set on having you visit us as soon as you can and for as long a time as you will. It is too bad that you are not nearer us.

I wrote Wedderburn that Carmichael had sent an article to the *Annals* on "Non Homogeneous Linear Equations with an Infinite Number of Variables." Here Carmichael presents as new the very first method developed in the solution of such equations, namely that of von Kötteritzch; I am indebted to Bôcher for this reference. Whether there is anything left to the paper now, I do not know but I have sent the reference to Carmichael and asked for a statement of the relation between his work and that of von Kötteritzch. It struck me as almost self-evident that the work could not be new, but evidently Carmichael did not feel the same way or he would have searched the literature.[8]

What do you think of E. B. Wilson's work on Relativity?[9] I positively refuse to be interested in any speculations here, unless they are made by men of the very first rank.

It will be good to have a talk with you again, and I hope that the opportunity for a number of them is near at hand.

Cordially yours,
George D. Birkhoff

Veblen Papers, Library of Congress

1. BIRKHOFF (1884–1944) obtained a Ph.D. in mathematics under E. H. Moore. At this date he had just left Princeton for Harvard where he remained.

2. EMIL BOREL (1871–1956) of the École Normale in Paris.

3. JULIAN LOWELL COOLIDGE (1873–1954).
4. WILLIAM FOGG OSGOOD (1864–1943).
5. VITO VOLTERRA (1860–1940) of the University of Rome.
6. "Jules Henri Poincaré," *Proceedings of the American Philosophical Society* 51 (1912): iii–ix.
7. LEONHARD EULER (1707–1783) and K. G. J. JACOBI (1804–1851).
8. This is an example of the refereeing process in science. JOSEPH H. M. WEDDERBURN (1882–1948), born in Scotland, was at Princeton (1909–1945). He, Veblen, and Birkhoff were involved in editing *Annals of Mathematics*. ROBERT DANIEL CARMICHAEL (1879–1967), who was then at Indiana and later went to Illinois (1915–47) (dean of the Graduate School 1933–1947), had submitted this article which was evidently rejected because of its lack of recognition of the work of ERNST THEODOR KÖTTERITZSCH (1841–[?]), hardly a household name even among historians of mathematics. A Jena Ph.D. 1867, he taught in German secondary schools, retiring in 1882. But there were American mathematicians who took umbrage at the lack of proper citation of a predecessor. Carmichael gave the paper at the 1913 meeting of the American Mathematical Society, and it was printed in the *American Journal of Mathematics* 36 (1914): 13–20, with proper acknowledgment of Kötteritzsch. Carmichael was a Princeton Ph.D. 1911, which perhaps got him close scrutiny from Veblen and Birkhoff. MAXIME BÔCHER (1867–1918) was an American mathematician at Harvard from 1891 to 1918.
9. EDWIN BIDWELL WILSON (1879–1964) was a student of J. Willard Gibbs. Wilson in 1901 presented his teacher's lectures on *Vector Analysis*. After a stay at the Massachusetts Institute of Technology, in 1922 Wilson became professor of vital statistics at the Harvard School of Public Health. His work crossed into many areas. For many years, he was the editor of the *Proceedings* of the National Academy of Sciences. The reference here may be to the article he wrote with G. N. Lewis, "The Space-Time Manifold of Relativity: The Non-Euclidean Geometry of Mechanics and Electromagnetism," in the American Academy of Arts and Sciences *Proceedings* 48 (1912): 387–507.

Norbert Wiener to Leo Wiener, September 20, 1913

Dear father,

Many thanks to Constance for her letter. I am taking good care of my appearance. I shave regularly. I took a sponge-bath in my washbasin Sunday evening, and I am having one pair of shoes reheeled.

I have begun my work. Following Russell's advice, I bought Goursats' "Cours d'analyse," yesterday, and have covered more than 30 pages *thoroughly* already. It is a big book, in two quarto volumes of about 605 pages each. I paid 12/0 per volume for a second-hand copy. Its first-hand price is 16/0 per volume. I had quite a time finding it in the bookstores, as most of them were out of it. Excuse my spending so much money, but, as Russell told me, it is a book I should own.

Russell, as you know, invited me to his room Saturday evening. He had in his room at the same time another Fellow of the College, a mathematician. Between them, they made it very unpleasant for me.

Half the time they were talking entirely between themselves; the other half, they were casting aspersions upon my mathematical knowledge. One would say, "I wonder whether Mr. Wiener's mathematical knowledge is sufficient to enable him to take up this course of study with profit?" The other would answer, "But this line of work requires very little real mathematical preparation, you know." Then the first would say, "But it isn't so much the amount one has studied as the way one has been trained that counts." And so on for an hour. I have read a good deal about the studied insolence and conceit of the English University man, but this is the first time I have had the misfortune to encounter it.

As to my research-work, Russell's attitude seems to be one of utter indifference, mingled with contempt. I gave him my thesis to read, and he said he was so busy that he could not tell when he would get through, whereas you know that he told us that he was free *every* evening. I told him about the particular work I was interested in, but asked him for advice as to what particular work to take up, but he said, "Our method of doing research work differs from the German and American methods in that we let the students find their own problems, instead of assigning problems to them, and I think our method is better."

Russell has invited me to his rooms, without, however, naming any particular date, further than saying that Thursday is his especial "evening at home." However, I think that I shall be quite content with what I shall see of him at lectures.

I hope that you have had a pleasant trip, and that everyone is in tip-top shape. What have you decided to do with Constance, Bertha and Fritz? Is mother rested yet? Please tell me about all these things in your next letter.

Your loving son,
Norbert

P.S. When should I write to Schmidt and Perry?

P.S. Don't imagine that I have made any break before Russell, for I haven't.

Wiener Papers, Massachusetts Institute of Technology

Norbert Wiener to Leo Wiener, October 1, 1913

Dear father,

I am afraid that I have sadly misunderstood Mr. Russell. Yesterday, as I was walking around back of Trinity, I came upon him, and he invited me to tea. He gave me back my paper, with a list of criticisms carefully made out (though, I believe, mostly invalid), and said, that as a technical piece of work it was very good, and showed a thorough acquaintance with the use of symbols. He said that he looked forward to my making things interesting in his course, and was, in general, very pleasant to me. He also gave me a copy of Vol. III of the "Principia."

By the way, it is very likely that Santayana[1] will be here in a fortnight or thereabouts. From what Russell tells me, he has been staying in Paris, and not, as the legend at Harvard runs, at Madrid. I shall most certainly try to meet him if he comes.

It is Wednesday now, and I have had no letter from you yet. Please tell me all about your trip.

Your loving son,
Norbert

Wiener Papers, Massachusetts Institute of Technology

1. GEORGE SANTAYANA (1863–1952), the philosopher, formerly at Harvard.

Norbert Wiener to Leo Wiener, October 4, 1913

Dear father:

I must confess that I feel guiltily lazy nowadays. You see, though term began on the first, lectures do not commence until the thirteenth, and I have as yet no access to the library. I am trying to make myself study up my "Cours d'Analyse," but then, I have gone far enough to persuade myself that the subject is well within the range of my mathematical comprehension, and no matter how far I do go, we shall cover all that work in class anyhow, so I feel myself justified in not putting over an hour or two a day into it.

As to my research work, I am really at a loss what to do. Russell at first protested that a theory of types without an e-relation was impossible. I carried my work far enough along to convince him that it is possible, and that Schröder is consistent on this point, thereby substantiating the results of my thesis. So far, so good, but what am I to do next? I thought, from what I had read in Russell's books, that he tries to base his philosophical dogmas on his mathematical work, but apparently it was the other way round. Now, this removes much of the significance of the line of work I was intending to pursue, which was simply directed at undermining the claims of his mathematical work to philosophical importance. This leaves me in grave doubts as to whether I should continue this line of investigation, especially as R. keeps urging me to take up work that does not simply deal with other books. And I haven't any other problems as yet.

I tried the other day to write a philosophical essay, but I found before I got through with the first sheet of paper that a philosophical essay can only be written in a philosophical library. The very first sentence would need to be verified by references to half a dozen works to which I had no access. So I gave that up as a bad job.

I went to bring my results to R. Thursday evening. As I said before, I succeeded in establishing my point. R. invited me to a sort of meeting of students he held later on that evening, and as I wanted to meet students, I went. One of the things the discussion turned on was the Cambridge Union. I was surprised to learn that it played a much less important place in student life than the Harvard Union. In fact, from all I can gather, there is less genuine "college life" here than at Harvard.

Yesterday afternoon I went to a moving-picture show. I paid 3d for my seat. It was rather interesting for me to note that all but one of the pictures represented some phase or other of American life. I saw a statement in the Transcript last year quoted from some English periodical that the "movies" have had a tremendous influence in Americanizing England. I can easily understand that now.

I am glad that all of you are settled. How are the children learning German? Of course Constance has passed her exam. Please tell me soon about everything.

I got my suit from London the other day. I just got my laundry this morning. It was 2/10 1/2. I paid my rent yesterday. I practically spent the whole of my first cheque by Thursday, but it was all on things I needed for the whole year. I was forced to get some collars to tide over the time from Wednesday [to] Saturday.

Cambridge is an awfully sleepy little place,—far sleepier than our Cambridge. I really don't like it a bit, notwithstanding its beauty and associations. But then, of course I haven't got used to it yet. I'll try.

Your loving son,
Norbert

P.S. I got a registration blank the other day from Harvard, and sent it off.

P.S.$_2$ I believe I can live all told (except for books and University fees) for 31s. a week, or about $7.50. If this is too much, tell me so, and I can cut it down to 29s., or about $7.00.

Wiener Papers, Massachusetts Institute of Technology

Norbert Wiener to Bertha Wiener, October 22, 1913

Dear ma—

I am taking good care of my appearance in every way. I had my pants pressed yesterday, and tomorrow I shall have a haircut. I take a good, thorough sponge-bath twice a week. I not only look, but *feel* far more self-respecting than I did at Cornell. I feel confident that you would feel quite satisfied with me if you saw me. I shave every morning.

T'other day I got into a little philosophical tiff with G. E. Moore,[1] one of the big bugs hereabouts. Of course I got neatly squelched. However, as a student that was with me at the time remarked, when G. E. is really disgusted with a person's philosophical ability, he gets very rude, while he was entirely polite to me.

Friday the Moral Sciences club meets in Russell's rooms, and he reads a paper. I'm invited to butt in. Friday afternoon I go to tea with a Scotch phil. student, who will have some of his philosophical friends in also.

My math. work isn't so bad as I thought it was; I apologized to Baker[2] for the fewness and incorrectness of my problems, but he said that he had no fault to find with me, and that for the scantiness of my previous technical training and the time that had elapsed since I went over my work last, I was doing very well. My work with Hardy[3] is flowing along smoothly enough, while, so far, I have no work to do in Russell's courses, for I have covered all the ground myself.

I licked the work I did last summer into some sort of shape, and showed it to Hardy and Russell. Hardy seemed rather indifferent about it, but R. is highly interested and pleased by it. I think, as things look now, that I may get something of genuine mathematical value out of it—that I may develop, perhaps, a theory of circular ordinal numbers from it. I couldn't have done anything with it last summer, however, without books. But here, now that I have gotten into the swing of my work, I feel confident of accomplishing something.

My life here is about the same mixture of loafing and work that it was at Harvard. On the one hand, it is neither possible nor profitable for me to put in several consecutive hours of highly concentrated effort—it pays me rather to work, and then knock off work for a short while, and to work again, and knock off work again, and so on. On the other hand, in my last 2 years at Harvard, I have formed habits of work and concentration that make it quite impossible for me ever to fritter my time away again, as I did at Cornell. I am, I hope, far more of a man now than I was then. I feel that I am accomplishing something during my stay here.

You must remember that in the lecture-rooms my fellow-students are the pick of the mathematicians of the pick of the mathematical colleges of the pick of the mathematical universities of the world. And I feel that I am quite holding my own in these courses, though my technical knowledge is much less than theirs.

I hope everybody, and you particularly, Mother, is well. I am sorry you must go to the inconvenience of moving again. Write to me soon, and tell me of everyone from Fritz to father.

Your devoted son,
Norbert

Wiener Papers, Massachusetts Institute of Technology

1. The British philosopher G. E. MOORE (1873–1958).
2. HENRY FREDERICK BAKER (1866–1956) was best noted for his work in geometry. From 1914 to 1936 he was Lowndean Professor of Astronomy and Geometry.
3. G. H. HARDY (1877–1947) was the leading British pure mathematician of the era; author of a notable autobiography, *A Mathematician's Apology* (Cambridge, 1940).

Norbert Wiener to Leo Wiener, October 25, 1913

Dear father:

My work is progressing fairly well in general, except that in Baker's course we go so fast that I can neither take notes nor follow it completely without taking notes. The student that sits next to me in that course is one of the most damnable cads I have ever met. The other day he made several entirely unnecessary remarks in derogation of American students, and today, when I brought in some problems which were not rigorously done, he went out of his way to insult me about them.

Yesterday I visited a Scotchman, Dorward. He is rather a pleasant fellow,—the Scotch, in general, are, I believe, more sociable than the English. Last night I attended the meeting of the Moral Sciences club at Russell's rooms. Russell read a paper. The whole thing was very formal—the club is too large and unwieldy to be anything but formal. Today I met an American—a Yale man—who invited me to his rooms. The Anglicized American combines all the faults of the Englishman and the American, with the good qualities left out. I went to see a football game this afternoon. Of course, I could not completely understand it, but it seemed far less interesting than the American game, and the spectators seemed rather bored with it.

The Cambridge student, damn him, prides himself on a blasé attitude to everything in general. It is considered bad form to talk to a man at the Union without an introduction, and even students sitting beside one another in class all year long, or living next room to one another do not in general know one another. It is the deadest, most desolate place in the world.

I have a great dislike for Russell; I cannot explain it completely, but I feel a detestation for the man. As far as any sympathy with me, or with anyone else, I believe, he is an iceberg. His mind impresses one as a keen, cold, narrow logical machine, that cuts the universe into neat little packets, that measure as it were, just three inches each way. His type of mathematical analysis he applies as a sort of Procrustean bed to the facts, and those that contain more than his system provides for, he lops short, and those that contain less, he draws out. He is, nevertheless, within his limitations, a wonderfully accurate thinker.

Hardy is a typical Englishman; he plays cricket; lectures on his subject in a remarkably lucid manner; is, however, utterly indifferent to the students under him; and mispronounces his French and his German in a particularly atrocious manner.

Baker lectures at such a rate that few can follow him, but seems far more interested in his students than either Russell or Hardy. He has a marvellous mind—I can hardly understand a thing he says.

One particularly provoking thing is to hear when you speak to Hardy, say, about some of your plans in which he has not shown the slightest outward sign of interest. I have just spoken to Russell about that!

I hope it will not be necessary for me to stay another term in Cambridge.

I am keeping my personal appearance up; I had a haircut yesterday.

I got my card from the Union[1] today. By mistake they made it out in your name.

I'll send a post-card to Fritz tomorrow.

Goodbye,
Norbert

Wiener Papers, Massachusetts Institute of Technology

1. The Union Society of Cambridge.

Norbert Wiener to Leo Wiener, April 30, 1914

Dear dad:

I am all settled here, and my work has begun. I take two lecture-courses on philosophy from Husserl,[1] a course on the Theory of Groups from Landau,[2] & a course on Differential Equations from Hilbert.[3] I have called on all three. Hilbert is the typical German mathematician in appearance, & Husserl the typical german philosopher. Husserl's chief business is trying, as all good German philosophers do, to evolve new sciences à priori. He would be perfectly ready to evolve from his inner consciousness a pure Phenomenology of the camel (You remember your favorite yarn). Hilbert, the mathematician, has taken up almost every branch of Mathematics in his life, from the theory of aggregates & symbolic logic to the principles of statistical mechanics. At his invitation, I visited the mathematical club here. I am scheduled to take part this semester in a Sammelreferat on the *Principia* of W. & R.[4] Landau, who had a letter from Hardy about me, was very pleasant to me. He has the reputation here of being a stickler for logical rigor. I heard his first lecture in the introductory course on the calculus today (I came an hour early by mistake, thinking it was his course on Groups), & it certainly was interesting to hear how he spoke of the lack of logical rigor in the mathematics of the German gymnasia, for instance in the treatment of irrational numbers. It is a relief to know that the U.S.A. isn't the only place where the secondary schools have their faults. He said that hardly any of the Gymnasium graduates are able to prove rigorously that $\sqrt{2}$ is an irrational number!

I am exchanging English for German with a student here. I now wish I hadn't made the arrangement, for it simply means that I talk German only half the time instead of all the time, & he takes up an awful lot of my time that I would rather put in to work. It is true we go for walks together, but I really do wish I hadn't made the agreement with him. I don't know how to break it politely. Heaven only knows my German is weak enough, but I can't learn any German when he insists on our talking English all the time.

I am getting along O.K. here. I occasionally go to the movies, or the theater etc.

Good bye,
Your loving son,
Norbert

Wiener Papers, Massachusetts Institute of Technology

1. EDMUND HUSSERL (1859–1938).
2. EDMUND LANDAU (1877–1938).
3. DAVID HILBERT (1862–1943). See Constance Reid, *Hilbert* (Berlin, 1970).
4. The three-volume *Principia Mathematica* of Whitehead and Russell had just appeared (1910–1913).

6
Biology to 1915: A General View

Like other scientific areas in the United States around 1900, the biological sciences were undergoing both growth and intellectual ferment. Parallelisms to the physical sciences existed, but only a misleading tour de force could construe these as a significant overlap or as evidences of reduction of the phenomena of life to the laws of physics and chemistry. Despite strenuous efforts, social settings and intellectual content combined to give these fields a distinctive character. Even in our era when molecular biology and related specializations are so visible, the life sciences somehow elude the confident predictions of so many nineteenth-century philosophers and scientists.[1]

From a bird's-eye view of the intellectual landscape of 1900, biology was a widespread, rather amorphous activity. Even some of its ardent practitioners decried the absence of the structure and purposefulness of the physical sciences. The old natural history tradition had not quite passed away. Amateurs still contributed to field studies and to taxonomy, while the cult of nature study enlisted some members of the middle and upper classes as supporters of research. Differing viewpoints created strains that sometimes produced ruptures between professional and amateur. The Marine Biological Laboratory at Woods Hole, Massachusetts, originally provided instruction in nature study for schoolteachers and interested amateurs. Led by C. O. Whitman, the director, the younger professional scientists gained control from the older trustees for the express purpose of running the laboratory by themselves for their principal concern, original investigation. The split was not simply a matter of professionals versus amateurs. Supporters of the old order included such individuals as C. S. Minot, a competent embryologist at Harvard.

A source of strength for biology was the rather recent spectacular success of Darwin and Darwinism in its various forms. Biology now had a success story and a great rival to such men of physical science as Newton and Faraday. Even scientists with reservations about natural selection benefited from the intellectual and popular activities that arose in the wake of *The Origin of Species*. For the first time since the seventeenth century, intellectuals of all stripes could use an organic metaphor, rather than a physical one, in organizing their effusions. Biological subjects had a kind of topicality which penetrated even into ivory towers. All kinds of biological research expanded, many not obviously Darwinian, and the new metaphor became widely applied. Because most of these attempts failed, many biologists stepped back, as it were, into the models provided by the physical sciences. Some sought improved rigor in the forms and practices of those disciplines, others wanted to reduce biology to the laws of physics and chemistry.

As a reasonable continuation of the concerns of natural history, an active morphological tradition expanded, influenced by the success of Darwin. The reaction to this was a concern for a different kind of biology characterized by experimentation and quantification. Most notable were the investigations in embryology and in cytology. These gave rise to the very important school of classical genetics which was created largely by THOMAS HUNT MORGAN (1866–1945) and his students. Like his friend, the cytologist EDMUND BEECHER WILSON (1856–1939), Morgan was a graduate of Johns Hopkins University in its "golden age" during the latter decades of the last century. They were both influenced by their teachers, the morphologist WILLIAM KEITH BROOKS (1848–1908), and the English physiologist HENRY NEWALL MARTIN (1848–1896) who was at Hopkins from 1876 to 1893. Viewed institutionally, what Morgan, Wilson, and like-minded peers were trying to do was to establish an academic biology based in the liberal arts college and the graduate school independent of both the older field biology and the rising pressures from the medical school.

Although retrospection accentuates these pressures, a more visible source of support of biology was agriculture. By 1900 both the U.S. Department of Agriculture and the experiment stations in the states were expanding considerably, although unevenly in terms of quality. Increasingly, the work in the Department and in the states had come under the control of professional scientists. They brought to their positions the standards and attitudes acquired at graduate schools and at foreign universities. The leading administrators of agricultural research bore a generic resemblance to leaders of geology—men like T. C. Chamberlin. By adroit appeals stressing utility, they obtained support at local and national levels for both practical and abstruse research. Until the coming of a genuinely new order in World War II, agriculture was the preeminent example of governmental science. The various secretaries of the Department and the chiefs of the scientific bureaus were the successors to such men as Henry, Bache, and Powell. In World War I, George Ellery Hale's move to perpetuate the National Research Council had to pass the watchful eye of Secretary David F. Houston. Early in the New Deal it required approval from Secretary Henry A. Wallace to form a short-lived national Science Advisory Board.

Agriculture is not treated in this volume beyond occasional references because the focus is on the scientific community as such and the principal intellectual and administrative concerns of its leaders. The fields in which the agricultural sector excelled were not central to these concerns. From their basic science orientation, agriculture seemed to the leading scientific figures of this era to have a wrong mix of theoretical and practical investigations—too much of the utilitarian and not enough of the abstruse. They preferred to point with pride to, for example, Whitney's General Electric Research Laboratory.

By its very existence, the agricultural research sector created pressures on national policy in science persisting to the present. It was a successful example of a strategy contrary to the views of Hale and his allies. Instead of concentrating resources at a few key institutions or individuals, the agricultural research program called for a wide diffusion of support on the basis of geography and population. It assumed growth of research based on local support, and growth meant consciously accepting an element of risk in disbursing funds (unlike Woodward at the Carnegie Institution of Washington). Also present was a recognition of a social function to the goal of advancing theoretical understanding.

Medicine was different. Although physiology, the application of chemistry to biology, and microscopy were established portions of the medical curriculum, the medical school and its graduates lacked a certain obvious generality of influence early in this century. Like agriculture, medical research appeared outside the central intellectual concerns and policy issues of the scientific community. Unlike agriculture, medicine did not represent any obvious alternative to the policies of the leaders of the scientific community. Unlike agriculture, medical scientists and medical institutions had strategic roles from which they influenced research policies, perhaps even more than the emerging world of industrial research. By 1900 bacteriology had spectacularly demonstrated the potential of scientific research. Because health involved every one in an emotion-laden manner agriculture could not rival, medical research had extraordinary potential for attracting support. Unlike industrial research, supporting medical research appeared untainted by sordid concerns. That view disregarded the role of scientific medicine in maintaining the social and economic position of medical practitioners. In the post–World War II period, medicine rivaled national defense and industrial research in determining national science policies.

What follows is an introduction to two specialized sections on genetics and on scientific medicine. It first briefly introduces an encounter between biologists of the old and new styles, one at the Department of Agriculture, the other at a rising graduate university. The next theme is the growth of scientific journals. A scattering of letters display aspects of research in the medical orbit.

At this point in the documents, JACQUES LOEB (1859–1924) appears. German-born and educated, Loeb, a physiologist, came to the United States in 1891. After teaching at Bryn Mawr where he met Thomas Hunt Morgan, Loeb moved to the University of Chicago, then to the University of California at Berkeley. In 1910, he became a member of the Rockefeller Institute. Loeb had great influence both symbolically and by the impact of his work. A convinced believer in scientific materialism, he was a vigorous propagandist, greatly influencing both scientists and the lay public. Apparently, he never realized that his basic position was as untestable as the opposing viewpoints he attacked. Sinclair Lewis used Loeb as one of the two sources for Gottlieb, the archetype of the scientist in *Arrowsmith*. (The University of Michigan bacteriologist, FREDERICK GEORGE NOVY [1864–1957], was the other.) Loeb performed striking influential experiments on behavior, artificial parthenogenesis, and the chemistry of living matter. His successes convinced him and those like-minded about the ultimate reduction of living processes to the laws of physics and chemistry. Loeb had strong feelings about the nature of science and a genuine concern for social issues. Although he believed in the ivory tower, he also stood up to be counted. Here, among other topics, he writes about universities, medical schools, and research institutes. Like Morgan, Loeb wanted a biology independent of both the older natural history and the influences of medicine.

1. Perhaps the best introduction is Garland Allen, *Life Science in the Twentieth Century* (New York, 1975), which unwittingly displays this tension between past prophecy and present fulfillment.

Edwin G. Conklin to Charles Wardell Stiles,[1] February 25, 1902

My dear Dr. Stiles:

We have from year to year students who are taking both graduate and undergraduate degrees with us who desire to enter some of the government bureaus where they may be able to do biological work, and I should be greatly obliged to you if you could inform me whether there are any opportunities of this sort which are open to women as well as men, and if so, about what their nature is? If there is any demand for trained biological assistants in any of the government bureaus we should be very glad to fit such of our students as care to look forward to such a career, for this work provided we knew just what might be required of them. I should be under obligation to you if you can give me any information on this subject.

Very truly yours,
[*Edwin G. Conklin*]

Conklin Papers, Princeton University

1. STILES (1867–1941) was an entomologist at the Department of Agriculture until 1902. In that year he went to the Hygienic Laboratory, the predecessor of the National Institutes of Health. Stiles is best known for his work leading to the eradication of hookworm disease by the Rockefeller Sanitary Commission starting in 1909.

Charles Wardell Stiles to Edwin G. Conklin, March 1, 1902

My Dear Doctor Conklin:

I am in receipt of your letter of February 25, inquiring into the chances of Government positions for biologists, and in particular for the opportunities open to ladies in Government work.

Regarding the ladies I would state that we esteem them very highly, but it is not the policy of the Government to appoint them as laboratory assistants. The reason for this is that we must have assistants whom we can send into the field, and from the standpoint of the Government it is not feasible or proper to order a young lady out on the plains to live alone among the cowboys for three or four months, or to send her to a slaughter-house to collect material. If ladies were appointed as assistants it would necessarily throw an extra amount of field work upon the men, and this would immediately lead to dissatisfaction in the service. If you have any young lady students who are anticipating Government life as scientific workers I think you will do them a great kindness to tell them frankly that they are on the wrong track.[1]

The best opening for young ladies in Government work is in the line of librarians, translators, and in technical work as private secretaries. These positions are all Civil Service positions, and the next examinations are to be held either in March or April. You can obtain a list of the examinations by addressing the Civil Service Commission, Washington, D.C.

As for opportunities for young men, we are very loath to accept a person without first having had him under observation. It has been our experience that the biologists who have come to us from universities in the last ten years have been good cytologists and have been up on the literature of speculative zoology, but comparatively few of them have any idea that zoology is a practical science.[2] My advice to any young man anticipating entering the Government service would be to volunteer his services without pay for one of his summer vacations. He should pick out a laboratory either in zoology or botany as he prefers, and make regular application to work as a regular assistant without pay. This will give him an insight into Government methods, and this insight may prove to him that he does not wish to take a permanent position, or it may prove to the person in charge of the division that the volunteer is not a person who would do well in Government work. All of the positions of assistants are on the Civil Service list, and we are powerless to appoint a person except from that list.

There is a position known as scientific aid which pays $40 a month, and which is intended practically as a fellowship, enabling a man to spend one or two years in becoming acquainted with the practical application of science to public affairs, and in order to give him an opportunity to try his hand at original investigation. Most of the men taking these positions in biological lines utilize the opportunity to attend the night lectures at Columbian Medical College. These positions are open without a regular Civil Service examination. The applicant must file his papers showing that he is a graduate of a land-grant college; he must also file an essay on some subject, and give full information regarding himself. If he has had a year's teaching or post-graduate course, that counts in his favor. Full details in regard to these positions can be had by applying to the Civil Service Commission.

There is another position known as student assistant. This is not a Civil Service position, and pays only $25 a month. Students in various colleges here in the city occupy these positions and help themselves along in their education. They are given an opportunity to do considerable studying during their office hours.

Regarding your proposition to train men for Government work, I would state that several other colleges have approached me on the same subject, asking for suggestions as to what should be done. In this line I would say that experience has shown it necessary for the Government to train its own men. The swing of the pendulum among the colleges the last ten years has been to the extreme away from practical biology, and there is a very prevalent opinion among the biological students who have come to us that biology is a study of the useless, as once defined by C. H. Hurst,[3] of Manchester. There is also an unfortunate tendency among these men who call upon us inquiring regarding positions, to look down upon systematic work which is absolutely necessary from our standpoint of quarantine and practical dealing with diseases of animals and plants. The first suggestion I would therefore make is that the colleges should preach the doctrine that the well-known saying of "knowledge for knowledge's sake" is not entirely unlike the idea of money for money's sake, and that while biology has its strong points as a college study in general training and culture, it also has its practical application.

I would not advise any radical change in the college curriculum by which systematic zoology should be substituted for the present courses, but I would urge the introduction of an elective in systematic zoology. Most of the college students would not care to take this, but it is very advisable that every man entering the field of applied zoology shall know the principles of classification and the principles and customs of nomenclature.

I would also suggest that any college expecting to fit men to work in applied zoology should institute a chair of applied zoology and fill that chair with a man who has had practical experience in this line. We find such at present chiefly in the agricultural colleges, and I am perfectly frank to say that were I looking for an assistant I should prefer to have a man from one of the agricultural colleges, or from Cornell, rather than from any other college or university in the country. When I first came to Washington I felt that the Washington men were rather severe in their criticisms of the college men who come here to work, but I have now had to examine large numbers of Civil Service papers written by college students from all over the country and from all of the prominent colleges and universities, and I have come to the conclusion that my former criticism of our Government biologists in this regard was unjustified.

I do not consider that biology as taught in colleges today is entirely free from criticism, nor do I consider it the most practical course that can be given as a preliminary to the study of medicine or as a preliminary to official work. While the Government represents one extreme and the colleges represent another, I believe we must endeavor to find a happy medium, and as I do not think it possible for one man to cover both fields, my solution of the problem would be the appointment in the first class universities of professors of applied zoology, who should supplement the courses, as at present given, in the desired direction. I do not think it at all feasible for the biological department of any university to train men in applied zoology without an applied zoologist on the faculty. A comparison would be the proposition to graduate men in medicine without providing clinical facilities for them, or without having men on the medical faculty who had had actual experience with cases of disease.

This letter is a little lengthy, but as you have asked my ideas on the subject, I have complied with your request. I tried not to be extreme in the views expressed, and hope that the information may be of some service to you.

With best regards,
Very sincerely yours,
Ch. Wardell Stiles

Conklin Papers, Princeton University

1. Note the assumption that the work is in the field, not in the laboratory. It reflects an older natural history tradition as well as attitudes toward women.

2. Again note that the newer field is defined as pure in the sense of useless but that the older taxonomy is clearly out of the realm of theory. See the further comments of Stiles.

3. CHARLES HUBERT HURST (1856[?]–1898).

Herbert Spencer Jennings[1] to Edwin G. Conklin, January 17, 1903

My dear Professor Conklin,

I have been glad to receive your letter of Jan. 3. My election for a term of three years did not appear peculiar to me, as the Assistant Professors at the University of Michigan are elected for the same period.

I should rather like to see tried the plan of running a course in beginning zoology with the frog taken up first. I have always carried on courses the other way, and it has worked very well, but at Ann Arbor the beginning course is much heavier than yours, each student putting in six hours a week in the laboratory in zoology, and at the same time an equivalent amount in botany, the latter carried on in the same way, beginning with the cell. With so heavy a course we could get them to working pretty well with the microscope before unicellular organisms were left. But the plan had decided drawbacks even there, and with only three hours a week, I should think it might make serious difficulty. I shall therefore be glad to assist in running a course the other way. In that case I suppose my work in the general physiology part of the course will come in later.

I suppose you must have been very busy with the scientific meetings at Philadelphia this year. I wish I could have been present. I am very glad to hear of the new Journal of Experimental Zoology, and am certainly pleased to stand as one of the editors. We certainly needed a journal badly in America. I am particularly glad to hear that there is some financial backing to keep the journal going till it is well on its feet, as it is rather mortifying to see so many American journals run for a short time and then suspend. With this new journal we shall have the field fairly well covered in America, except for purely morphological work (in the old sense). I wonder if there is any possibility that the Journal of Morphology will start anew? I should like to have sent in my paper on Amoeba for the new journal, but was bound to send it to the Carnegie Institution, and in any case I judge that it would have been too long for the journal, as I had to go over the matter from the ground up. It is remarkable that matters could have been in such a condition for an organism so much studied as Amoeba, and one on which so much work has been done by men who stand high both in descriptive and experimental lines. There has been too much guidance from theoretical considerations, here as in many other places.

We are enjoying life at Naples very much, and keeping well now, after some tribulations at the beginning.

With best wishes,

Yours truly,
H. S. Jennings
Conklin Papers, Princeton University

1. JENNINGS (1868–1947) was then at the University of Pennsylvania. From 1906 to 1938, he was at Hopkins. He specialized in the study of the behavior of microorganisms.

Franklin P. Mall[1] to Jacques Loeb, April 25, 1903

Dear Loeb:

I regret very much to hear that you have had so much sickness this winter when starting in your new home for it may make you feel that you had better remained in Chicago. Now that the worst is over you can begin work with the untold number of new hearts which will come flocking to you all the way from China. No doubt fortune will come your way in the Pacific as well as [it] did in the Atlantic. But I am rejoiced that MacCallum[2] has found health in you, your climate & that you have found him worthy & congenial. We are all attached to him so very much and all of us are happy to know that he also is happy. The change has meant every thing to him for he now can be a decided addition to the forces in the country.

I do not think that you are right about our gifts to science coming too slowly to do good. If poverty is what is desired we certainly had it a century ago but it did us little good. To be sure we produced a Franklin but he would have been "produced" to-day just as well had he lived. The increased opportunity will necessarily produce all kinds of new "managers," impressarios as I call them but they cannot destroy all opportunity. Docility may be encouraged by them but it cannot carry the day. The new opportunity aided to clear the way for Richards, Flexner, Hilfricht,[3] Theobold Smith as well as for yourself, together not a bad showing for one year. Also we now have 20 young men starting in scientific work now for one ten years ago. It appears to me that the present is good and the outlook for the future excellent. The development will reach a high state when the recognized leaders are the men who really work instead of those who manage & do not work,—the impressarios. Maybe it will be found that the singing birds are not dependent upon the cages nor upon managers.

Let a crowd of scientific men do the managing & they will make a sad showing of it at present. Their best exhibition is made in our various scientific societies. If endowment gives talented men leisure & salary, the scientific men must do the rest. They make a scientific career dignified, not the opposite. So I feel that your presence & work is more necessary than the Carnegie but both should work together.

The country at large will never appreciate science until it is productive. That it show appreciation before it is productive is practically out of the question. Maybe we will see the time when part of this country is upon the high culture level of Germany, but even this has its drawbacks.

It is always a great pleasure to see something of enthusiastic and appreciative people like Wheeler, Flint[4] & others & I sincerely hope that you will be with them often.

We shall sail from here in a month to be gone until Sept. 15. I think a month in each, Germany, Holland, & England will make me long for home again. You know I am not ambitious but lazy.

With best wishes to Mrs. Loeb

Sincerely Yours
F. P. Mall

Loeb Papers, Library of Congress

1. MALL (1862–1917) was first professor of anatomy at Johns Hopkins. His collection of human embryos resulted in the formation of CIW's Department of Embryology which Mall headed from 1913 to his death.

2. The physiologist JOHN BRUCE MACCALLUM (1876–[?]), a Toronto undergraduate, received the M.D. from Johns Hopkins in 1900. He left Berkeley in 1906. He died of tuberculosis shortly afterward.

3. This is a conjectural reading. No "Hilfricht" is known to us.

4. JOSEPH MARSHALL FLINT (1872–1944) was an anatomist who went to a post at Yale in 1907. A "Wheeler" in the sciences at Berkeley remains unknown to us. Perhaps a reference to President Wheeler of the University of California, Berkeley.

Ross G. Harrison[1] to Franklin P. Mall, August 17, 1904

Dear Dr. Mall,

I find your letter of May 19 has remained unduly long unanswered. However my time was so taken up with my work while in Bonn that I came to very little else. I am much more than satisfied with the results of the season's work, for I feel sure now that I have settled the question of the histogenesis of the peripheral nerves beyond the shadow of a doubt. By cutting out the neural crest of young frog embryos, I succeeded in getting motor nerves to develop on which there are no cells of Schwann. One can follow the naked motor nerve fibers all the way to the remotest parts of the abdominal musculature. In such embryos the spinal ganglia and sensory nerves of course fail to develop. The experiment proves that the nerves develop entirely independently of the cells of Schwann, and also that these cells are ectodermal and not mesodermal as it is supposed. I have also a mass of other observations which entirely support the process theory. I gave a short account of the work to the Niederrheinische Gesellschaft which is now being printed in their proceedings. With the exception of his view regarding the origin of the cells it is entirely right. Personally I have never had very much doubt about this, but the large numbers of workers who have recently flocked to Dohm and Bethe rendered it necessary that the question should be definitely settled. I packed up my things in Bonn last week, and started off for Switzerland via the Black Forest. Mrs. Harrison has gone with the children to her home and will remain there until we sail. The International Zoological Congress, which is now in session here, is very well attended. I have met a number of men, who have interested me very much, amongst whom are Forel, Kronecker, Spemann[2] etc. Minot, Bowditch, Bigelow, Osborne[3] and a number of other Americans are here so that we are fairly well represented. It has been decided to hold the next congress (1907) in Boston, and there will be an effort made to secure the next International Physiological Congress for the same time and place.

After the close of the Congress I am going to the Bernese Oberland for a couple of week's tramping. This will put me into good shape. At present I feel somewhat used up after the summer's work. We expect to sail on the "Chemnitz" for Baltimore direct on Sept. 15, due Sept. 29. The examinations will be held, I suppose, on Oct. 1 & 3. The osteological preparations were all finished before I left Bonn and will make quite an addition to our collection. The total amount is a little over six hundred marks. I have had packed with the skeletons the new microscope and all of the laboratory property, which I brought with me, i.e. Lewis's microscope, my slides and notes, a number of odds and ends. Instruments, which belong to me personally, I am carrying myself. I have made out a consular invoice including everything and there should be no trouble at the other end. The boxes are being sent from Bonn by way of Bremen. I could not get them ready in time to be sent to Leitz as you suggested. Several weeks ago we had a very pleasant, though very short, visit from the Barkers. Barker seems to be well pleased with his summer's work. He looked well and happy and said you were enjoying yourself at Chebeagere. After Switzerland it is a pretty good place.

With best greetings von Haus zu Haus

Yours sincerely
R. G. Harrison

Mall Papers, Department of Embryology, Carnegie Institution of Washington

1. HARRISON (1870–1959) was an experimental embryologist who studied at Hopkins with W. K. Brooks as did T. H. Morgan and E. G. Conklin. From 1907 to 1939 he was at Yale. Harrison is best known for his development of tissue-culture techniques.

2. AUGUSTE-HENRI FOREL (1848–1931) was a Swiss physician who did neurological work, studied the brain, and was an expert on ants. He was active in reform movements. HUGO KRONECKER (1839–1914) was a physiologist at Bonn. HANS SPEMANN (1869–1941), an embryologist, received the Nobel Prize in 1935, then went to Wurzburg.

3. HENRY PICKERING BOWDITCH (1840–1911) of Harvard was a physiologist. The zoologist ROBERT PAINE BIGELOW (1870–1955) was at M.I.T. The biochemist THOMAS BURR OSBORNE (1859–1929) was at Yale.

Benjamin I. Wheeler to Jacques Loeb, May 1, 1905

Dear Professor Loeb:

I do not think the exclusion of the state universities from the Carnegie beneficence will hurt them at all. The allowances to retired professors at this University are already as large as they are likely to be made by the Carnegie gift in the private institutions. I think you are right in supposing that Mr. Carnegie has been influenced by ruling powers in the private institutions to take the step he has. Undoubtedly President Harper and President Butler are very actively engaged as against the growing prestige of the state universities. Funnily enough our own faculty recently fell into line and aided them when its Graduate Council refused to vote in favor of certain state universities for membership in the Association of American Universities. President Eliot and myself nominated these state universities for membership; Harper, Butler, and others vigorously opposed, and our Graduate Council voted with them, or at least the few members who were present on the occasion of the meeting voted that way. We must begin at home and convert our own faculty to a faith in the state university. We have a certain number of weak-kneed graduates of Harvard and Yale who after all, I fear, vaguely believe there is no future for the state university. They are totally wrong; the type of the state university is the ultimate American type I am sure.

Very sincerely yours,
Benj. I. Wheeler

Loeb Papers, Library of Congress

Jacques Loeb to Theodore Roosevelt, February 8, 1909

Dear and Honored President:

At the request of Professor John Graham Brooks,[1] I take the liberty of repeating two ideas which I expressed in a lecture on "The Physiological Basis of Altruism."

1st: It seems to me—and I believe the success of your administration proves it—that real statesmanship consists in the rendering accessible to the whole nation the benefits to be derived from the discoveries of fundamental laws in pure science. To give one illustration: the law of conservation of energy was discovered less than seventy years ago. Had the statesmen at that time realized the bearing of this law upon the economic welfare of the nation, they would have saved the sources of energy of the sun stored up in water power, coal and oil fields, forests etc., to the whole nation, instead of allowing individuals to appropriate them, and thus partially enslaving all coming generations.

I believe it to be no accident that England, which produces the largest number of real statesmen, also leads in the number of eminent representatives of the experimental sciences, especially of Physics. Nowhere are statesmen more closely in touch and intimately acquainted with the fundamental sciences (Physics, Chemistry and experimental Biology) than in England; e.g., Salisbury and Balfour.

The progressive character of Emperor William's government is due to his appreciation of science and his close contact with the leading German scientists.

The transition from the past to modern statesmanship is marked by the substitution of fertile science for sterile bureaucracy and jurisprudence for the limited horizon of the business men.

2nd: My work on the analysis of reflex actions and instincts has lead me to the opinion that the "sense of justice" is an instinct equally elemental with the social instinct and the instincts of mating and taking up food; and that it differs from the other instincts only in regard to the time during which it can remain latent. In this instinctive character of the "sense of justice" or the desire for a "square deal" lies, in my opinion, the basis for an optimistic forecast of the future of society, since no predatory influence, no corrupt press, no bureaucratic and juristic narrowness can permanently overpower this instinct any more than they can overpower the social instinct or the instinct of mating or eating.

It may at times appear as if the instinct of justice had died out in a community. Those watching the attitude of the population of San Francisco toward the graft prosecution were at times driven to such a pessimistic view; yet the attempt on Heney's[2] life demonstrated clearly that this instinct still existed in San Francisco in its elemental force, and that it only needed a spark to kindle it into a flame.

The recognition of the sense of justice as an elemental instinct by the statesman can invest him with unusual power of benefitting his nation. For it can be shown by experiments on lower animals that any instinct can be caused to become pre-eminent by a certain treatment, and in the case of a human instinct this can be done by fostering and embellishing it; and likewise, any instinct can be weakened by the fostering and embellishing of other instincts. No other instinct, however, is more important for the peaceful and steady development of a nation than the instinct of justice. Moreover, the statesman who makes it his policy to foster this instinct in the masses will win deeper and more permanent gratitude than the one who works primarily for the gratification of other instincts.

I question whether these statements contain anything new to you, since your administration served me as a model for their illustration.

Believe me to be, with high regard,
Your obedient servant,
[*Jacques Loeb*]

Loeb Papers, Library of Congress

1. BROOKS (1846–1938) left the Unitarian ministry for a career as a sociologist and reformer. He was particularly concerned with labor-employer relations.

2. FRANCIS JOSEPH HENEY (1859–1937), a lawyer, was then investigating political corruption in San Francisco.

Theodore Roosevelt to Jacques Loeb, February 13, 1909

My dear Professor Loeb:

Your letter interested me much, perhaps especially the sentence in which you say that "modern statesmanship is marked off from the statesmanship of the past, by the substitution of fertile science for sterile bureaucracy and jurisprudence or the limited horizon of the business man."

I entirely agree with you. I am particularly pleased also by your scientific basis for the belief that we are justified in our optimism.

With regard, believe me,

Sincerely yours,
Theodore Roosevelt

Loeb Papers, Library of Congress

Herbert Spencer Jennings to Edwin G. Conklin, April 27, 1911

My dear Dr. Conklin:—

I am extremely sorry I didn't get a chance to see you at Philadelphia. I found myself indisposed Saturday afternoon, and had to cut out the dinner, to my great regret, and go home.

I wanted to talk with you; and among other things, about the program for the American Society of Naturalists. You know we have two sessions: one forming a discussion on some subject of general interest;

the other a more special program of papers giving recent results etc. in Genetics. For one thing, I want to get track of any good pieces of work along the latter line, so that an invitation may be given that they be presented before the Society. I would be glad if you would let me know of anything of the sort.

But the main thing I wanted to talk with you about was the subject for the General Session. There has been some tendency at the last meeting or two for this to become rather special in character: to become a mere extension of the special Genetics program. This needs to be avoided if we are to maintain the Naturalists as a sort of central biological society. At the same time I would be glad on the whole if we could take up some subject of general interest that was connected with Genetics, though I do not feel that this is essential. I would be very glad for suggestions from you on this matter.

I have thought of a number of possible subjects for discussion; and talked them over with a few people. One of them is a subject, in connection with which the paper that (I believe) you *didn't* give at the Amer. Philos. Soc. would be valuable for us. I had thought of proposing a discussion of the Mechanism and Material Basis of Heredity and Development, or something of that sort. It seems to me that there has been considerable progress, or at least change of view, on this matter, of late; and that it is a subject on which many have ideas; and one on which there might be sufficient divergence of view to make it interesting. Morgan has recently changed his ideas on the matter; apparently Montgomery[1] would be interested in the matter; perhaps Wilson would be ready to give a summary of the way the thing appears now to him. There are certain men at work from the chemical standpoint: it would be most interesting if we could get something from such. And your paper, dealing with the matter from the standpoint of the student of embryological development, would be perhaps the most important of all.

I should think there might be papers—

(1) Treating the matter from the standpoint of cytology: Wilson?—Montgomery—(Miss Stevens?)[2] (others?)

(2) From the study of embryological development—Conklin.

(3) Material basis for sex limited inheritance, or the like—Morgan.

(4) Treating it from the chemical standpoint.

(5) Possibly we could get some straight physiologist to give his views on the matter. It would be specially interesting if we could get someone who is "down on" chromosomes to tell why, and what the opposing views are.

(6) From the standpoint of the breeder, or student of inheritance.

What do you think of this? Would you be willing to go into it, if that subject should be taken up?

I thought of certain other subjects. Bateson[3] in his recent book remarks that it "becomes necessary to revise the whole structure of the evolution theory in view of Mendelism." The possibility occurred to me of having a sort of critical review and discussion on such a question as—How has the recent experimental and analytical study of genetic problems affected our ideas of the methods and factors of evolution?

Such a discussion might have papers on—How our views of *heredity* have changed as a result of this study; our ideas of *variation;* its relation to *natural selection;* how far have we *seen* or *produced* actual steps in evolution, that is,—Has any work been done that deserves the name "experimental evolution"? Some special points that might come out are—Davis[4] promises later a contribution on how his results with Oenothera affect the views of Mutation; the question whether all inheritance is Mendelian (a point in which I have heard much interest and desire for light expressed by persons interested in the general problems of biology); What are the positive achievements of the statistical study of these problems (Harris)?[5] etc. etc.

I am inclined to think that a discussion on this question would be a little more difficult to arrange satisfactorily, though if it *were* properly carried out, it might be very interesting.

Other possible subjects:

What work in genetics is needed, in order to throw light on the problems of evolution? (Suggested by Montgomery.) Seems to me *good.*

What is the significance and what are the results, of biparental propagation (as compared with what we should have if propagation were uniparental). That is, the significance and consequence of amphimixes.

The scientific basis for Eugenics; or a general discussion of Eugenics.

I am afraid this long letter will frighten you; but if you could indicate to me your views even very briefly it would help much.

Yours truly,
H. S. Jennings

Conklin Papers, Princeton University

1. The zoologist THOMAS HARRISON MONTGOMERY (1873–1912) had a Berlin Ph.D. (1894). He was a cytologist and did important work on the chromosomes. At this date he was at the University of Pennsylvania.

2. NETTIE MARIA STEVENS (1861–1912) of Bryn Mawr had worked with Morgan on regeneration. Independent of Wilson, she had shown in a beetle species that sex was determined by part of a chromosome.

3. WILLIAM BATESON (1861–1926) was one of the earliest Mendelians in Britain; he has been credited with coining the term "genetics." At this date he was director of the

John Innes Horticultural Institute at Cambridge. He became a strong opponent of the chromosome theory of heredity.

4. The botanist BRADLEY MOORE DAVIS (1871–1957) was at the University of Pennsylvania.

5. The biologist JAMES ARTHUR HARRIS (1880–1930) was then at Cold Spring Harbor.

Jacques Loeb to Henry Edward Armstrong,[1] November 3, 1911

My dear Prof. Armstrong:

I hope you will forgive me for not having answered your letter of July 25th. It arrived at a time when I was about to sail for Europe and was mislaid in the rush of my preparations. I went to Europe to attend the Congress of Monists, and I delivered an address there on "Life" which I am afraid would not find the approval of Sir Oliver Lodge.[2]

The change from California to New York was in some respects a necessary one, since conditions out there had become rather intolerable. The American university organization, with an omnipotent head for each institution, is one of the worst drawbacks to the development of science in this country. America excells in astronomy and biology simply for the reason that these two lines of work are carried on in observatories and marine stations which are outside of reach of the autocratic president. I mention this fact because you are interested in educational problems, and I some times have the fear that American methods might creep into your universities too.

In Berkeley the university had lost practically all its investigators and I felt very lonely. In this respect New York is much more satisfactory, although, needless to say, I miss the beauty and the climate of California.

As far as an article to SCIENCE PROGRESS is concerned, I have not forgotten my promise, and I shall keep it. I have in mind an article on the role of acids in life phenomena or something of this kind, but I am working on this topic and I should like to develop it a little more before I am ready to write on it. If you would prefer an article on fertilization I could let you have such an article at any time.

I remain with kindest regards,

Yours very sincerely,
Jacques Loeb

Loeb Papers, Library of Congress

1. ARMSTRONG (1848–1937), a British chemist, was an editor of *Science Progress*. "Das Leben" appeared in English the next year as "The Mechanistic Conception of Life" (*Popular Science Monthly* 80 [1912]: 5–21).

2. SIR OLIVER JOSEPH LODGE (1851–1940), a physicist, was an avid proponent of spiritualism.

Edmund Beecher Wilson to Charles B. Davenport, December 27, 1913

Dear Davenport:

In reply to your inquiry, no, I do not remember ever to have seen the statement that America leads in cytology. I could, on the other hand, refer you to some statements quite to the contrary, including some rather pungent remarks concerning the characteristics of the "American School." Perhaps I ought to add that the most spicy of these comments emanated from a man whom I consider the d——dst liar in Europe (!) If you have not looked into the matter you would be astounded at the violence and (as I think) the dishonesty with which some of our work in this country has been contested by some of our friends in Germany. The fact seems to be that these people can not forgive us for having discovered certain things that they failed to see. Were it worth while it would be good fun some day to bring together some of the things that have been said about us.

With best wishes and holiday greetings

Yours sincerely,
Edm. B. Wilson

Davenport Papers, American Philosophical Society

Jacques Loeb to Hans Driesch,[1] May 12, 1914

My dear Professor Driesch:

I have just received your little book on "Problems of Individuality."[2] Needless to say that it interests me very much and that I shall read it. I think you have touched the very basis of biology, perhaps of physics too, but I am not sure that I fully agree with your conclusions. I am inclined to believe that the mechanistic analysis will be able to explain these things completely.

In dictating this letter my eyes rest on the sentence "The facts of active adaptation are very numerous." I have been busy trying to find such cases and I am almost in despair because I find just the reverse: they are so limited. I have been working on the case of adaptation to higher concentrations but I think there is almost nothing of that kind, though some papers state the contrary. I think I sent you my publication on that topic a little while ago.

I have not heard from Oxford, and of course shall not; McDougall[3] surely would not want me there and in a certain way I don't regret it since I am very busy at present and hardly know how to get around with my work.

With kindest regards to all of yours and Herbst.[4]

Yours very sincerely,
[*Jacques Loeb*]

Loeb Papers, Library of Congress

1. DRIESCH (1867–1941) was a notable German biologist whose position on vitalism was quite contrary to Loeb's beliefs. See *Problems of Individuality* (London, 1914).

2. Based on lectures delivered at Oxford.

3. WILLIAM MCDOUGALL (1871–1938), psychologist, was at Harvard 1921–27 and at Duke University 1927–1938.

4. CURT ALFRED HERBST (1866–1946) was an embryologist at Heidelberg at this date.

Jacques Loeb to Elias Potter Lyon,[1] November 8, 1915

My dear Dr. Lyon:

I was glad to receive your letter of November 3rd and since I have had perhaps more experience in regard to research positions than the majority of my colleagues, I will gladly tell you what I think about the desirability of developing research independently at Rochester[2] or at the University of Minnesota. You probably will be surprised to hear that for the time being I candidly recommend that Rochester be given every possible opportunity to develop into a place for pure research even if this happens apparently to the detriment of the University, and my reason is as follows: I was called to the University of California as a research professor, I had not asked for a pure research position but the administration offered it to me as an inducement. The outcome was that some or many of my colleagues, and very soon the community and the newspapers, quickly resented the idea that I should receive full pay and do little teaching. The people at large felt that they sent their children to the university to learn something and there was a parasitic professor who drew a salary without rendering services. The administration, while resisting this for some time, could not indefinitely ignore the voice of the people and then the conflict arose. While the administration did not go back on its word, yet I was made to feel that I was really a superfluous element in the university.

With all good intentions on your part and on the part of Vincent,[3] the same situation is bound to arise if you have research positions in the university. In a democracy today there is as yet no room in a state university for pure research. It may be done on the sly, but public pressure is against it. What I tell you I have talked over with Morgan and Wilson who are in very favorable positions in regard to research at Columbia. They both tell me that they believe that the Rockefeller and the Carnegie Institute are saving research in the universities of the east. Morgan once told me, what keeps the university presidents from harassing research men more than they do is merely the fact that they are afraid their good research men will be taken away from them by the Rockefeller Institute or by the Carnegie people. It is my firm conviction that before the state universities can have efficient research work established they will have to go through the intermediate stage of having a research institution independent of the university administration and the pressure of a democracy, and in the State of Minnesota the Mayo place is the only chance as far as I can see for the present. Therefore, if these people are really bent on creating an atmosphere of pure research, I think the university could make no greater mistake than to try to draw that away from Rochester to Minneapolis.

In the State University you may be able to have young men do research work under some professor; this will be tolerated because as a rule they do not pay any attention to the men on smaller salaries. But you can not get a man with a full professorship and full salary to do research without arousing opposition. The number of students increases so rapidly in state universities (on account of the fact that it is considered a matter of social prestige to have a university degree) that all the means the administration possesses will be needed to supply the demand for instruction and the taxpayers in general sympathize with this policy and with no other.

It is a severe indictment against democracy and it pains me, since you know that I am a good democrat, but it is a fact that a research man is really safe only at present in a privately endowed institution, while he can not feel safe in a teaching institution.

I remain with kindest regards,
Yours very sincerely,
[*Jacques Loeb*]

Loeb Papers, Library of Congress

1. The physiologist LYON (1867–1937) was dean of the University of Minnesota Medical School.

2. The Mayo Clinic in Rochester, Minnesota, had encouraged biomedical research. In 1915 the Mayo Foundation affiliated with the University of Minnesota to launch a graduate school for clinical medicine. See Helen Clapesattle, *The Doctors Mayo,* 2d ed. (Minneapolis, 1954), pp. 303–306, 320–332.

3. GEORGE E. VINCENT (1864–1941), a sociologist, was president of the University of Minnesota 1911–1917 and president of the Rockefeller Foundation 1917–1929.

7

Genetics:[1] The Classical Stage

Darwin had a problem. He did not know how variations came into being nor how these were transmitted through the generations. Nor did he understand the mechanism for the development of the adult organism. This lack of knowledge resulted in unsuccessful speculations on his part and engendered a sequence of explanations based on varying degrees of reliable evidence. Problems of heredity and development understandably attracted all manner of ambitious young scientists toward the end of the last century.

Not only had Darwin given the world an impressive organic metaphor, he also had a number of other metaphors guiding the thoughts and behavior of humans. Some resolute actions were fought by devout laymen and scientists who were comfortable with a natural theology in which science and religion were one. Darwin had affected the idea of the Deity, the place of man in the universe, and, most important, he had placed the future of mankind in the shadows of chance. The possibility of a science of heredity raised other disquieting questions.

While evolution had a far-off cosmic air—the time spans were so huge—a science of heredity could effect everyone and in a most immediate personal manner. What had before been a loose collection of commonsense beliefs slightly embellished by reliable empirical evidence might hereafter apply rigorously to plants and animals. And like evolution, the real sensitive question was mankind: Why and how was humanity as it was? And if a science was applicable to plants and animals, to improve them for the use of Homo sapiens, what about the improver? The issues ultimately were intellectual and moral, not at all narrowly technical.

As genetics came into being in the first four decades of this century, the potential of the organic metaphor—the implication of biology—became very clear. It was roughly parallel to the case of Darwinian evolution. Because of the implications for humanity, the prevailing pecking order of the scientific disciplines was at stake. During the prior century, the degree of generality was crucial. Physics clearly won. (Mathematicians thought otherwise, but theirs was not simply another science.) And physics underwent an enormous intellectual revolution in this century. Despite all the advances in the biological fields, physics held the laurel for generality buttressed by impressive applications. The less spectacular revolutions in biology could not match physics in generality. Nevertheless, the overall trend could very well indicate a change in the pecking order, and one based upon implications for humanity. Genetics has a strong claim as the first biological specialty to attain a status rivaling anything in the physical sciences for intellectual quality and potential.

The most significant developments came from individuals working in cytology and embryology. The parallel with the successful strategies of physical sciences was deliberate. Instead of complex organisms or entities, single cells, relatively simple organ systems, easily manipulable species became objects for experimentation. Investigators working with complex higher organisms had serious problems and lesser degrees of success. In a later phase of genetics, investigators would turn to protozoa and viruses. E. B. Wilson and his students helped establish the role of the chromosome in heredity, setting the stage for Thomas Hunt Morgan and his students.

Morgan was initially skeptical about the claims of the Mendelians after the rediscovery of the nearly forgotten work of the Abbot of Brno. Among the letters to follow are examples not only of his early correspondence with Wilson but also expressions of his doubts. Reading Morgan's letters and his publications places a strain on a historian's obligation to retain a critical distance. He is very interesting and impressive. There was in this man a great drive for critical rigor, applied to himself as well as to others. Only after his 1910 experiments with Drosophila did Morgan become a Mendelian; after that he and his students at Columbia University "fly room" gave genetics its classical form.

It was an experimental science in which numerical data counted. Observing that many of his predecessors had glibly postulated all manner of complex entities and processes not susceptible to experimentation, Morgan adopted a strategy of almost geometric chasteness. The gene was undefined as to structure; it was simply a point on a line (the chromosome). Morgan knew and later said that some chemical and chemical processes were involved, and he simply refused to speculate beyond his experimental resources. But Morgan could observe processes in the cell involving chromosomes and could treat these in a simple kind of geometry and algebra to create testable hypotheses. It was the spirit of Euclid peering down a microscope.

Like many of his contemporaries, Morgan had reservations about natural selection as a specific factor in evolution. There were too many undemonstrated assumptions. As in the case of heredity, the successors of Darwin simply did not meet his standards of rigor. Their formulations were not science but philosophy. Worse, they were "sociology," unscientific attempts to foist ideologies in the name of science. As late as 1917 Morgan was asking quietly about the relation of the hard findings of genetics to the grand sweep of evolutionary theory. That question was answered when population genetics was developed in the time between the two world wars. This development generated the modern synthesis of genetics and evolution. A sense of Morgan's position appears in his 1913 response to George Ellery Hale's proposed evolution lectures.

Another theme is his growing antipathy to the claims and actions of the eugenicist. Perhaps, like many of his generation, Morgan started out with rather vague beliefs in the existence of heredity, and a willingness to consider modest eugenics proposals as being in a reasonable reformist tradition. Many Progressive Era figures had such stances. But as his knowledge grew, the expanding activities of the eugenicists became clearly an inadmissible "sociology."

The case of CHARLES B. DAVENPORT (1866–1944)[2] illustrates this trend. Davenport was a Harvard-trained biologist who convinced the trustees of the Carnegie Institution that he was an exceptional man. In 1904 they established a Station for Experimental Evolution at Cold Spring Harbor, New York, for Davenport. His 1903 letter to Billings speaks for itself. The voluminous Davenport Papers at the American Philosophical Society contain many examples of his cogent remarks regarding the hazards of drawing conclusions about man from dubious experiments and shaky statistics. Unfortunately, they themselves also contain many such instances, sometimes in the very letters warning of the hazards. A convinced believer in the racial inferiority of many groups, Davenport in 1910 founded a Eugenics Record Office which became part of the Carnegie Institution in 1918. From that office came the "authoritative scientific evidence" behind the immigration quotas of the twenties. Before then Morgan, in accordance with his desire to be involved only with pure science, had parted from the eugenicists. Shortly before World War II, an embarrassed Vannevar Bush, a successor of Woodward's at CIW, ended that episode of the institution's history.

A happier application was the development of hybrid corn. Some of the basic work occurred before Morgan did his classical work. Burbank's observer, George Harrison Shull, did the investigation which ultimately led to the concept of hybrid vigor. As his letters show, Shull was quite certain of the significance of this development. Shull and his contemporary E. M. East[3] were not doing simply pure research, as some later literature indicates. That is, the case of hybrid corn is cited as an example of how pure disinterested science yielded a great economic payoff. The development into the 1920s involved a mixture of skills and motivations. Both East and Shull were aware of work on improving corn by scientific and commercial plant breeders. Their findings were picked up by both groups, who carried the work forward to the point where a commercial seedsman like Henry A. Wallace could make hybrid corn on a regular commercial basis.

1. See the following articles by Garland Allen: "Thomas Hunt Morgan and the Problem of Sex Determination, 1903–1910," *Proceedings of the American Philosophical Society* 110 (1966): 49–57; "Thomas Hunt Morgan and the Problem of Natural Selection," *Journal of the History of Biology* 1 (1964): 113–139; "Thomas Hunt Morgan and the Emergence of a New American Biology," *Quarterly Review of Biology* 44 (1969): 168–188; "The Introduction of *Drosophila* into the Study of Heredity and Evolution: 1900–1910," *Isis*

66 (1975): 322–333. See also Charles Rosenberg, "Factors in the Development of Genetics in the United States: Some Suggestions," *Journal of the History of Medicine and Allied Sciences* 22 (1967): 27–46.

2. See Charles Rosenberg, "Charles Benedict Davenport and the Beginnings of Human Genetics," *Bulletin of the History of Medicine* 35 (1961): 266–276; and Mark Haller, *Eugenics: Hereditarian Attitudes in American Thought* (New Brunswick, N.J., 1963).

3. EDWARD MURRAY EAST (1879–1938) was a Ph.D. from the University of Illinois, 1907. By 1905 he was already investigating the effects of inbreeding in corn at the Illinois Agricultural Experiment Station. He continued the work from 1905 to 1909 at the Connecticut Station, and from 1909 at Harvard. Like most of his professional colleagues, he was a believer in eugenics. East will reappear later in this volume espousing his view that crosses between divergent races of man are biologically detrimental.

Charles B. Davenport to John Shaw Billings, May 3, 1903

My dear Dr. Billings:

In my report I have indicated several lines of investigation that should be undertaken in a Station for Experimental Evolution. But there is one that deserves study beyond all others and I want in this letter to speak of it more fully.

I need not dilate here upon the importance of a more exact knowledge of evolution upon mankind. We have only to consider the influence that the acceptance of the doctrine of evolution, made possible by Darwin's work, has had upon our ideas of human responsibility and of man's origin and destiny. Evolution has replaced the idea that man is apart from the rest of creation having been made of a superior type by a special dispensation of the Creator by the idea of man's origin out of some thing lower by lawful, orderly, processes that are still at work raising him to a more perfect manhood. From his very source among the lower animals, and especially from the imperfectly social animals, he carries with him certain feeling and desires which, when uncontrolled, tend to bring disaster into the social order that man is trying to maintain. As an animal he may, when through arrested development or disease his strictly human qualities are lacking, act like an animal, and this without responsibility on his part. The history of man's evolution in the past gives us the more courage to believe that he can evolve to something still higher by the same processes. But what are these processes by which man has evolved and which we should know in order to make use of them in hastening his further evolution? The answer to this question must seem astounding: We do not know the processes of evolution: they have never been studied.

From a hundred proofs we know that there is some relation between environment and evolution. All animals are adapted to the environment in which they live. But how has this adaptation been brought about? We do not know.

We have hypotheses, to be sure. Darwin believed that a change in environment will act directly upon the organism and make it different and that this difference will be inherited. If the change is a disadvantageous one the animal will be killed off by the competition of its more favored brothers. If the change is an advantageous one the individual will have a better chance of surviving and, transmitting the advantageous quality, will become the ancestor of an improved race.

The practically unanimous "opinion" of naturalists, led by the sharp *logic* of Weismann, is that this explanation is all wrong. It is wrong, they say, because the modification wrought on the organism by the changed environment cannot be transmitted to the next generation. The second generation begins at precisely the same level with the first.

Let us examine the application of this "opinion" to man. You can improve a man by putting him in a good environment. His early training teaches him elementary manners and morals and the church continues and extends those teachings. Good books are placed in his hands, his ideals are raised, his imagination kindled, his ambition to do his best aroused. Schools show him how to make use of his powers and show him the direction in which he can work to the best advantage. Through all these influences a person born in the slums can be made a useful man. According to the prevailing opinion, however, the influence of all this care bestowed on the individual is not inherited by his offspring. They begin at precisely the same level that he did and receive no dowry of a finer mental stuff from all his intellectual accumulations. The only advantage his children get is the better training they receive in their early years.

Now, if the contrary is true we ought to know it. For what greater encouragement can there be for the teacher, the philanthropist or the man himself than to be assured that the improvement of the individual means an improvement of the race through a permanent bettering of the germ-plasma: that our children will, on the average, reap an advantage and be born on a higher plane because of all our strivings. If it is not true we want to know it and realize that a permanent improvement of the race can only be brought about by breeding the best.

We have in this country the grave problem of the negro—a race whose mental development is, on the average, far below the average of the Caucasian. Is there a prospect that we may through the education of the individual produce an improved race so that we may hope at last that the negro mind shall be as teachable, as elastic, as original, and as fruitful as the Caucasian's? Or must future generations, indefinitely, start from the same low plane and yield the same meagre results?

We do not know; we have no data. Prevailing "opinion" says we must face the latter alternative. If this were so, it would be best to export the black race at once.

There is only one way to settle this matter and that is by experiment. Are the effects of environment, education, and so on, inherited? Few adequate experiments have been made, but within the last five years two crucial experiments on insects make it probable that certain acquired characters are inherited; and if some then certainly others and perhaps the results of education.

The experiments to which I allude are those of Standfuss and Fischer.[1] They have shown that when moths are reared at a high temperature the color markings of the wings are altered. The offspring of such modified parents, when reared at ordinary temperatures, inherit the modified markings of the parents. Consequently, some acquired characters are inherited in moths; I have seen the results of the experiments and can testify to this.

Now the way is clearly marked out for us. We must experiment with animals by altering their conditions of life and thus altering their form, color, and, through training, their instincts and behavior. Rear offspring from these modified parents and see in what cases there is an inheritance of the altered qualities even when the second generation is reared under the original, unaltered conditions of life. Learn what kinds of acquired qualities are inherited.

The details must depend upon available material for experimentation. It is desirable, for one thing, to repeat the experiments of moths. Rooms for keeping butterflies, moths, and other insects at a constant high or low temperature are required. Many insects are modified structurally by changing the food plant. Caterpillars must be reared on strange plants and their offspring compared with the grandparents. Experiments should be begun, intended to last for many years, on rapidly breeding wild animals, such as field mice, to see whether daily training will make the successive generations more and more tractable. This matter must be studied quantitatively. If in the first generation 100 trials are required to learn a trick which is thereafter practiced till maturity, will the offspring, on the average, learn in 80 trials? Then if these children are all kept in training until maturity will their offspring learn the trick in 60 trials, and so on. Not one pair but a score or more should be experimented with. This experiment is feasible and will yield results one way or the other.

The problem that I have suggested will be recognised as important. But why has it not been solved already? Because it is practically impossible to get such a series of experiments carried on in connection with a University. A retired place, a variety of environmental conditions, and above all an opportunity for the investigators to give their whole time for a series of years—all are demanded. And there is at present no University far sighted enough to provide the means for such an investigation. There are scores of men of science who are hoping that the Carnegie Institution may be able to do this work and I am only acting as their spokesman.

In regard to place and money requirements, I have already written. There is hardly a better conceivable place for this work than Cold Spring Harbor. A building will be necessary with constant temperature rooms and breeding rooms. I have drawn plans for such a building which I will submit if desired.

As to a person to carry out the proposed work, I am ready at the present moment to abandon all other plans for this. If in the future it would be impossible for me, I should be glad to propose a suitable man.

Respectfully yours,
Chas. B. Davenport

Billings Papers, New York Public Library

1. MAXIMILLIAN RUDOLPH STANDFUSS (1854–1917); EDUARD FISCHER (1861–1939). Both are Swiss.

Edmund Beecher Wilson to Daniel Coit Gilman, October 5, 1903

Dear Sir:

Replying to your letter of September 21st: I sent to Secretary Walcott last July a brief statement of the work which I have carried on with the aid of a Carnegie grant during the past winter, adding that I would send a more detailed statement if desired. I do not know whether such a statement is necessary, but will again review the main points.

I employed the grant received from the Carnegie Institution to defray the expenses of a visit to Naples Zoological Station, extending from the early part of February to the end of July, and during the whole of this time I was actively engaged on the studies in experimental embryology, for which I went to Naples. It was my purpose, first, to search for available material for the experimental analysis of the early developmental stages in mollusks and annelids, and if such material could be found, to make as exhaustive a study as possible of this subject, which possesses a high theoretical interest in its bearing on the general problem of differentiation. I am glad to report a large measure of success in these aims. After a somewhat prolonged period of fruitless search and experiment, I found two excellent objects for my research, and made as exhaustive an analysis of them as the time would allow. Without entering too far upon the technical details, I will state that I have succeeded in demonstrating in a conclusive way the mosaic character of development in the molluscan egg, and have obtained most striking evidence of the self-differentiation and specification of embryonic cells. This result is interesting to me, not merely in its bearing on the problem of differentiation, but also, and perhaps in even greater degree, through the firm basis which it gives for the general method and point of view in studies in cellular embryology.

A second general division of my work has included an experimental study of prelocalization in the unsegmented egg, which has yielded results of no less interest than those relating to the cleavage stages. Of these, the most important relate to the embryonic basis of correlation and to the relation between quantitative and qualitative prelocalization in the germ.

While my results are, as I think, definitive on some of the questions involved, it is evident that an extremely wide field for further work of a far reaching character lies open in this direction. If I may be allowed to add a general comment on the nature of this work, its principal significance lies in its connection with recent studies on the cellular basis of inheritance and development, taken in connection with experimental studies on heredity such as those that have grown out of the rediscovery of the Mendelian law. I am fully persuaded that there is now a very good prospect of making an essential advance toward an understanding of the actual mechanism of hereditary transmission, and I venture to express the hope that studies in this direction may receive their due share of support from the Carnegie Institution. I may take this occasion also to express my appreciation of the great value of the opportunities afforded at Naples, and my hope that the Carnegie Institution may continue its support of the Naples Station.

Yours very truly,
Edm. B. Wilson

Archives, Carnegie Institution of Washington

Thomas Hunt Morgan to Edwin G. Conklin, June 8, 1905

Dear Conklin:

I hear you are back and will come later to Woods Hole where I hope to see much of you. Let me thank you for your big and equally valuable paper on the cell lineage of Cynthia and Curia. Little did I dream that day when I showed you first the eggs with the red cheeks that you would make so much out of them. I rejoice in the very modest share (which I should feel proud to boast of in the affair) that might have fallen to my lot had it been allowed.

Your second paper on Mosaic Development[1] I have just read. You have written Driesch on the hip, *yet I am not sure* but that even what you describe for the half-(lateral) larvae does not involve *something* of a regulation towards a whole. If I mistake not you will hear more of this later. In regard to your point that only people who have studied cell lineage should be allowed to do experimental work (with citations) is it not unfair to overlook *the* point, that they did not even realize the importance and interest of work of this sort until poor, untrained men, but men of ideas and genius, such as Roux and Driesch, drew down the fire from heaven to illuminate these fields. I have had, and you have had too, students no more than third rate *intellectually,* that could do more accurate book-keeping-cell-lineage than men like Loeb & Roux[2] & Driesch ever dreamt of. The world judges fairly I believe in giving first rank to the pioneers who open up the fields of research.

If you wanted to get a rise you see you have got it. Whether it will get to you remains to be seen; for I may relegate this to the waste basket tomorrow morning.

Meanwhile kindly greetings to Mrs. Conklin. I hope we shall see her this summer. Don't get drowned in the Bermudas, and for God Sake don't find out that there are no half larvae in Amphioxus.

Sincerely
T. H. Morgan

P.S. I showed Mrs. Morgan[3] the papers you had sent me and she promptly discovered you had sent the Mosaic paper to her. Why?

Conklin Papers, Princeton University

1. "Mosaic Development in Ascidian Eggs," *Journal of Experimental Zoology* 2 (1905): 145–223.

2. WILHELM ROUX (1850–1924), the embryologist.

3. Lillian Vaughan Sampson Morgan was a former student of T. H. Morgan at Bryn Mawr and a notable cytologist.

Edwin G. Conklin to Thomas Hunt Morgan, June 10, 1905

Dear Morgan,

Oh! what a beautiful rise I got out of you. Morgan you are impaled on a bare hook. I never said that experimental work should be in the hands of cell-lineage cranks, nor did I deny the great value of the work of Driesch and Roux. I grant everything which you have said about the intellectual stimulus which has come from these men, and yet I believe that Driesch made a mistake about the Ascidian.

As to the fact that there is regulation in the lateral half embryo, I readily grant it, but it does not lead to the formation of a complete embryo in *Cynthia*. This regulation is shown best in the formation of a complete Notochord and an atypical sense vesicle (see p. 207), but such regulation is not antagonistic to the Mosaic view; at least I cannot see that it is.

As for Amphioxus, Heaven protect me! I am resolved to find just what others have found, except possibly with respect to the development of isolated quadrants of the eggs. Otherwise I should be counted insane. But most probably I shall find nothing at all, as I did at Tortugas.

I am delighted to know that you and Mrs. Morgan are going to Woods Hole this summer. I shall be there about the middle of August and I hope that Mrs. Conklin and the children will also be there at that time. They are going to visit Mrs. Conklin's parents while I am in Bermuda.

I wanted to send Mrs. Morgan a reprint in acknowledgment of her fine paper on Hylodes, but I had forgotten your home address, so I sent it to you. If you would like to have a copy of that paper to tear to pieces, just say the word; I have more of them than I know what to do with.

I am very sorry that I did not mention the fact that you first called my attention to the Cynthia egg. I fully intended to do so but merely overlooked it. I give you all the credit for having started me on this work, for you will remember that I was of the opinion that such "measley" animals as Cynthia could not have good eggs.

Please accept my thanks for the many good papers which you have sent me since I saw you. You are still the same old fellow in spite of our doleful predictions before your marriage.

[*Edwin G. Conklin*]

Conklin Papers, Princeton University

Edmund Beecher Wilson to George Ellery Hale,[1] November 15, 1906

My dear Professor Hale:—

Regarding your inquiries in the matter of cooperative work in the field of biology, it seems to me that there are many subjects in which there is not only opportunity but also the greatest need of such cooperation on the broadest possible basis. One such subject very obviously is the general one of geographical distribution and the nearly related one of species and varieties as affected by climatic conditions and the like. No systematist can possibly master more than a few groups. He must cooperate with others in any general survey of a given area. Conclusions sufficiently broad and thorough to have much real philosophic value can only be attained by the united labors of many observers, extending over wide areas and long periods of time. Perhaps a still better example may be drawn from a field in which I am personally more directly interested, namely the correlation between the results of the experimental study of heredity and those derived from the cytological basis of heredity in the germ-cells. I can not think of any subject in the whole field of biology in the investigation of which cooperative work is more necessary or promises to yield more important results than in this one. The results we may hope for here are absolutely fundamental to the whole theory of heredity and evolution. They are results that can be attained; but only by the united labors of experimenters on the one hand and of microscopical observers on the other. Cooperation we have, in a certain sense, but not organized cooperation or any program that is generally agreed upon. It seems to me that any effort, by the National Academy or any other organized body of authority, in the direction of cooperative work in zoology would be a good work.

Very sincerely yours,
Edmund B. Wilson

Hale Papers, California Institute of Technology

1. This is part of Hale's campaign to promote cooperative work in the sciences, ideally under the aegis of the National Academy.

George Harrison Shull to John Merle Coulter[1] [February 11, 1911]

Dear Doctor Coulter:—

I consider the most important results of the recent work in genetics to be the complete triumph of Mendelism and of the unit-character idea. Several years ago many exceptions were presented by various investigators but further analysis of these supposedly exceptional cases have brought them into simple harmonious relations with recognizedly typical cases, until now there remains extremely few exceptions and some of these will undoubtedly yield to more careful study. One of the most famous exceptions—the case of the yellow mice—was recently worked out by Castle[2] who showed that yellow mice are always heterozygous, the one homozygote being incapable of developing, thus producing a ratio of 2:1, in the same manner as in Baur's variegated *Antirrhinums*.[3]

Next in importance I would place the development, and the demonstration of the correctness, of Johannsen's genotype-conception.[4]

This idea is directly opposed to the conception that a gradual change may be made in the fundamental living structure through selection. Each genotype is permanent and unchanged from the time of its origin, until the death of the last individual belonging to it. Any selection among the individuals of a single genotype produces no amelioration. When selection is effective it means that the work was begun with a mixture of genotypes.

The meeting at Ithaca devoted to this conception showed in a striking way what a mass of data has been accumulated by different investigators, for which no other satisfactory explanation is available.

The theoretical importance of both Mendelism and the genotype-conception are sufficiently obvious. The practical importance consists in demonstrating the necessity of utilizing the "isolation-principle" in plant-breeding. Old forms of mass-selection must necessarily result in greater or less delay in arriving at desired results. Two individuals which are externally identical may belong to different genotypes or may be heterozygous with respect to any one or more of its dominant characters, the other individual being homozygous. Isolation methods find at once the individual which belongs to the superior genotype or which is homozygous, and the finding of this individual forms the starting point of a new race, which is uniform and constant from the beginning, needing neither fixation nor purification. This latter process is held by all old-fashioned breeders to require about six years, and in some cases longer.

I think that I may be justified in suggesting that my work with Indian corn will take rank with the theoretically and practically important results of the last few years, particularly in its relation to the genotype-conception. Previous studies on the genotype have been limited to self-fertilized and vegetatively reproduced plants. I have shown that in this cross-bred plant many genotypes are intermingled in the form of complex hybrids and that the physiological vigor and yield are apparently dependent upon the degree of hybridity. The separation of pure genotypes and their combination in the production of F_1 hybrids on a commercial scale promises to yield larger crops and more uniform with respect to desired qualities than can be produced by any other method.

Another extremely important result has been the demonstration during the last four years that sex is a Mendelian character and that one sex—most generally the female, but in some cases the male—is homozygous, the other sex heterozygous. Homozygous females are found in Bryonia and Lychnis, in some insects, worms, spiders, and in man, but heterozygous females occur in some moths (Abraxas), in domestic fowls, and possibly in sea-urchins.

Another result which deserves mention because of its practical importance is that disease-resistance may be inherited as a Mendelian character, thus allowing the easy production of disease-resistant strains.

For the future it may be pointed out that all the unit-character phenomena need a fuller and more critical analysis, especially from the stand-point of chemistry and experimental cytology. The definition of unit-characters in terms of chemistry must gradually replace the methods now in vogue, of using external morphological characters. More minute analysis by the breeder will bring to light many features previously overlooked, which will give a fuller insight into the Mendelian behavior. Daily, instances are being found of such complications in material which has been considered fairly simple. For example, Prof. Emerson,[5] at Ithaca, gave a very interesting case of "spurious allelomorphism" in the color of the cobs of corn, which a number of previous investigators, among them himself, had found as a result of many crosses, to give the simple ratio of 3:1.

In practical plant breeding the problems are still much the same as they have been for a thousand years or more, with the exception that now the methods which have long been applied to the production of garden plants, or rather the recent refinements of those methods, must be used in the improvement of common field crops, and especially in the production of crops more resistant to drouth, cold, disease, and other untoward conditions. This work is already in progress at many of the State Experimental Stations.

I fear that these discursive suggestions will be of little value to you, but as each statement would require a chapter or two in order to set it in its proper relations to the rest, I will let them go in their present form, and hope that I have approximately fulfilled your request.

Geo. H. Shull

Genetics Department (University of California, Berkeley)
Papers, American Philosophical Society

1. COULTER (1851–1928) was a botanist at the University of Chicago. A pupil of Asa Gray, Coulter was editor of the *Botanical Gazette*.

2. WILLIAM ERNEST CASTLE (1867–1962) was a geneticist who was at Harvard during the years 1897–1936.

3. ERWIN BAUR (1875–1933) later was head of the Kaiser Wilhelm Institute for genetics at Dahlem.

4. The Danish biologist WILHELM LUDWIG JOHANNSEN (1857–1927) introduced the genotype-phenotype distinction into genetics.

5. ROLLINS ADAMS EMERSON (1873–1947) was then at the University of Nebraska. From 1914 to 1942 he was at Cornell University.

Thomas Hunt Morgan to Jacques Loeb, March 16, 1911

Dear Loeb,

Last week I gave my preliminary paper on wing mutations to Cattell, and it should appear in Science next week. The paper on eye mutations that is more fully worked out will appear the week after.[1] As I told you last summer all my wing mutations go back to my flies treated with radium, as do also *at least* two of the eye mutations. In regard to short wings all types show Mendelian segregation—two types are sex-limited, according to Mendelian expectation; the other two types show segregation beyond a doubt, but do not give the Mendelian proportion *for one pair of characters*. Evidently these cases are Mendelian with more than one factor involved. I have been at work on the latter types for just a year, and have many more results than will appear in my preliminary note. Three of us are working at it pretty strenuously still. We begin to see light at the other end.

This fly work has kept me so busy this year that I haven't yet been over to see you in your new quarters, but I still have hope. Are you going to Washington? I have sent in a paper entitled, "The Disintegration of a Species and Its Reconstruction by artificial Combinations." Doesn't that make you want to go.?! With best wishes—

Sincerely
T. H. Morgan

Loeb Papers, Library of Congress

1. "The Origins of Nine Wing Mutations in Drosophila," *Science*, n.s. 33 (1911): 496–499; "The Origin of Five Mutations in Eye Colour in Drosophila and Their Modes of Inheritance," *Science*, n.s. 33 (1911): 534–537.

George Harrison Shull to Ernest Brown Babcock,[1] July 26, 1911

Dear Professor Babcock:—

Your letter of the 18th has come to hand and I appreciate very much your kindness in seeing a member of the Committee having in charge the arrangements for the Hitchcock lectures, and also for referring my letter to President Wheeler. Shortly after writing you I also took occasion to write direct to President Wheeler but have not yet heard from him. I feel sure everybody will be well pleased if Doctor Johannsen should be chosen for your lectures. I regret very much to learn that Doctor Hale is still so indisposed, and hope he will speedily regain his health.

I lectured on my inbreeding experiments with corn before the students of the summer school here night before last, and found a great deal of interest manifested. Many of those from the "corn states" stopped after the lecture to discuss how to get my methods into general use in their home districts. Such a desired end is doubtless far in the future, for all of this interest can be at best only the leaven in the lump, and there is no way of knowing how rapidly the lump can be leavened. I confidently expect that some form of hybridization method will sooner or later supplant the selection method now in general use.

I have called the attention of a young friend in Santa Rosa, Mr. Howard E. Gilkey, to your contemplated course in Plant Breeding. He has much talent for horticulture and I believe he has been recently canvassing Berkeley and taking orders for "America" Gladioli. He worked for a year or two for Mr. Burbank, and I believe he has a future in this line of work. I have influenced him to complete his High-School course, which he has finished this year. I think it would be unfortunate if he should by any short-sightedness lose the opportunity which a college career will make secure. I hope you will get hold of him and encourage him to go on patiently to the end.

Mrs. Shull joins me in kindest regards,

Sincerely yours,
Geo. H. Shull

Genetics Department (University of California, Berkeley) Papers, American Philosophical Society

1. BABCOCK (1877–1954) was a geneticist at the University of California, Berkeley.

George Ellery Hale to Edmund Beecher Wilson, October 14, 1913

Dear Professor Wilson:—

After a careful study of various plans for the William Ellery Hale lectures, I wish to submit to the members of the Committee the following scheme for consideration and criticism.

In order that the lectures may add as largely as possible to the interest of the Academy meetings, I propose that three be given each year at the annual meeting in Washington, and three at the autumn meeting. A larger number on any one occasion would hardly seem feasible, on account of the limitations of time.

To inaugurate the course, I would further suggest that a series of lectures on evolution, covering a period of several years, be provided. The outline of such a series might be approximately as follows:

Three lectures each on

(1) The evolution of matter.

(2) The evolution of the earth.

(3) The evolution of the earth's surface.
(4) Paleontological evidences of organic evolution.
(5) Variation and evolution in botany.
(6) Variation and evolution in zoology.
(7) The evolution of man.
(8) Early civilizations of the Orient.

I am not satisfied with either the sub-division or titles suggested, but the outline will serve for purposes of criticism. The object I have in mind is to secure continuity and some degree of homogeneity in the series, the treatment to be thoroughly scientific, but not so technical as to be beyond the easy comprehension of members unacquainted with the lecturer's special field. The lectures on the evolution of matter might open, for example, with a description of the various forms of matter and their chief physical and chemical relationships. Then would follow an account of the results of the researches in radio-activity and a discussion of the electron theory of matter. This would furnish the raw material, as it were, for the astronomical course, which might begin with a description of the heavenly bodies and the structure of the universe, describe the sun as a typical star and sketch the probable stages of stellar evolution, and conclude with an account of the most promising theories of the origin of the earth and moon.

I will not attempt to go into the remaining courses, but trust you will criticize them, as well as those which precede, in the freest manner, providing that you agree in thinking that the general plan of a series on evolution is sound. I believe that a brief account of the formation of rocks should be included, and think it advantageous to say something about the chief phenomena of terrestrial magnetism and meteorology, if this is feasible in the short space of these lectures. It is evident that the broadest possible treatment will be necessary in all cases, and that the temptation to dwell on minor matters must be strenuously resisted by the lecturers.

As for the courses (4), (5), (6) and (7), I am much in doubt, especially as I am not acquainted with the ability of possible lecturers to handle broader topics. Is it desirable, for example, to include a separate course on theories of organic evolution, or will these be best described, as I have assumed, in connection with the courses named? It might be better to unite (5) and (6), but could we get a man to deal with the combination? As for (8), it would certainly seem both interesting and advantageous to bring the series down to modern times by a short account of the earliest civilizations, treated from the standpoint of evolution.

In addition to your criticism of the proposed plan, please send your choice of lecturers in each subject. As the available fund is one thousand dollars per year, I suggest that we give say $700 to European lecturers, and $300 to American lecturers, to cover the greater expenses of the former. We might also plan, in general, to invite European lecturers for the courses to be given at the annual meetings and American lecturers for the autumn courses. I would also suggest that the first course be given next April, preferably by Rutherford (it is probable that J. J. Thomson could not come).

An early reply will be appreciated, in order that an invitation may be sent in the near future to the first lecturer selected.

Very sincerely yours,
[*George Ellery Hale*]
Chairman,
Committee on William E. Hale
Lectures

Hale Papers, California Institute of Technology

Thomas Hunt Morgan to Edmund Beecher Wilson, October 24, 1913

Dear Wilson:

In regard to Professor Hale's interesting proposals relating to the William Ellery Hale Lectures, I should like to ask whether you do not think that a more up to date arrangement of the topics under Nos. 5 and 6 might prove more interesting for two reasons: First, because there is no legitimate distinction between variation and evolution in zoology and in botany; and second, because the matter as here stated has been gone over so often that it has lost interest, especially in the light of the more recent advances. I suggest, therefore, instead of Nos. 5 and 6, two of the following subjects be proposed as substitutes:

1. The modern study of genetics and its bearing on organic evolution.
2. Darwin's theory of natural selection in the light of modern criticism.
3. The Mutation Theory and its bearing on organic evolution.
4. Organic evolution as a chance product of variation or as a direct adaptive response (in Bergson's[1] sense) to the environment.
5. The materialistic and vitalistic interpretations of organic evolution.
6. The control of the course of organic evolution by man.

Sincerely yours,
T. H. Morgan

Hale Papers, California Institute of Technology

1. HENRI BERGSON (1859–1941), French philosopher, was a believer in creative evolution.

Thomas Hunt Morgan[1] to Robert S. Woodward, May 21, 1914

Dear Professor Woodward:

I have decided to appeal to the Carnegie Institution of Washington for assistance in an investigation, the nature of which is, I venture to hope, of sufficient importance to enlist the support of the Institution. The work has now progressed to such a stage that under favorable conditions a successful issue is certain, and the material on which the work is being done is unparalleled for the purposes to which we intend to put it.

Studies that have been carried on by myself and my students during the past three years on the heredity of seventy mutants of the fruit-fly, Drosophila, have shown that the seventy mutant-characters fall into *four* great groups. The characters in each group are inherited as though tied together. One group shows sex-linked inheritance, which means that the characters of this group follow the known distribution of the sex chromosomes and by implication are conditioned by these chromosomes. If the other linked groups can also be referred each to a special chromosome we have a ready explanation of the occurrence of such groups; for, it is a fact of extraordinary interest in this connection that there are just four pairs of chromosomes. The linkage within each group is, however, not absolute but relative in Drosophila, and in the study of the breaking up of the linkage relations we have an opportunity to attack one of the most important problems of biology, viz., the constitution (in the sense of a material configuration) of the hereditary germ-plasma in relation to heredity. It is for the study of this problem that I ask your support.

In order to conduct this work properly I need the assistance of two trained investigators. The salary of these assistants I should estimate to be about $1,800 and $1,500 a year respectively. The time required to bring the investigation to a reasonable condition of completion I place at *three* years. All other expenses connected with the work I am prepared to meet myself, including the salary of an artist, which amounts to $1,000 a year. The University will, I have reason to believe, house us and supply the necessary equipment, janitorial help, etc.

I shall submit to the Carnegie Institution for publication the results of the work, either during the progress or at the end of the specified time, with the understanding that the Institution is in no wise committing itself to the acceptance of such results for publication, but is left free to accept or reject them when presented.

The two men whom I should like to engage for this work will not be free until a year from June, 1914. One of the men, Mr. A. H. Sturtevant,[2] who has worked on the problem of linkage for two years, and who has shown aptitude of a very high order for this kind of investigation, goes abroad this year on a travelling fellowship, and will visit the chief centres of genetic work in Europe. The other man, Mr. C. B. Bridges,[3] is at present a Fellow, taking his degree a year hence. He has been my assistant (on part time) for two years. He is resourceful in meeting difficulties and well prepared to take part in these studies. The published work of these two men is submitted for examination. I also submit as much of my own work on the problem at issue as has appeared.

Respectfully,
T. H. Morgan
Archives, Carnegie Institution of Washington

1. This application for CIW support resulted in Morgan becoming a research associate. CIW continued to support his group at the California Institute of Technology even after his death.

2. ALFRED H. STURTEVANT (1891–1971) was one of the original group in the Fly Room at Columbia and a codeveloper of classical genetics with Morgan. He followed his teacher to the California Institute of Technology.

3. CALVIN B. BRIDGES (1889–1938), was another of the original group in the Fly Room and a codeveloper of classical genetics. He joined Sturtevant and Morgan in the move to California.

Thomas Hunt Morgan to Charles B. Davenport, January 18, 1915

Dear Davenport:

I have just written to Mr. Popenoe,[1] resigning from the Committee on Animal Breeding. I am sending you just a line to give a further explanation of why I have done so. For some time I have been entirely out of sympathy with their method of procedure. The pretentious title for one thing, the reckless statements, and the unreliability of a good deal that is said in the Journal, are perhaps sufficient reasons for not wishing to appear as an active member of their proceedings by having one's name appear on the Journal. If they want to do this sort of thing, well and good; I have no objection. It may be they reach the kind of people they want to in this way, but I think it is just as well for some of us to set a better standard, and not appear as participators in the show. I have no desire to make any fuss or to discuss the matter, but personally I would rather be out of it and remain a simple member of the Association for the sake of the Journal.[2]

I enclose a letter which may interest you and also another dealing with the Redfield[3] propaganda. It seems to me that if a man of the latter stamp is allowed to make use of the name of the Association and of the Committee in charge of eugenics, that one is better out of the business. You will, of course, understand that I am merely explaining to you personally and confidentially why I have resigned because I should like you to know the real reasons back of the step that I have taken.

Sincerely yours,
T. H. Morgan

Davenport Papers, American Philosophical Society

1. PAUL POPENOE (1888–1979) edited the *Journal of Heredity* from 1913 to 1917. He had no training as a biologist, nor was he trained in any cognate field. In a long career, Popenoe kept himself busy promoting sundry good causes.

2. The *Journal of Heredity* was published by the American Genetic Association. Prior to 1913, the organization was named the American Breeders Association. It consisted of scientists and people interested in breeding animals both for the farm and as pets. The association had a decided eugenicist set. Morgan is reacting both against amateurism and the content of the association's work.

3. CASPAR LAVATAR REDFIELD (1853–1943) described himself as a patent attorney and evolutionist. He was an engineer who devised quantitative confirmations of his eugenic beliefs.

Charles B. Davenport to Thomas Hunt Morgan, January 21, 1915

Dear Morgan:—

Your letter of January 18th just at hand. I am glad to understand your reasons for resigning from the Committee on Animal Breeding. I am afraid the Journal of Heredity does represent American biological science in this field very poorly; indeed so as to make us ashamed at times, but, yet, what can one do about it? I suppose the editor runs short of material and puts in some worthless stuff. He especially urges the need of photographs or pictures of some sort with descriptive text. One of the worst features is a bad habit that the editor has contracted of writing up the reports of committees himself and publishing in the Journal. This is what apparently happened in the recent report of the committee on eugenics which came out while I was in Australia. On the other hand, I had hoped that the Journal might serve to interest intelligent people in an inherently interesting part of biological work, the necessity of science; the esoteric does not so strongly appeal to me.

I spent some time once in working in cooperation with Mr. Redfield on the influence of the behavior of parents on milk yield of offspring. The results came out negative and Mr. Redfield said that probably the subject for investigation had been poorly chosen. I have felt a little skeptical about Mr. Redfield since that time but I do feel we know very little about the subject and I do think it is rather absurd to print such a letter as that which Mr. Redfield sent to the editor of the Journal in large type.

I don't know what to do about the Journal. As you see I am not on the Council and, on the whole, with the exception of a very few names the Council does not strike me as liable to be very critical. I have often thought of resigning as secretary of the Eugenics Committee. There is no reason why I should keep on excepting an historical one and, as I said, it seemed to me possible that it might be well, if the results of biological work in this field could be presented in a way which they would reach a larger audience of intelligent persons.

Sincerely yours,
[*Charles B. Davenport*]
Davenport Papers, American Philosophical Society

George Harrison Shull to Ernest Brown Babcock, September 13, 1915

Dear Professor Babcock:

Your letter of September 8 is at hand and I am glad to know that you received my criticisms of your paper in the spirit in which they were given. Regarding loss of factors, you should remember that I have been from the start a very ardent advocate of the *non-committal* attitude toward the *real nature* of hereditary factors. I certainly would not think of the loss of the factors as *necessarily* the loss of material things from the germ-plasm. I would be inclined rather to define the loss of factors as the loss of the capacity for a given reaction (under given conditions of course). The cause of this loss of *capacity* may conceivably be of such a nature that the loss may be regained by a new mutation. Even the loss of a *material particle* or of a *chemical substance* does not make *inconceivable* the regaining of such a particle, or of such a substance, by a mutational process. Doctor Holmes[1] lays too much stress upon the reversal of mutations as a difficulty for "particle hypotheses" of heredity.

Regarding the double publication of scientific articles, I may say that the plan you suggest would be *possible,* though I think scarcely desirable. As there is no intention to copyright papers which are published in "Genetics"[2] you would be at perfect liberty to publish again in your local series. Generally, it seems like a waste of precious facilities for publication, to duplicate in this manner. I am naturally anxious, however, to gather into "Genetics" as large a portion of the fundamentally important American papers as possible. I am pleased to know that you have been able to secure fertile seeds from Hemizonia. If I remember correctly these are matings of individuals in the same, supposedly pure, strain, or at least between individuals having closely similar characteristics.

We are now in our new home, which is at the above address, but I will be unable to begin my investigations as early as usual, because my greenhouses are not yet completed; but excellent progress is being made and in three weeks I shall probably be able to start by Bursa cultures, and thus will probably be able to carry through about two-thirds of my usual stint.

Mrs. Shull joins me in kindest regards to you and yours,

Very sincerely yours,

Geo. H. Shull

Genetics Department (University of California, Berkeley)

Papers, American Philosophical Society

1. SAMUEL JACKSON HOLMES (1868–1964) was a zoologist and geneticist. From 1912 to 1939 he was at the University of California, Berkeley.

2. *Genetics* came into existence in 1916, in part as a reaction against the *Journal of Heredity*.

8
Scientific Medicine and Its Perils

The practice of medicine in the United States in the first half of the nineteenth century often lacked the unquestioned esteem the physicians considered its just due. Licensure laws were weak and often unenforced. Enthusiasts promoted other forms of medical practice, for example, homeopathy. The organized medical community recognized only their own practices as based upon rational knowledge. They considered other types of practitioners to be sectarians. In response, the sects called for a democratic, laissez-faire policy, and they accused the regulars of also being a sect—allopaths—seeking a profitable monopoly. Because the practice of medicine had only a modest scientific basis, the irregulars could point out its flaws while insisting that they filled gaps in the services offered by the medical community. Although echoes of the Jacksonian rhetoric of the sects persisted into the Progressive Era, long before 1900 the allopaths had clearly triumphed. Advances in scientific medicine, particularly bacteriology, helped decide the issue.[1]

During the Progressive Era two aspects of medicine—education and research—received the forms which persist, with very few modifications, today. Important participants in this area were the Flexner brothers. ABRAHAM FLEXNER (1866–1959) was an educator who received his training at Gilman's Johns Hopkins University where his experience provided the template for much of his later work. As a staff member of the Carnegie Foundation for the Advancement of Teaching, Abraham Flexner produced in 1910 a highly influential report on medical education in the United States. Later, he served with the Rockefeller philanthropies and then founded the Institute for Advanced Study at Princeton, New Jersey.[2]

The 1910 report was a devastating, factual account based upon personal inspection of each and every existing medical school. (Flexner was a scholar of the old school who shunned questionnaires and the paraphernalia of the social sciences.) Its effect was to close the marginal proprietary schools and any others which were clearly of dubious quality. Flexner was hastening a process already underway, one resulting in consolidation of medical schools plus affiliation with educational institutions. Eliminating marginal institutions reduced the number of practitioners. Those who now graduated bore, tacitly, the stamp of scientific expertise and an assurance of high income. But the reduction in numbers affected both the type and quality of medical care given the nation. The marginal practitioners, after all, did provide services in areas shunned by the more established. And not all medical care requires the pretensions of scientific medicine.

Abraham Flexner applied a fairly strict standard—his interpretation of the Johns Hopkins experience and of the German medical school. The consequences of his reforms in medical education are still effective. In retrospect, his reading of the past can be faulted. But it prevailed because it was believed in—and acted upon. Because he assumed a static, rather simple situation, the net affect was to suppress any tendencies toward a healthy pluralism and severely limit educational experimentation for decades. The training supposedly yielded a "scientific" physician, one capable of applying science to practice or to research. Rarely did that occur. The education was often far too scientific and technical for the realities of most medical practices and, at the same time, often inadequate for research in the real frontier areas of biomedical research. Physicians usually settled into a defined routine; to advance beyond limited clinical studies, researchers had to educate themselves or get special training.

SIMON FLEXNER (1863–1946) became the first head of the Rockefeller Institute, a great research organization with many contributions to its credit. When the federal government expanded its role in medicine, the National Institutes of Health clearly showed the imprint of his achievements. Aspects of the institute's founding and early years are given below. Simon Flexner astutely combined laboratories with basic research functions with a hospital devoted to clinical investigations in the scientific spirit. It far surpassed Johns Hopkins as a model, and the medical schools incorporated aspects of his organization. (Note the parallel with university departments and the research institutes of the Carnegie Institution.) Flexner himself favored separation of teaching and research.

Mall and Loeb appear again commenting on medical education and on the effects of private philanthropy. A future Nobel Laureate, Joseph Erlanger, contrasts the state and private universities. Unspoken are the effects of popular pressures on medical practice and research. The 1911 letter of Loeb to Upton Sinclair, the novelist, illustrates one aspect of how organized medicine benefited from its association with science. Note Loeb's statements about medical education and practitioners and how he identifies the physician with the scientist. Sinclair might very well be in error, but Loeb is also concerned with limiting the right of the public to question experts.

Another clash over the rights of experts and the concerned public—more extensive and weighty—occurs in the antivivisection dispute between William James and Walter Cannon. It is essentially an argument about the autonomy of experts from the concerns of outsiders. James and Cannon, colleagues at Harvard, try to avoid the extremes and emotionalism of the arguments made for and against experimentation with animals. Somehow, the issues are not quite joined. They remain in different frames of reference. When the issue of human experimentation became acute after World War II, the issue was placed in a far broader perspective.

1. See Joseph F. Kett, *The Formation of the American Medical Profession* (New Haven, 1968); William G. Rothstein, *American Physicians in the Nineteenth Century: From Sects to Science* (Baltimore, 1972).

2. See his autobiography, *I Remember* (New York, 1940).

Franklin P. Mall to Wilhelm His, January 5, 1900

My dear Prof. His:

Your kind letter as well as your paper on "Protoplasm" reached me a few months ago, and in a very busy state. My Wirkungskreis has now broadened so much that I am allowing work to accumulate and must soon reorganize my daily routine so that I may become a more worthy pupil of you.

The autumn an excess of work with some sickness in my family (which is now two) brought my own work to a standstill which only today I have been able to begin instead of on Oct 1, as in the past.

We are yet discussing Medical Education and I have made great headway during the past year. In medical education we have progressed more in 1899 than during the previous 20 years. The prospects are now that we shall soon have a dozen Medical Faculties of good rank in the near future. My own efforts on behalf of reform in the programme of studies are meeting with much greater success than I ever anticipated. In America we do things with great enthusiasm and then undo them with the same enthusiasm.

I venture to inclose a plan for cold storage for the preservation of anatomical material. It is a modification of my own apparatus which is in use in Philadelphia. The plan is to cool the vault through the direct expansion of the ammonia in the ceiling pipes as well as in pipes within the brine tanks. By this system it is easy to keep a vault −15° C by operating a machine 6 hours a day. During the past year we have also greatly improved the CO_2 freezing microtome and I hope to send you one in the course of a year. The perfected instrument is now being manufactured.

I have just gotten a perfect 3½ and 5 weeks embryo. The latter is like your KO and this stage we so much need in our work on the development of the muscles. Both will soon go to the microtome. I am now getting too many embryos as it keeps me busy in looking after them. Over 50 pathological specimens I have just prepared for publication in the Welch Festschrift.

I attended a meeting of the Anatomical Society for the first time last week. Wilder[1] was president and talked over 6 hours on anatomical terms. Your mistakes of course were pointed out. He also gave notice that he would continue to talk on terms as long as he lived. I think we can stop the talk soon as all are heartily sick of it. At the Anatomical Society I saw much of Minot who is so much interested in embryology that he is writing another book on it. Through him also the old fashioned teaching at Harvard has been overthrown.

With best wishes for the New Year, Very sincerely yours

F. P. Mall

Mall Papers, Department of Embryology, Carnegie Institution of Washington

1. BURT GREEN WILDER (1841–1925) was professor of neurology and vertebrate zoology at Cornell.

Luther Emmett Holt[1] to Simon Flexner, May 11, 1901

My dear Doctor:

You may possibly have heard of the projected Research Laboratory to be established through the benefaction of Mr. John D. Rockefeller. Two meetings have been held and the work of organization is going forward rapidly.

At a meeting held in New York last evening, you were duly elected a member of the Board of Directors. The following gentlemen at present compose the Board: Dr. Welch, President, Drs. Prudden, Biggs, Theobald Smith, Herter and myself. The next meeting will be held in New York on Saturday evening, May 25th. The hour and place will be given later.

It was decided last night that for the present the fund at the disposal of the Board be used for investigation in different laboratories throughout the country and you are accordingly requested to come to the meeting prepared with answers to the following questions:

I. Is there any problem at present along the general line indicated which you are specially interested in carrying out?

II. What facilities have you and what would you require to accomplish this?

III. Have you any well trained men in view who could give all or half of their time to this work.

Very truly yours,
L. Emmett Holt

Simon Flexner Papers, American Philosophical Society

1. HOLT (1855–1924) was an eminent pediatrician, for many years a trustee and secretary of the Rockefeller Institute.

Luther Emmett Holt to Simon Flexner, May 15, 1901

Dear Dr. Flexner:

The details of the organization of the Pathological Laboratory I will explain to you more fully at our next meeting. For the present the laboratory will have no endowment, although it is expected that ultimately this will be the case. Mr. Rockefeller has promised to begin with $20,000 a year, to be expended under the direction of the board.

The scope of the work is to be purely of a research character and it was thought by the members at our last meeting advisable to attack two kinds of problems. One of the more simple, practical character, which might lead to definite results in a comparatively short time, i.e., in a year or so; the other of the problems of a more obscure character and more difficult, such for example as the search for the micro-organisms of scarlet fever, or measles, which would doubtless require a long series of careful experiments. At our meeting to be held next week it is hoped that there will be presented for consideration half a dozen problems of the simpler type, upon which we might profitably expend say one-half of our sum in keeping two or three men at work, reserving the balance for the more obscure ones.

The Board of Directors will have the authority to manage fully the scientific work. I trust that you will surely come to the meeting and I know that you will come prepared with practical suggestions, so that something may be begun even during the present summer.

Very sincerely yours,
L. Emmett Holt

Simon Flexner Papers, American Philosophical Society

Luther Emmett Holt to Simon Flexner, May 14, 1902

Dear Dr. Flexner:

You will be pleased to know that Dr. Herter and I had a long interview with Mr. Rockefeller last evening and went over with him the ground covered by our report and recommendations. He is quite in favor of the general proposition, but wishes to know what proposal we have to make for immediate action. This brings up the question of the site, the size of the lot required for the fully developed Institute and what building or buildings should be constructed at once in order that work might be begun a year from now. I presume he will make a reply to the Board after receiving the formal report and after going over the matter with his father. The opinion expressed last night is perhaps to be regarded as a provisional one but I think that there is little doubt that this will be his father's attitude also.

He has already talked with Mr. Carnegie regarding the prospects and scope of the Carnegie Institute and had been told by him that the Carnegie Institute did not wish to enter the medical field at all. This I think clears the ground for the development of our Institute.

I am sending these facts to each of the members of the Board in order that we may come together after having given some thought to the questions that are likely to be discussed at our June meeting.

Yours sincerely,
L. Emmett Holt
Secretary

Simon Flexner Papers, American Philosophical Society

John D. Rockefeller, Jr., to John D. Rockefeller, Sr., June 13, 1902 (telegram)

Rockefeller Institute directors after year of careful thought and planning, are now ready to take forward step. The scheme as outlined of which I spoke to you casually before your departure, contemplates eventually, for land, buildings and endowment about five millions. Present requirements, land sufficient for entire scheme and one building for present requirements probable cost between three and four hundred thousand dollars. Annual expense for salaries, maintenance, etc.; from 40 to 60 thousand dollars. I strongly recommend a pledge of one million dollars, to be drawn at the option of the Board during ten years. Of this, three to four hundred thousand dollars will be for land and buildings balance for current expenses. Would also allow original pledge of $200,000 during ten years to stand in addition. This amount will be used for investigations thro'out the Country as at present, while new pledge will be for work centered in New York. I should be glad of intimation of your feelings on the question by tomorrow morning. Board meets tomorrow night and I leave for Providence at noon. I feel confident this is conservative and wise. Have been considering matter with individuals of board for some weeks.

J. D. Rockefeller, Jr.

Simon Flexner Papers, American Philosophical Society

John D. Rockefeller, Sr., to John D. Rockefeller, Jr., June 14, 1902 (telegram)

To J.D.R. Jr.

Yesterday's telegram received in reference to Rockefeller Institute. As you so earnestly recommend you may pledge One Million Dollars to be distributed through next ten years. If it were left as you suggest to be drawn at option of the Board, they might take a large portion in the early part of the ten years. We cannot say anything about five millions now.

JDR

Simon Flexner Papers, American Philosophical Society

Franklin P. Mall to Wilhelm His, October 29, 1902

Dear Professor His:

I was so much gratified by what you said about organization that I sent your letter to the President of the Carnegie Institution. I have also sent him your pamphlet on Wissenschaftliche Institute, which he is having translated to be added to his report. President Gilman seems to be much of your opinion in this matter and I think that we shall probably have a series of Institutes thus organized. If the preliminary organisation meets with any general approval in the scientific world we shall have untold millions to carry the work on. We do not think that we can change dollars into "grey matter" but if we fail not we can encourage talent by making it independent. At present we are very familiar with the "Streber" and for some time at least we shall not be troubled with him. Our endowments for research are in the hands of Gilman, Billings, Weir Mitchell and Welch and we have the greatest confidence in their ability to manage them.

Before our Institutes are organized the leading universities are beginning to meet the new conditions by granting much greater freedom to some of the professors. At Harvard there are two professorships for research only and California has created such a chair for Dr. Loeb of Chicago.

Had we only the men our outlook would be most brilliant, but as it is we are not given to worry much over our own shortcomings.

I have made all arrangements to go abroad next summer and will take my family to Dresden. We would of course pass a few weeks in Leipzig and I hope to travel about somewhat. I fear sometimes that the business side of education occupies me too much so that from time to time it is necessary to return to the fountain. In addition to my troubles of that kind I am greatly overrun with students having 175 in gross anatomy this year. Dr. Myers and Miss Sabin have been added to my staff but I should like to get half a dozen more.

We have been back from Maine for over a month and feel now as if we had never been away. The autumn with us is always very beautiful and mild.

Believe me, Very sincerely yours,

F. P. Mall

Mall Papers, Department of Embryology Carnegie Institution of Washington

Frederick Taylor Gates[1] to Simon Flexner, December 6, 1907

Dear Dr. Flexner:

I have just completed reading the sheets of your article. The proof is much stronger than you, in your extreme conservatism, had led me to expect. Judging only from the statistics of recovery in the number of cases treated, the proof would be conclusive to a layman, but when there is added to this the almost uniform drop in temperature and subsidence of symptoms after each injection, these injections numbering hundreds, the heightening of the opsonic index, the destructive effect of the serum on the staining qualities and the germinal or reproductive life of the bacilli, surely these amount, when taken with the reversed ratio of cures to deaths, to a demonstration.

My first impulse was to wish you had put forth stronger claims and had eliminated the sentence in which you decline to declare your case proved, but on reflection I think you are right. An assertion that you have proved your case, even though I think you have indeed proved it, would be likely to provoke public challenge, or at least doubt, hesitation, interrogation, depreciation, but when you put forth this demonstrative evidence and make no claim for it at all, and indeed distinctly repudiate any claim of demonstration (which might be put forward by your friends), you conciliate the profession in advance and will receive encouraging and favorable comment rather than provoke criticism. Yes, you are right in maintaining this attitude of reserve even though you may be inwardly much better assured than you are willing to admit.[2]

I thank you for your considerateness in forwarding me these sheets. Mr. Murphy[3] is now reading them. I will keep them as a souvenir unless you wish them returned.

Very truly yours,
F. T. Gates

Simon Flexner Papers, American Philosophical Society

1. GATES (1855–1929) was a Baptist minister who became principal advisor in philanthropy to John D. Rockefeller, Sr.

2. Flexner's caution, for whatever reason, was wise. The results proved not as rosy as Gates had stated. He was a remarkable man. See Robert S. Morison, "Frederick T. Gates and the Rockefeller Institute for Medical Research," pp. 3–12 in *Trends in Biomedical Research (1901–1976)* (Pocantico Hills, N.Y.: Rockefeller Archive Center, 1977). (Proceedings of the Second Rockefeller Archive Center Conference, December 10, 1976.)

3. STARR JOCELYN MURPHY (1860–1921) was an attorney who worked with Rockefeller, Sr., in the philanthropic area.

Frederick Taylor Gates to Simon Flexner, January 14, 1908

Dear Dr. Flexner:

I sent your last letter on the Serum, to Mr. Rockefeller, Sr. with some words as to the significance of your work. I cannot refrain from giving you a glimpse of the delight which your work is affording him. In replying he uses the following words—"How can we be thankful enough for Dr. Flexner's discovery of the serum for cerebro spinal meningitis! Let the good work go on. It makes Mrs. Rockefeller and me so happy to know about it. Surely, life is worth the living."

Very truly yours,
F. T. Gates

Simon Flexner Papers, American Philosophical Society

Walter Cannon to Charles W. Eliot, May 5, 1909

Dear Mr. Eliot:

In answer to your request for specific instances of the more complex relations between research in sciences not directly concerned with problems of disease, and practical results in medicine and surgery, I offer the following:—

I. From Physics:—

1. The contriving of the compound microscope and the introduction of the oil immersion lense (which increased illumination and di-

minished spherical and chromatic aberration) determined the possibility of discovering the infective agents which cause disease and studying the minute changes which they produce in living tissues. These improvements in the microscope have been due to investigations of the way in which light is refracted in passing through media of different density—physical research not in the slightest degree actuated by concern for human suffering.

2. The ophthalmoscope, called by Helmholtz[1] his "optical toy," was another application of physical principles which has had great value in testing not only for ocular disease but for abnormal conditions in the brain.
3. The x-rays have proved highly important for detecting the presence of bone fractures, calculi, pulmonary tuberculosis, aneurisms, foreign bodies, and many other pathological states; investigating the functions of internal organs, as the stomach and intestines; and curing superficial cancer (epithelioma). The x-rays were discovered in the course of a research on electrical discharge through attenuated gases—with no idea of resulting value to medicine.
4. Researches on oxidation and combustion, and the proof of the conservation of energy, were at first purely physical interests. The application of these conceptions to food values, to respiration and animal heat, and the demonstration that animal bodies obey these physical laws have been fundamental to a rational attitude toward bodily activities both in health and disease.

II. From Chemistry:—

1. The discovery of ordinary (ethyl) ether (Valerius Cordius, 1540),[2] and the discovery of chloroform (Soubeiran, 1831; Liebig, 1832)[3] likewise, were made without the slightest hint of the unspeakable blessing they were to confer on mankind, both in the direct abolition of pain, and in the benefits from medical and surgical research on the lower animals, made possible by these anaesthetics.
2. The industrial importance of chemical research on the aniline dyes is well known. No less important is their use in bacteriology. The basic aniline dyes have a specific affinity for bacteria, and thus they render easily visible these minute organisms. Certain bacteria (for example, the tubercle bacillus) are differentiated from other bacteria in diagnosis by peculiar staining reactions to these aniline dyes—a use quite unsuspected in the original chemical investigations.
3. The development of the science of pharmacology has been largely dependent on advances in organic chemistry. These advances

permitted syntheses of related compounds; the pharmacologist then studied the physiological actions of these related substances with the purpose of securing more exact control of bodily processes. Thus through the application of methods of organic chemistry new local anaesthetics (cocaine, eucaine), new soporifics (sulphonal, trional, tetronal), new drugs for reducing fever (antipyrine, phenacetin, acetanilide), and other important agencies of great medical and surgical utility have been made available.

4. The growth of Pasteur's ideas revealed in a remarkable manner how research of prime interest only to the individual investigator may bring results of highest value to all mankind. Starting as a chemist he was struck by the fact that penicillium glaucum destroyed dextro-tartaric acid and did not affect levo-tartaric acid in the same solution. From that time forward, through studies on lactic acid fermentation, disease of beer, silk worm disease, anthrax, boils and puerperal fever, chicken cholera, and rabies, he, and others stimulated by his researches, were led step by step to a demonstration of the microbic nature of infectious diseases, to methods of conferring immunity, and to the triumphs of aseptic surgery.

III. From Biology:—

1. Although the ingestion of foreign bodies by living cells had been suggested as a means of overcoming invading organisms, Metchnikoff,[4] the zoologist, first observed this phenomenon (phagocytosis) in a small crustacean (daphnia). This observation was fundamental in understanding the relation of leucocytes to immunity and in treating disease by stimulation of phagocytosis.
2. One of the most striking conquests of disease in recent times is the practical abolition of yellow fever in civilized communities, made possible by the research of Walter Reed and his comrades in Cuba. It should not be forgotten that the immediate effectiveness of their work was dependent on entomologists who had been interested in the scales and veins on the wings, and the scales and hairs on the bodies of mosquitoes, and thus had classified them, and had studied their methods of breeding. (This last instance I have already used in a paper "The Opposition to Medical Research," published last year.)

These are only a few of the more striking examples of the way in which facts obtained in a disinterested search for truth in the biological and physical sciences have proved of highest importance in medical and surgical practice. One need cite only the discussion between Volta and Galvani[5] regarding the action of dissimilar metals on frog legs, and Pfeffer's[6] observations on turgescence of plant cells, to show also how biology has contributed to physics and physical chemistry. The isolated groups of fitted blocks in a puzzle have their place in making the final picture.

I hope that you may be able to use these instances of the results of free research in your argument for freedom.

Yours sincerely,
[*Walter Cannon*]

Cannon Papers, Harvard University Medical School

1. One of the great figures of nineteenth century science was the German physicist and physiologist HERMAN LUDWIG FERDINAND VON HELMHOLTZ (1821–1894).
2. CORDIUS (1515–1544) was a German physician and botanist.
3. EUGENE SOUBEIRAN (1797–1858) was a French pharmacologist. JUSTUS VON LIEBIG (1803–1873) was one of the seminal figures in the growth of chemistry in Germany.
4. ELIE METCHNIKOFF (1845–1916), a Russian best known for his work on the function of the white blood cells, was director of the Pasteur Institute 1895–1916. He received a Nobel Prize in 1906.
5. ALLESSANDRO G. A. A. VOLTA (1745–1827) and LUIGI GALVANI (1737–1798).
6. The German botanist WILHELM PFEFFER (1845–1920).

Walter Cannon to William James,[1] May 25, 1909

My dear Professor James,

Your letter to the Vivisection Reform Society, reprinted from the New York Evening Post in last night's Transcript, so valiantly strikes at both parties in the antivivisection agitation that perhaps it is best for each to be satisfied with the raps given the other side. There are some statements in the letter, however, which I am sure your fair-mindedness would not have permitted you to make if you had not so long been unacquainted with the conditions of medical research. Even I, who have lived in a medical laboratory for nearly ten years, have been surprised by the results of a recent inquiry into the circumstances of animal experimentation in the laboratories of this Country.

In most of the large laboratories regulations governing animal experimentation have been posted and enforced for many years—in one laboratory for more than thirty years. These rules have recently been collected and combined, with the following result:

RULES REGARDING ANIMALS

I. Vagrant dogs and cats brought to this Laboratory and purchased here shall be held at least as long as at the city pound, and shall be returned to their owners if claimed and identified.

II. Animals in the Laboratory shall receive every consideration for their bodily comfort; they shall be kindly treated, properly fed, and their surroundings kept in the best possible sanitary condition.

III. No operations on animals shall be made except with the sanction of the Director of the Laboratory, who holds himself responsible for the importance of the problems studied and for the propriety of the procedures used in the solution of these problems.

IV. In any operation likely to cause greater discomfort than that attending anesthetization, the animal shall first be rendered incapable of perceiving pain and shall be maintained in that condition until the operation is ended.

Exceptions to this rule will be made by the Director alone and then only when anesthesia would defeat the object of the experiment. In such cases an anesthetic shall be used so far as possible and may be discontinued only so long as is absolutely essential for the necessary observations.

V. At the conclusion of the experiment the animal shall be killed painlessly.

Exceptions to this rule will be made only when continuance of the animal's life is necessary to determine the result of the experiment. In that case, the same aseptic precautions shall be observed during the operation and so far as possible the same care shall be taken to minimize discomforts during the convalescence as in a hospital for human beings.

The foregoing regulations have in some instances replaced the older regulations; they have been adopted in other places to such extent that now practically every laboratory in which animal experimentation is extensively practiced is under such government.

From the foregoing information you will understand that the defenders of animal experimentation do not, as you state, protest "against any regulation," nor do they agree with you that it is "no one's business what happens to an animal, so long as the individual who is handling it can plead that to increase science is his aim," nor have they "disclaimed corporate responsibility." You will be interested to know that the Medical School in which you were trained has a Committee on Animals, with plenary powers to control the conditions of research and to discharge instantly any employee guilty of inhumanity to the laboratory animals. A representative of this Committee daily inspects the rooms in which animals are kept and is under orders to report any case of neglect.

Have you not confused a protest by medical men against special legislation directed at them and an assumption commonly made that they object to any control of their work? They do indeed object to special legislation. Such legislation, they contend, is unwarranted. General legislation against cruelty has in two states during the past year proved to be adequate to punish cruel dissection of animals. Special legislation in England has led only to distrust of the inspectors, and to increasing violence of agitation. And furthermore, the Vivisection Reform Society is unique in demanding only legal restriction; the fifteen antivivisection societies of England, and at least four in this Country, though crying for special legislation, exist for the abolition of the means of medical progress.

The American Medical Association, the national body of medical men of this Country, has appointed a Council on the Defense of Medical Research. That Council has secured reports of the conditions of animal experimentation in nearly all the laboratories of the United States. The regulations given above are one of the results of its labors. It believes that placing the well-being of experimental animals in charge of the responsible director of a laboratory, constantly at hand, will ensure a greater degree of attention and consideration, indeed a greater degree of humanity, than would be provided by the political appointment of a government inspector, with unknown qualifications, who could make only occasional visits.

We have been convinced by careful examination that our house is in order. We are planning to give publicity not only to the conditions of animal experimentation in laboratories of medical research, but also to the way in which the important results for the relief of man's estate have been derived by such experimentation. We shall perhaps be unable to convince our opponents that we are not callous monsters, rejoicing in pain, for they take the attitude which you take—that we "cannot be trusted to be truthful or moral when under fire."

Yours sincerely,
Walter B. Cannon
William James Papers, Harvard University

1. JAMES (1842–1910) was a philosopher and psychologist. To the point is his medical degree (1869) and early work in physiology. One of the great figures of American intellectual history. One suspects Cannon knew this of his former teacher which underlines Cannon's efforts to win over James.

William James to Walter Cannon, May 28, 1909

Dear Cannon:

When I wrote my letter to Mr. Taber, I well knew that I should give pain to friends like you and H. P. B.[1] (by whom, however, I hope it may remain unread, as he ought now to have graduated from the polemics of this world). The regulations of which you inform me are something the wide extent of which I am glad to learn, but altho I stand condemned thereby of a certain amount of ignorance, the plea I make is in principle only confirmed. The laboratory directors are singly acknowledging the ethical responsibility, and the medical association is beginning to respond to the attacks in the way which I call for. If now the American Physiological Association will officially formulate an ethical code with a censorship with power of expulsion attached, it will have something dignified to meet the public with. *That* is the kind of initiative that ought to come from them, rather than the initiative of stirring up the American Naturalists to pass vague denunciatory votes which sound to the reader only like expressions of club feeling and claims of class privileges. That, I mean, is the *general* impression that all the pro-vivisection utterances give. Instead of frankly admitting that experiments on live animals are an atrocious necessity, and confessing the duty of the utmost economy of misery and brutality, they habitually single out what is manifestly weak and preposterous in the attack, reply solely to that, ignore everything else, seem to claim that a laboratory is a sort of Garden of Eden, and leave on the reader an impression of prevarication, not to say mendacity, which but strengthens the anti-vivisectionist cause. I don't except Porter's manifesto of some years ago, or even yours, dear Cannon of some months ago, which I turned to with expectations of something radical and vital on the whole question, only to be disappointed. You turned only a narrow edge of defence, to certain narrow attacks. But why not deal with the whole matter from the bottom, and constructively and radically, not as a class under fire, screening itself by a certain policy of retort? Why swell the public impression that it is a class-cause that is being defended, with all the lack of candor that in such cases prevails? I wouldn't guarantee our College faculty (individually extraordinarily conscientious men) not to *lie* thru thick and thin, to ward off attacks from without. I couldn't trust our philosophic department, myself included, not to lie when assaulted by the laity. Deny everything provisionally; and meanwhile get the house in order so as to reply more truly to the next attack! Don't you think that the anti viv. agitation has had much to do with the establishment of all those laboratory rules which your committee has collected? I imagine that but for it, few of them would now exist. That *that* is the effective line of defence is my whole contention; only it ought to be deepened and consolidated. I miss, for instance, any discountenancing of lecture demonstrations the absolute vanity of most of which is one of the strongest impressions I took away from my student days. It isn't merely the cruelty and coarseness, it is the *stupidity,* before an audience whose intellects you are supposed to be addressing, of whipping up the flagging mind by an appeal to action and sensation (like the newspapers with their scareheads and portraits) in spite of the fact that what the experiment *is,* no one sees at all. The ether, blood, etc. are a *stimulus,* that is all! I hope that this is diminisht since my [time?]. I blush myself at the memory of certain cats vainly used by me before my classes in the '70's, and quite as much do I blush at the brainless frogs and pigeons which I used to show yearly to my students in psychology, not because of the pain, which was certainly absent in the case of the pigeons, but because the essential refinement of the facts I was trying to illustrate was entirely submerged in the coarse laughter which the show provoked—the room used to take it only comically. Our conceptualizing faculty has after all to be depended on for most of the things we learn; and where the vivid touch added by sensation has to be purchased by the life of a fellow creature, I think it ought to be part of our responsible moral equipment to practise economy, and to see that something genuinely intellectual (as distinguisht from more retinal stimulation) is gained. I know that this will have a sick-room and squeamish sound; yet I am sure that this whole side of the business is the one that has to be articulately *emphasized as the ideal,* for men's native callousness and tendency to abuse of power can be trusted to weigh the other side sufficiently, I am sure.

There, dear Cannon, I have written at too great length. You are in a position to do more good, and to do *more to meet attack efficiently,* I am persuaded, if you will take the course I suggest, than in any other way. What individual physiologists are most in need of, I am sure, is more moral courage in face of their own club opinion. Why can't you yourself become a more aggressive leader along that line? I think you will be surprised at the adhesion you will get from persons now mute and inert within the profession, if you become an energetic advocate of reform from within, and let outside attacks be answered *only in that way.* The really judicious public is, I believe, equally [sick] of the idiocy of the one side and of the *suppressio veri* of the other.

Yours faithfully,
Wm. James

William James Papers, Harvard University

1. HENRY PICKERING BOWDITCH (1840–1911) was Cannon's predecessor in physiology at Harvard.

Walter Cannon to William James, June 4, 1909

My dear Professor James:

You give pain to your friends in such a tolerable manner that they may easily forgive you. Please do not think, however, that I have taken your strictures on physiologists in any personal sense; I mentioned your belief that they as a class are incapable of telling the truth when attacked because I consider your attitude, common enough, an example of one of the most desperate forms of social injustice. It belongs to the class of unreasonable prejudgments of race and condition, whereby men are huddled, classified and stamped, stamped with doctrinaire notions of what they must be and what lives they must lead. We crowd the negro inexorably into menial positions and then cry, "Behold, he is incapable of rising." In a not dissimilar way the antivivisectionist attacks, with prejudice and imagined evil, the medical investigators, and if the investigators defend themselves they are at once branded as liars. Why? Because no body of men, when attacked, are truthful or moral. To render them perpetually untruthful and immoral, therefore, merely continue the attack! Then no one will believe what *they* say, and the cause is won!

Is not that the essence of unfairness and injustice? Your attitude automatically makes it impossible for investigators to defend their position. They might do what you propose—formulate an ethical code, with censorship, and penalty of social degradation—but on your own statements must not you yourself believe that they would set up these devices with a sly wink at one another and a nod towards the Public? Just what degree of honor in such matters should you expect from a body of untruthful and immoral men?

Can you be patient with me while I state again my view? Certainly I should assert the ethical responsibility of the experimenter both to his community and to the animals experimented upon; but I should equally insist on a fair hearing and on judgment based upon the facts, and not on the fancies of ignorant accusers. It is right that the community which receives the benefit, should decide how far the employment of animal life for its own gain is justified, and naturally the experimenter should abide by that decision; but it is equally right that they who know the conditions of medical research should see to it that these conditions are not misrepresented.

I like to regard this question as I regard any complex question—as one in which the evidence must first of all be established. What evidence is there that investigators are heartlessly cruel, or even careless and indifferent in the use of animals for experimental purposes? When you write that there is a *suppressio veri* on our part, what is your evidence for that statement? When you mention the "unspeakable possibilities of callousness, wantonness and meanness of human nature," what testiomony leads you to suppose that these qualities prevail among men of science? I desire evidence, not imaginings.

This was the attitude I took in my address last June; and since that time, in connection with the Council for the Defense of Medical Research, I have been searching for such evidence myself. Thus far my search has been vain. The conditions of animal experimentation throughout the Country are, I am convinced, thoroughly respectable. Nevertheless as chairman of that Council I have gone on record with the promise that if ever I learn of an instance of wanton cruelty to any animal in a laboratory there will be no more ardent advocate than I, of just punishment of the offender through existing laws. Can you have the will to believe that your vivisectionist friend is capable of burning with moral indignation at cruelty to animals?

So far as *evidence* of the practices in laboratories is concerned I can only recall again the regulations cited in my previous letter. These regulations doubtless were originally formulated in response to a recognition of public interest and responsibility. Personally I do not believe that they have in any respect altered the treatment of animals in the laboratories; they merely state for general information the conditions of animal experimentation. If you or any other sober-minded persons do not believe that these regulations are seriously regarded by the experimenters, I can only reiterate the invitations that have repeatedly been given that you come to the laboratories at any time and see for yourselves.

My previous letter was intended to be informative and not controversial, and I therefore did not mention the use of animals for demonstrations. Certainly on that point we must immediately admit disagreement. If I were teaching philosophy and not a medical science, concerned with the facts and methods of observation, I should perhaps feel as you do in this matter. If my students looked on demonstrations as a joke, and greeted them with laughter, as you state your students used to do, I should agree that demonstrations were worse than useless. If the demonstrations were seen by only a few men in the front row, again I should admit that they were a failure. So far as the "waste of animal life" is concerned, that argument does not appeal to me. About 15,000 cats and 6000 dogs are killed annually in Boston merely to be rid of the excess. Perhaps two dozen of such cats are used annually for demonstrations. The total number of dogs and cats killed in New York City alone was, in 1907, about twenty times as many as were used for all purposes in all the large medical schools of the Country. To my mind an animal on its way to death by etherization is much less "wasted," if used for instruction, than if directly killed. A person can make himself sick with the thought of such use of an animal just as he can turn his gorge by vivid realization of the fact that he has frequently sunk his teeth into pieces of the carcass of a cow. But if instead of observing a demonstration in this "sick-room and squeamish" manner, we contemplate it, as I believe most of my students contemplate it, as an exhibition of one of the marvels of bodily functioning, there can be aroused only deep interest and serious respect. Why assume that these exhibitions are given to "help out the professor's dullness"? There is a more reasonable explanation in the desire to establish a knowledge of the subject, as in all teaching of natural phenomena, on the relatively firm basis of sensation.

Just one further point and I will cease abusing your patience. You have written, as many of the older antivivisectionists properly wrote, as if the physiologists were sole offenders in animal experimentation. If you should visit medical laboratories today, however, you would find that this important method of medical progress was being used by bacteriologists, pathologists, pharmacologists, hygienists, biological chemists, by surgeons in surgical laboratories, and by physicians in laboratories of clinical research.

We physiologists take our place among these active investigators, but we are not pre-eminent, except possibly in our service as scape-goats.

Please do not feel under the necessity of replying to this long letter. I felt that I must explain myself to you or be tormented by an unhappy sense of being misunderstood.

With cordial best wishes,

Yours sincerely,
Walter B. Cannon
William James Papers, Harvard University

William James to Walter Cannon, June 7, 1909

Yours of the 4th finds me up here, dear Cannon, and makes me grieve that I should have imposed the trouble on you of writing another long letter. In such discussions the cancelling process can rarely be applied to the pleadings *pro* and *con,* so as to get a clear residual balance. The only effect is an often inarticulable practical hitch in either direction wrought in the disputants. If you take the least hitch forward in favor of some positive *ethical expression of a mandatory sort* on the part of your committee (as contrasted with mere repelling of accusations, &c.) my letter will have only too promptly served its purpose,—that being to me the one really promising line of defence against the anti-vivisectionist attacks.

My calling most of these "idiotic" sufficiently clears me of complicity with the great raft of them. And I am only too happy to be corrected by you as to the fact of absence in our present laboratories of the indifference to the animals which left so strong an impression in my mind 30 and more years ago, both in the laboratories which I saw, and in the lectures which I heard. I cannot help suspecting that the improvement has in part been due to the anti-viv. agitation, for I don't wholly share your good opinion of scientific human nature. Class for class, the scientific class is no doubt superior, & original geniuses in biology must have free rein, but is every idea-less imitative student a consecrated soul? The motives of individual scientific men may [be] as vulgar and egoistic as anyone's. And I have never seen any reason for supposing that the scientific class, if given privileges, would be a mite less unscrupulous in defending them than any other class would. Men are faulty, and no class of us can safely be left protected against outside criticism. But I gather from your last letter that your committee is even now working on the lines which my letter suggested, so there remains no practical ground for dispute. The total ethical situation is the *hauptsache,* and ought not to be obscured by quarrels about the accuracy, either way, of this or that detail.

Yours always truly
Wm. James
William James Papers, Harvard University

Walter Cannon to William James, February 11, 1910

My dear Professor James:

I told the reporter last night that what you had stated was an incorrect inference from what I had written to you. You will note on reference to my letter of May 25 that I presented regulations regarding animal experimentation which were a combination and revision of rules found to have been long enforced in some of the larger laboratories of the country. I stated that the acceptance and enforcement of these rules indicated that experimenters are not, as your letter to Mr. Taber declared, indifferent to "what happens to an animal, so long as the individual handling it can plead that to increase science is his aim," indeed that much care is given to the comfort of animals used in research.

In my letter the American Medical Association was mentioned only with reference to the investigation conducted by its Council on the Defense of Medical Research—an investigation which revealed the rules previously cited. I wrote that the Council believed that placing the responsibility for experimental animals on the director of the laboratory was more likely to ensure the results we all desire than arranging for occasional government inspection.

This morning's (Friday) New York Herald states, "Professor William James of Harvard University yesterday still maintained that he had been told by a member of the American Medical Association that that body was about to admit the principle of ethical responsibility in vivisection and would probably publish disciplinary rules on the subject.''

Just what is meant by ''ethical responsibility'' might be long debated. If it means that no one should wantonly or carelessly cause pain to animals I should say that the point has never been in question among experimenters. The rules above mentioned were a spontaneous expression, in widely separated places, of the caution every decent man naturally takes in experimenting on animals. The only part played by the American Medical Association was the appointing of the Council. The Council has summarized the rules which it discovered and has distributed them in places where they did not exist. I should say that these rules were not in any sense ''disciplinary'' in character. They merely express to new-comers in laboratories and to the interested public the manner and the spirit in which animals are used for experimental purposes.

I should not write to you further about this matter did I not feel that your statement gives a wrong impression of the attitude of the American Medical Association towards the management of laboratories (I have been telegraphed to both yesterday and today about this matter), an attitude which the Association is in fact incapable of maintaining in any sense of absolute control. And as I am involved with you in the matter I submit to you the question whether my statement justified your inference.

You will perhaps be interested to know that in an antivivisection ''chamber of horrors'' in New York City, your name is posted with others in large letters on the wall as having expressed opposition to vivisection. The society managing this exhibition is that presided over by Mrs. Balais.

I get very much discouraged at times about this fight. We are all so prone to misinterpretations that I wonder if mutual understanding will every come. Surely I should turn quietly to my investigations if I did not feel that grave problems of human welfare were dependent for their solution on freedom of medical research.

Yours sincerely,
W. B. Cannon

William James Papers, Harvard University

Walter Cannon to William James, February 13, 1910

My dear Professor James,—

I wish to thank you for your letter which came this morning. You are not more strongly in favor of the control of the methods of medical research from within than I am. It is quite possible that the exercise of class-pressure, which you have advocated, may prove to be the most efficacious means of assuring uniformly humane treatment of animals. At present, however, we are trying out the influence of a public declaration of principles, with the trust that any person tending to be careless may be kept in the path of considerateness because those principles have been formulated and published. And the enforcement of the principles in any case is left with the director of the laboratory—the most trustworthy and responsible person whom we could select.

If the plan now in trial does not operate satisfactorily for the humane care of experimental animals we must devise something more drastic. I hope that because the Am. Med. Assn. hasn't gone the full length of your recommendations you will not be discouraged. The situation is very complex, and we must move with full consideration for all the factors involved. My protest was not against the desirability of corporate responsibility and discipline, but against the statement that the Association now takes that attitude. You have admitted that your inference that the Association now takes that attitude is incorrect. I may some day have to announce to you that an association has been formed which has adopted the "James plan"! Certainly I shall use my influence to have the laboratory procedures arranged to spare in every way the discomfort of animals that are rendering to man their supreme service. The "James plan" may be the best means of attaining that end.

We agree at least in our dislike of writing letters. I hope that this may be the last we need to exchange on this distressing subject.

With best wishes,

Yours sincerely
W. B. Cannon

William James Papers, Harvard University

Jacques Loeb to Upton Sinclair,[1] March 13, 1911

My dear Mr. Sinclair:

I have received your novel and I am reading it with the greatest interest. I shall write you more fully my impressions when I have finished it. As far as I have gone, I am inclined to believe that it is the best work you have done yet.

While I am in full sympathy with Upton Sinclair, the Novelist, I am afraid I cannot say the same with Mr. Sinclair, the Healer, and for the simple reason that discoveries in physics, chemistry and biology are not made by mere inspirations. Syphillis is caused by a microbe organism, by spirochaete pallida; and is cured by the killing of all the spirochaete in the body. This can be done in most cases by mercury, which is more of a poison to microbe organism and second, more effectively by Ehrlich's new arsenic compound. If you want to prove that fasting kills it, the way to do it is by experimenting on animals which have been made syphilitic on a large scale, as has been done in regard to mercury and to 606. Moreover, this can be done only by experts who are familiar with the organism and the tests for its presence, and can no more be done by a layman than the determinations of the number of molecules in a cubic centimeter of gas can be left to a layman not familiar with the theories and methods for such a determination.

I do not know whether you realize it, but, you are trying to substitute for the expert scientist the inspired layman, and in this attempt I cannot sympathize with you, since it means the negation of all civilization. You mean to fight physicians, in reality you fight the real principles which make progress of science and of civilization possible.

I hope you will forgive me this lecture. I should like to come to the meeting but I am afraid I can not since I have a lecture on Friday which I have not yet written out.

I remain,

Yours very sincerely,
[*Jacques Loeb*]

Loeb Papers, Library of Congress

1. SINCLAIR (1878–1968), the novelist and muckraker. This letter probably refers to his recently published book, *The Fasting Cure* (New York, 1911).

Joseph Erlanger[1] to Walter Cannon, March 10, 1913

Dear Walter:

I gladly make the attempt to compare for you the spirit of Hopkins and Wisconsin, though I believe that the comparison you ask of me can scarcely be considered typical, in that Wisconsin ranks with the foremost of the western state universities.

One difference between the two types of institutions consists in the attitude of the authorities with respect to the method of attacking problems. At Hopkins the research activities are directed, one might say, by purely academic considerations; the direct aim is the advancement of knowledge. Should the problem happen to have a local flavoring and should the result thereof be applicable to the improvement of local conditions, so much the better. In the state university the authorities prefer that investigators interest themselves in practicable local problems, though as a rule no pressure is brought to bear upon them to bring this about. In this connection I have heard President Van Hise say in effect that it is the duty of the state university to interest itself in local problems, and that in the long run such interest would be of advantage, not alone to the state, but to science as a whole.

The staff of the Medical Department of Wisconsin is not overburdened with teaching. Though they happen to do considerably more than at Hopkins, with all due respect to Harvard, I am inclined to believe that more teaching is done in the latter place than at Wisconsin. However I do not believe that this is the case in the other colleges at Wisconsin. Conditions in the Medical School there are largely attributable to the fact that with but one exception, all of the chairs are filled by Hopkins men who have striven to transplant Hopkins conditions.

The main difference between the two institutions lies, I believe, in the tendency in the case of state universities of interference by the legislature. Legislative bodies are rarely in full sympathy with the spirit of research, though probably they are not entirely to blame for this attitude. They are the representatives of the people, and I am inclined to believe that were the question "Shall a tax of fifty cents per year be levied on every person, man, woman and child, in the State of Wisconsin for the support of research at the University?" be put to the people of the state, it would certainly fail to carry. Yet it has been estimated that this actually is the amount contributed by the state to investigation, directly and indirectly, at Wisconsin. Within the last few years Wisconsin has twice been subjected to a legislative investigation. As a result the University of Wisconsin can claim priority over Harvard for the institution of a scheme for determining the relative amount of time devoted by instructors to teaching and to research. Each semester a blank is issued to instructors with spaces for this information.

In institutions like Hopkins the spirit that obtains is determined primarily by the president, although the trustees actually have the power to determine questions of this kind, they, as a matter of fact, never interfere.

It is scarcely necessary for me to remind you that there are many large private eastern universities which, as regards spirit, are no better off then the smallest of the western state universities.

I hope I have made myself clear and that I have furnished you the information you desire. If not, I will be glad to make another try at it.

The annual problem of deciding where to spend the summer is with us. Should you happen to hear of a place in your vicinity which you think would be suitable, I would be glad to hear from you in regard to it, but please do not put yourself to any trouble for us.

With kind regards,

Sincerely yours,
[*Joseph Erlanger*]

Erlanger Papers, Washington University

1. The physiologist ERLANGER (1874–1965) received the Nobel Prize in 1944. Most of his career was at the Washington University School of Medicine.

9
The Reform of the National Academy or Hale's Vision

Almost from the start of George Ellery Hale's election to the National Academy of Sciences in 1902, he worked for its revitalization. More than the reform of one organization was involved; Hale saw his moves as part of a grand strategy for all of the sciences—even all of learning—within the United States. Complete success eluded him, but what came about has had important consequences to this day. To know what Hale wanted and why his vision did not become reality is important in order to understand the thinking of much of the leadership of the scientific community as well as the role of scientific knowledge in American society.

Objective observers in 1902 might wonder at Hale's choice of the National Academy as the vehicle for reform. Established in 1863, the academy never lived up to the hopes of its founders. It did not become the recognized apex of the scientific community with authority, tacit and otherwise, to set standards and to determine policy. The government occasionally called it for advice, not necessarily on weighty issues. By the time Hale appeared, membership was mildly honorific. The academicians were, by and large, a rather staid elderly group, limited in size and in their relationship to newer developments. But even among the academicians there were stirrings for change.

Hale recognized one great asset in the National Academy, an asset which it retains to this day. It was the institutional embodiment of the great tradition of science. On paper, the academy was the peer of the noted societies and academies of Europe traditionally linked with the creators of modern science. Part of the ideology of science stresses the innovativeness of its activity, its forward-looking character. From the standpoint of professional historians or of the humanities, scientists sometimes are indifferent or hostile to the past—even quite ignorant of it. When they turned to the history of their disciplines, they often displayed—as they do even today—an extraordinary naivete. Not that they did not care; the pose of indifference obscured an important role ascribed to history. To define the nature of the scientific tradition and to legitimate forms of behavior, the scientists evolved a heroic account of their past, a folk mythology of an intellectual elite. Opening remarks in textbooks, commencement addresses, presidential effusions to societies, and innumerable hortatory essays all mentioned the Royal Society, the Paris Academy, etc., connecting these with Newton and other great figures. References to past events also crop up in private correspondence; many examples occur in this volume. It mattered little to Hale and the other scientists that their "history" was often erroneous, misleading, or incomplete. They were not historians of science; it was their truth which accorded with their visions. An American sociologist noted, sometime in this period, that things believed to be real are real in their consequences. He was correct, but Hale's case indicates the need for a corollary: if beliefs depart greatly from reality, consequences will not match expectations.

Because of his work on the sun, Hale stressed international cooperation. The revitalization of the National Academy at first had the aspect of simply providing a better national locus for joint ventures with foreign scientists. But Hale wanted his academy to be as prestigious as were the older ones of Europe. That required more than internal reform and greater visibility for the organization. If the grand tradition was to be perpetuated, the prestige of the nation in the field of science had to be increased. Underlying Hale's actions was the apprehension not only that the United States would not attain scientific greatness but that it would go its own way oblivious to the pattern of the past. By 1913 Hale was ready to act.[1]

First he sought advice from the elders: Henry Smith Pritchett representing the scientific community, and the attorney Elihu Root.[2] (If there was an establishment in the United States in 1913, Root was its closest embodiment.) Clearly, Hale already knew what he wanted. Next, Hale wrote to leaders of particular disciplines asking each one to rate the prospects of his field. In November he gave a talk on his planned program for the academy at its annual meeting; copies went to all the members asking for their opinions. Replies were analyzed and tallied. Hale then revised his talk, which appeared in two installments in *Science*.

Two extrascientific factors are immediately visible in the documents that follow. First, Hale explicitly thought in nationalistic terms, both domestically and in the foreign arena. His reforms were not simply the imposition of a traditional form for the furthering of abstruse research. Second, he postulated a new scientific culture to combat the increasing fragmentation of knowledge and society. Although formidably specialized himself, Hale looked back with nostalgia to a simpler era characterized by a greater degree of unity. His vision was conservative both in social and scientific policies. Evolution was to be the central connecting element in this new scientific culture, although the model for its exposition in education was not Darwin's writings but the encyclopedic *Kosmos* of Alexander von Humboldt, a work already anachronistic on its publication in five volumes in the years 1845–1862.

Hale's proposals for institutional and educational reform aimed to undercut the growing importance of specialized disciplinary societies, the universities, local societies, and other organizations—all in favor of the hegemony of the National Academy. Ever bold in his conceptions, Hale favored extending National Academy membership to more engineers, more social scientists, and even to the humanities. Moves in that direction had occurred even before Hale spoke in November 1913. Although continental precedents in Paris and Berlin were cited by Hale, his motives were very much in accord with the spirit of his day. The reformed National Academy would constitute a monopoly of talent for all fields of learning.

But the respondents (75 of a total membership of 130) were not jubilant over Hale's ideas. Only 15 were in total agreement and two in complete disagreement. The remainder had qualms on specific points. The official tally reported a 64% neutral response, hardly a landslide. Although only three academicians were against including the humanities, only eight bothered to vote in favor. The rest remained silent. This point fell, but despite the tepid reaction Hale and his allies continued to press for his program. And many of its aspects came into being.

The responses on the status of fields must have presented a dilemma. Hale probably expected a continuation of the familiar laments about American indifference to basic research, a theme persisting down to the present. All but one respondent thought American work equaled or surpassed the best in Europe. Whether that judgment was true or not, it reflected the vigor and ebullience of the expanding research effort in universities, the government, and industry as well as the infusion of funds from the great philanthropies. The one naysayer was the physicist J. S. Ames[3] of Johns Hopkins. If a significant number of scientific fields were publicly admitted to be on a par with the best in Europe, crying poor would not work, as in the past. To Carnegie, in private, in the letter given here, Hale admitted the real strength of many fields. Such strength implied that lagging fields might advance in older modes, casting doubt on the need for Hale's program. Hale did not publicly avow what his correspondents asserted.

For the governance of the scientific community, Hale was promoting an academy model. A self-perpetuating elite of elites, the National Academy's stature and effectiveness were to arise from the achievements of the members. Collectively, the academy was to exemplify research and to symbolize the greatness of the scientific tradition of western civilization. Hale even wanted a "Faraday" to serve as Home Secretary of the Academy. By setting standards, the academy could encourage and reward the promising young. Research eminence of the national scientific community would enable the academy to deal with similar national bodies on equal terms. Domestically, a revitalized academy could represent science as an autonomous force in negotiations with governments, industry, and various professional groups. And it could negotiate as a peer with these other segments of the society.

While Hale worked with passion for his vision, Cattell continued his persistent efforts to make the American Association for the Advancement of Science the governing body of learning in the United States. His mechanism was the Committee of One Hundred, a device typical of Progressive

Era reform efforts. It was a serious threat to Hale. Unlike Hale, Cattell favored using the multiplicity of societies in a kind of federal system. He actually used a political analogy: the association would serve as the lower house and the academy as a powerless House of Lords. Cattell was also an academician; his committee enlisted some of the very people Hale relied on for implementation of the reform of the academy. Cattell also included a wide range of other interests, notably in the social sciences and humanities. Like Hale, he wanted a federation of like-minded intellectuals from all fields. Unlike Hale, Cattell had an obvious relation to reformist currents. Cattell wanted to work with government; Hale, a genuine conservative, mistrusted the political process, preferring to build the strength of his community before getting involved. Hale eventually bested Cattell in the struggle, but the Committee of One Hundred attracted the attention of the Rockefeller Foundation, which had great consequences in later decades.

Other assumptions undergirded Hale's actions. This was a vision of a small number of great men and near-great men generating new facts and new theories. Facts and theories then percolated down to the level of applications. It did not shun utility; rather, practical consequence came best in a certain path, under the aegis of the best scientists. Implied was the notion of deference due the best scientists. Crucial here is the tacit belief in the small size of the community—essentially an expanded National Academy, a corps of neophytes, and a ragged battalion of investigators who had not quite made the level of the near great.

The realities were quite otherwise, insuring the frustration of many of Hale's bolder visions. The United States by then had a widely diffused and pluralistic culture of the sciences, pure and applied. There were brigades of applied practitioners—teachers, engineers, doctors—far from the level of the leading researchers. An expanding number of institutions tolerated or actively promoted all kinds of investigations. In fact, the United States had a growing scientific culture which was not basic-science centered but research centered. This country has always been hospitable to research. The sharp distinction between basic and other kinds of research is a relatively recent one. In the United States the tendency was to blur the distinction and treat basic and applied as an identity under one rubric—research. It was an older viewpoint elsewhere obliterated by scientific advances and professional pride. A great part of the history of the sciences in twentieth-century America consists of the frustrating efforts of the scientific community to get the larger society to honor a distinction alien to much of the national style. Even Hale sometimes lapsed into the comfortable rhetoric of the last century, confusing theory with practice in the name of scientific progress.

1. For discussions of Hale's motivations and actions, see Daniel Kevles, "George Ellery Hale, the First World War, and the Advancement of Science in America," *Isis* 59 (1968): 427–437; Nathan Reingold, "The Case of the Disappearing Laboratory," *Amer-*

ican Quarterly 29 (1977): 79–101; Ronald C. Tobey, *The American Ideology of National Science, 1919–1930* (Pittsburgh, 1971).

2. ROOT (1845–1937) became an attorney, unlike his father and brother who taught mathematics at Hamilton College. The *Dictionary of American Biography* notes that his income of $5,000 in 1869 was five times that of his father. Root never forgot that in pushing for a higher status for science. He was secretary of war 1899–1903, secretary of state 1905–1909, in the U.S. Senate 1909–1915.

3. The physicist JOSEPH S. AMES (1864–1943) was at Johns Hopkins where he served as president 1929–1935. For many years he headed NACA, the predecessor of today's space agency.

Henry Smith Pritchett to George Ellery Hale, February 3, 1913

My dear Hale:

As I wrote you some days ago, I have gone over with much interest your memorandum concerning the National Academy of Sciences and venture now to send you some results of my reflections concerning it.

The essential question involved, in my mind, is, can the present National Academy be developed into an agency which shall touch the national life more directly, which shall help to dignify the calling of the scientific man in the eyes of the American people, and which shall also serve the interests of science at the same time?

The familiar example of the Paris Academy is the one to which we ordinarily refer when we think of such an agency in America. There is, however, a very great difference in the problem which such an agency would need to face in France or England and that which it would need to face in America. For example, it is hardly conceivable that weekly meetings of the National Academy could be held in Washington in the way in which such meetings are held in Paris. Our country, aside from the lack of interest in scientific work and the focusing of attention upon commercial matters, is too large geographically, its interests too diverse; the geographical distribution of members of the Academy would be too wide to permit such an agency to exist.

You mention two important ends to be sought: first, to provide the Academy with a permanent home; secondly, to teach the public to look to the Academy as the chief source and representative of science in America.

I am inclined to believe that a fitting home for the Academy in Washington is the first step towards its wider recognition. On the other hand, such a gift without an adequate endowment would be useless. The mere holding of annual or semiannual meetings in this building would have no appreciable effect on science or scientific recognition.

If a building is to be got, it should carry with it not only the endowment necessary to care for it, but also an endowment sufficiently great to carry on in this building certain activities. For example, the best known men of science might here give from month to month statements concerning the progress of their own work and of their own sciences, but even all this does not go very far toward accomplishing the purpose which you have in view. A suitable building with a good endowment and comparatively large income for such a purpose is, after all, an external thing. The gist of the matter, it seems to me, is contained in your second requirement, and to attain that the Academy itself must be the agent. There is no means, in my judgment, by which the public may be taught to look to the Academy as the chief source and representative of science in America except by the Academy itself, and this means in fact some sort of scientific leadership. It means men willing not only to work enthusiastically and heartily at science, but to consider themselves a part of the scientific brotherhood and to give themselves heartily and enthusiastically, like Faraday, to the work both of scientific investigation and of scientific leadership.

As I see it, therefore, the solution of the problem lies in developing in the Academy itself a leadership able, first, to impress itself upon a few men of means, later on upon Congress, and later on upon the country at large. Under such leadership I am sure a suitable building and good endowment can be secured. The development of such leadership will in the end bring about some sort of relationship between the National Academy of Sciences and the various local academies and associations throughout the country, but this relationship also can be brought about only by the development of an unselfish, enthusiastic, scientific leadership. The attitude of the Academy in the past has not generally been this. It has been a little inclined to take the position that Congress and the country should accept it and defer to it as the scientific center of influence, without at the same time giving to Congress and to the country any real leadership outside of the work which men do in their own homes and their own laboratories and which has counted, as a rule, for the individual influence rather than for the united influence. In fact, many individuals in the country have far more influence both upon public opinion and upon Congress than has the National Academy.

Whether the Academy can develop such leadership or not seems to me the question. In order to do it some man—perhaps some group of men—must be willing to live in Washington and to give their effort and their time unselfishly to the development of the idea. You know better than I whether such devotion and such leadership can be had, but without it I do not think that a building or even an endowment would be worth providing.

Yours sincerely,
Henry S. Pritchett
Hale Papers, California Institute of Technology

George Ellery Hale to Elihu Root, March 3, 1913

Dear Senator Root:—

I have thought it best to delay replying to your letter until I could give very careful consideration to a general plan for the National Academy which has been formulating itself in my mind for some years. I fully agree with what Doctor Pritchett said in his letter of February 3rd, and recognize the difficulties to which he refers. It seems to me that they are to be met by recognizing the fundamental differences between the conditions existing in the United States and in a more compact country like France, where centralization has gone so far as to concentrate most of the science of the nation in a single city. It is evident that the National Academy cannot hope to maintain weekly meetings attended by a considerable proportion of its members. For this reason it must lose one of the greatest advantages enjoyed by the Paris Academy or the Royal Society. Again, it cannot hope to produce a series of publications of equal standing with the Philosophical Transactions and the Proceedings of the Royal Society. These date back to the early days of science, before the special journals which now contain most of the technical papers had come into existence. At the beginning they represented the whole of science in England and acquired a prestige which could not be duplicated if the Royal Society were re-established to-day.

But such reasons are insufficient to cause us to despair of the possibility of developing greatly our own National Academy. At the weekly meetings of the Royal Society and the Paris Academy the scientific advances of the members are presented, and go out to the world under the name of the national body. There is nothing to prevent our accomplishing a similar result, provided funds for the purpose can be supplied, by publishing a journal which will contain the first announcements of the important results of all the members of the Academy. This need not interfere with the publication in detail of the various papers in the technical journals, where I think they belong. But it would have the advantage of affording a prompt means of announcing results through a national body, which would thus be recognized and rated as other national societies are, namely, on the basis of the work of its members.

Moreover, the members would thus have an opportunity to see the results of their colleagues in other fields than their own, which is impossible at present, because the multiplicity of special journals prevents investigators from reading those which do not lie in their own field. One of the most important functions of such a body as the National Academy is to widen the interest and sympathies of its members, and to lead them to take the broadest possible view of their own investigations. I believe that such a journal would be of great assitance in helping to accomplish this result. As for the possibility of publishing it, I have talked with several of the most prominent members of the Academy, who tell me that they would be glad to contribute the first announcements of their results, in the form of papers of moderate length, to a National Academy journal. It would thus resemble the Comptes Rendus of the Paris Academy, which does not contain long and detailed communications.

The question of the personal qualifications of the leaders of the National Academy, and their attitude toward the science of the country, is, as Doctor Pritchett justly states, the most important consideration with which we have to deal. It is true that some years ago the National Academy sometimes exhibited a narrow tendency and a disinclination to take a broad attitude toward the scientific interests of the country at large. There is reason to believe that some members were more interested in the exclusion of certain scientific men than in the election of new members to the Academy. At one period, for this or some other reason, only two or three members were elected per year, although the Constitution then allowed the addition of five new members annually. That this spirit has practically disappeared is shown by the change made in the Constitution several years ago, when the maximum number to be elected annually was increased from five to ten. It is probably undesirable, under the present conditions of American science, to add more rapidly to the membership of the Academy than is actually done under the revised Constitution.

The coming annual meeting is of great importance, not only because of the celebration of the fiftieth anniversary of the Academy's foundation, but also because the terms of office of the President and Home Secretary expire. It is essential to the successful conduct of the business of the Academy that both of these positions should be filled by the selection of Washington men, and I have no doubt that officers of the requisite liberal spirit, breadth of view, and willingness to devote themselves to the best interests of the Academy, can and will be found. In this connection let me repeat a suggestion which I made in a recent letter to Doctor Pritchett. In planning the future of the Academy, it seems to me that we should take advantage of any worthy examples which are found abroad. One of the most attractive of these is afforded by the Royal Institution, where Faraday carried on his work for so many years. In addition to his long series of investigations, which has made the Royal Institution so famous, he carried forward and developed the courses of lectures on scientific subjects which he had attended, as a bookbinder's apprentice, when Sir Humphry Davy held the Professorship of Chemistry. It is a literal fact, significant in its bearing on the possibilities of the Royal Institution's plan, that practically the entire scientific instruction of Faraday came through these lectures of Davy's. Courses given at the Royal Institution are not confined to the lectures of the Professor of Chemistry, or of others immediately connected with the staff, but invitations are extended to eminent men of science in various parts of the world, who think it an honor to describe their work before such a distinguished audience as is always gathered for the "Friday Evening Discourses." As a step in this direction, the children of the late William E. Hale have offered to the Academy the sum of $1,000 a year, for five years (with the expectation of continuing the same grant indefinitely), for the establishment of an annual course of lectures. If we could now secure, say as Home Secretary, a man who would be as nearly as possible a counterpart of Faraday, I feel confident that the future success of the National Academy as a useful American institution would be assured. Such a man must be, first of all, an investigator of high order, as the desired influence could not be exercised by anyone else. He must also be of attractive personality, and capable of describing in popular language the results of his work. To attract and hold an investigator, the principal essential is a well-equipped laboratory, with such assistance as might be required to carry on his researches in an effective manner. Therefore, if sufficient funds were available, I should make provision for such a laboratory in the proposed National Academy building. If we could create there such a tangible atmosphere of research as exists in the Royal Institution, where the original apparatus of Davy, Faraday, and other great investigators, is preserved and exhibited, and where present investigators may receive the extraordinary stimulus of lecturing at Faraday's own lecture-table, and of using some of his instruments to illustrate their work, I should feel that an immense step had been taken.

As for the possibility of finding the right kind of a man for this position, I do not think the difficulty insuperable. In fact, I have in mind two men, one a biologist, the other a physicist, either one of whom seems to me competent to accomplish what I have in mind. Both are carrying on important original research, both are attractive personally, and both are capable, I feel sure, of interesting others in the results of scientific investigations. Doubtless several other comparatively young men could be found with the requisite qualifications. In addition to the work of research and lecturing, the occupant of the position would serve as Home Secretary of the Academy, and act, perhaps in conjunction with a committee, in securing other lecturers for the proposed courses, which would be similar to those of the Royal Institution, including some of a semi-technical character, and others suited to the most popular audiences. If such lectures are to accomplish their full purpose, they should be freely illustrated by experiments, and for this purpose a collection of apparatus would be necessary. This need not involve any great initial expenditure, but the endowment should be sufficient to permit a sum to be spent annually to supply additional instruments and apparatus, so chosen as to serve, as far as practicable, for the work of research in the Academy laboratory and for use on the lecture table. In this connection I am sending you one of the circulars of the Royal Institution, which will serve to give an idea of the courses of lectures conducted there.

I think that provision should also be made for a series of exhibits, open to the public, of the current researches of the members of the Academy, as mentioned in a recent letter. Thus, through the publication in a suitable journal of the chief results obtained by the various members of the Academy, the periodical contribution to the press of brief abstracts of these papers, in clear and simple language; the exhibits of the work of the members, changed from time to time; the courses of lectures of the Home Secretary of various scientific men in this country and abroad; the exhibition of instruments and apparatus of historical interest used by former members of the Academy; and by various other agencies which could be developed if a building were available, the Academy should acquire in time a commanding influence of a broad and liberal character, favorable alike to the development of research and the public appreciation of science.

In addition to the above means of promoting the work of the Academy, I believe it should increase its participation in cooperative research, and strengthen its relations with the International Association of Academies. Through the influence of the National Academy, two international organizations for cooperation in research have already been established, and there is no reason to doubt that more can be accomplished in the same direction. Furthermore, I believe that in the course of time the Academy might well establish relations with local academies of science in various parts of the United States, and perhaps encourage their work in some such way as through the award of special prizes for the best papers of the year, or the participation of delegates from the local academies in certain sessions of the annual meetings in Washington.

It is somewhat difficult to make an estimate of cost, but the following rough figures may perhaps be of some service. A suitable building, with the addition of a laboratory, would probably cost from $600,000 to $700,000. The initial cost of the laboratory instruments and equipment would of course depend upon the character of the work to be done, as the instruments of a physical laboratory cost much more than those of a chemical or biological laboratory, on the average. Probably $50,000 would be a maximum figure, and perhaps not more than half this amount would be needed. Doubtless an initial expenditure of $10,000 for a working library, and an annual appropriation of $1,000 to maintain it, would be sufficient. The allowance made above for the building ought to be large enough to cover the cost of a heating plant, furniture, etc. As for the annual expense, I should suppose that the salary of the Director of the laboratory should be $5,000 or $6,000. He ought to have one Associate, at $3,000, and two assistants at perhaps from $1,200 to $1,500 each. There should also be one mechanician at $1,500, a stenographer at $900, a laboratory helper at $900, and a janitor at about $900. You will know the Washington scale of wages better than I do, however. If an allowance of $2,000 were made for heat, water, power and light, and $7,000 for apparatus, supplies and other expense, the total annual expenditure would amount to about $25,000. Probably this amount could be very materially reduced, but as you suggested a fairly ambitious scheme, I am supplying what appears to me to be liberal figures. These do not include the cost of publishing the proposed journal, which would amount to from $2,000 to $3,000 per year. They also do not include any allowance for lectures, other than those to be given by the Home Secretary and the income of the William E. Hale fund. Many of the lecturers would appreciate the opportunity to take part in such a course, and charge little or nothing for their services. Nevertheless, an additional sum of $1,000 a year for this purpose would be very desirable. There would also be some expenses involved in the installation of exhibits, etc. A total expenditure of $30,000 per year, however, ought to cover everything.

In addition to its trust funds, the Academy has a small endowment, regarding which you will find full details in the annual report. I believe, however, that the general expenses other than those indicated above will absorb all of the income, and that it would therefore be impossible to count on the application of the present income to the purposes outlined in this letter.

You will notice that I have not attempted to estimate the cost of the land, as I am not sufficiently familiar with Washington real estate values to do so.

I ought to add that the suggestions contained in this letter do not necessarily represent the official view of the Academy, as I have not had an opportunity to consult the members of the Council since receiving your last letter.

I shall be very glad to amplify the details here proposed, or to supply any other information you may desire.

Very sincerely yours,
[*George Ellery Hale*]

Hale Papers, California Institute of Technology

George Ellery Hale to Elihu Root, March 10th, 1913

Dear Senator Root:

Although my discussion of the National Academy is already a lengthy one, I trust you will permit me to write further regarding three points, as I am anxious to prove beyond doubt that there is a real opportunity for useful work. I refer to the participation of the Academy in international research, briefly mentioned in my last letter; the possibility of increasing the public appreciation of science, by directing attention to its cultural value; and its fundamental importance in the arts.

I am sending to you under separate cover the first and last volumes of the Transactions of the International Union for Cooperation in Solar Research, an alliance of twenty-two societies in England, France, Germany, Italy, Austria, Russia, Holland, Spain, Sweden, Switzerland, Belgium, Canada and the United States, formed through the initiative of the National Academy. At the last meeting of the Union, held on Mount Wilson in 1910, its scope was extended to include the whole subject of astrophysics. Interest in its work has steadily increased, and the advantages of international cooperation in this field of research have become more and more obvious. When inviting the principal European academies of science to send delegates to the preliminary meeting, held in St. Louis in 1904, the National Academy enjoyed the advantage of being a member of the International Association of Academies. The constituent bodies of this Association, which will hold its triennial meeting at St. Petersburg in May, are pledged to cooperate officially only with such projects as are recognized and endorsed by the Association. As the only American member, the Academy is thus in a position to introduce and support other undertakings in which international cooperation is desirable. The possession of a building would permit the Academy to invite the International Association to meet in Washington. Acceptance of the invitation would mean the selection of the National Academy as the "leading Academy" of the

Association for a period of three years—a position held successively by the Royal Society, the Paris Academy, the Vienna Academy, the Accademia dei Lincei of Rome, and at present by the St. Petersburg Academy.

The work of the International Committee on Paleontologic Correlation, initiated by the National Academy and conducted under its auspices, is another illustration of what can be accomplished in the field of cooperative research. Other opportunities of a similar nature lie open, and some of these should be embraced in the near future.

At the present time, science is rarely regarded as a cultural subject. It is not taught exclusively for practical ends, though the average student in a technical school is inclined to resent the courses in pure science which do not seem to apply directly in his chosen profession. These courses almost invariably present the details of a single branch of science, for a thorough knowledge of the subject can be obtained in no other way. But the average student is confused by the vast array of facts and the complexity of detail. He learns the formulae of chemistry or physics, but fails to grasp the general principles that underlie them. Even if he becomes a master in his chosen subject, he may not recognize its true place in the broad field of science. He thinks of astronomy and physiology as circumscribed by distinct boundaries, and the thought that a single idea unites them may rarely enter his mind.

The college student of the humanities, who takes a few science courses to broaden his horizon, is at a similar disadvantage. I have talked with such men and learned of their regret that they had never been given a picture of science as a whole. If they had been taught anything of evolution, it was in a single field—the evolution of the astronomer, the geologist, the biologist, or the student of languages. A comprehensive view of the development of the earth and its inhabitants had never been placed before them.

I am strongly of the opinion that every educational institution, from the high school up, should give a course in general evolution. Humboldt's "Cosmos," though written more than half a century ago, will never be out of date to the imaginative reader, who must be stirred by its all-embracing picture of nature. In many respects it might serve as a model for such a course of instruction as I have in mind. Add the broad generalizations of "The Origin of Species," and the discoveries of recent years, and we have the material for a course capable of inspiring thousands who find little of interest in the details of science. The cultural value of such instruction is beyond question. It would stimulate the imagination no less profoundly than the best works of art or literature. It would help to bridge the gap that still divides science from the humanities, by preparing the student to appreciate the meaning of evolution in every department of human activity. And it would aid in breaking down the provincial narrowness of the man whose mind is wholly occupied with the details of his own life. The United States must play a large part in the politics of the world, and take a deeper interest in foreign affairs. I believe that such instruction would be a useful means to this end.

The National Academy, better, perhaps, than any other institution, could set on foot a movement in this direction. Courses of lectures, amply illustrated by lantern-slides and experiments, would soon be repeated in educational institutions. A small model museum of evolution, containing comparatively few striking objects covering the entire history of the earth, would be imitated elsewhere. The persistent teaching of broad relationships in natural science, history and economics, could not be without influence on educators and on the public.

In still another respect, the Academy can perform a public service. If we are to compete successfully with Germany in foreign and domestic commerce, we must use science as effectively as she has done in the development of our industries. An exhibit in the Academy building, supplemented by lectures illustrating the benefits derived by manufacturers from the cultivation of pure science and its application to industry, would at once serve the interests of the investigators and stimulate the manufacturer to improve his product by research. The student of pure science is constantly asked as to the "practical value" of his work. If he could point to an exhibit which would show, for instance, how the modern dynamo goes directly back to the physical researches of Faraday, made with no thought of their application in the arts, he might ultimately see a change in the point of view of the public. Certainly the manufacturer, spurred on by German competition, and led to discover its source in the work of the investigator, should soon become the strongest supporter of scientific research. As Liebig once said in a letter to Faraday: "Practice alone can never lead to the discovery of a truth or a principle." Illustration of the values of principles as the basis of invention should lead the practical man to appreciate and aid the work of the National Academy.

It is almost hopeless, however, as Doctor Pritchett has remarked, to expect important results without the constant influence of a vigorous and inspiring personality. Here lies the necessity of securing a man, as nearly as possible like Faraday, to initiate and carry on the proposed work. I believe this to be the key to the whole problem. But without a well-equipped laboratory for his investigations, and a building especially adapted for the lectures and exhibits, no Home Secretary could be expected to contribute greatly to the standing of the Academy.

I should repeat that the responsibility for these suggestions lies with me. While I do not know whether the details of the plan will meet the approval of my colleagues on the Council of the Academy, to whom I am submitting them, I trust that its general features will receive their support.

Very sincerely yours,
George Ellery Hale

Signed in Mr. Hale's absence.

Hale Papers, California Institute of Technology

James McKean Cattell to William Henry Welch, April 26, 1913

Dear Welch:

It may be convenient if I send you some preliminary notes in advance of our conversation in regard to the problems confronting the National Academy. Permit me, however, to preface them with the remark that you are one of the few men, perhaps the only man, who, when I differ from him in judgment, I believe is more likely to be right than I—so egotistical are some of us. The problems before the Academy are difficult and wide reaching; their solution will require much time and various consideration. My own opinions, though I may express them dogmatically, are not fixed; indeed those who say they do not change their minds confess that they can not learn.

I need scarcely say that I was eager beyond measure that you should be elected to the presidency of the academy, for thus the respect, the admiration, and, I venture to add, the love which you command more than any other known to me, are made tributary to the advancement of science, for which I care more than for anything else, as the chief factor making for human welfare.

It seems to me that the theory of three coordinate departments of the government—the legislative, the executive and the judiciary—should be made to include a fourth department, the scientific and advisory. The government and the people need skilled guidance. The legislative representing under proper conditions the commonsense of the democracy requires competent advice prior to the enactment of laws. This advice must be given by men learned in past experience and fertile in new ideas. Such men are not to be entrusted with the enactment of laws, for by the nature of things their specialization interferes with their perspective. But their function is not less important than that of the legislature which enacts laws, the executive which puts them into effect or the courts which interpret and enforce them.

What then should be the fourth department of the government? Clearly a body similar to the National Academy, if the membership were composed of the men most competent. A self-perpetuating corporation electing its limited membership for life on the ground of scientific eminence can not be such a body. By the nature of things it must include many who are old, conservative, timid and inefficient. Scientific men are likely to be competent, sensible and cooperative. This holds especially in the natural sciences where these qualities are required for their work. In the mathematical sciences there is more likelihood of erratic genius. But the hundred men of the country most eminent for research would never be the hundred most competent to guide the advance of civilization.

If at our annual elections we eliminated by ballot the ten least competent members, it would be a truly dramatic vote, though by no means reflecting unfavorably on those who were dropped. This plan would not do; though it seems to me feasible to have a provision in the constitution to the effect that members who had not attended a meeting for three consecutive years and had presented no adequate excuse should be regarded as having resigned. But no method of elimination would give us a dominant body representing the scientific men and the scientific interests of the country and responsive to their will. At present the national scientific societies are the strongest organizations making for scientific progress. If they should elect, say for a term of six years and in proportion to their membership when brought to a common standard representatives to a national senate or council (with some others coopted or elected at large), this probably would be the strongest and most effective body we could obtain. The council of the American Association for the Advancement of Science is to a certain extent such a body, but it has only recently been organized in its present form and has never had the responsibility that leads to action. It has, however, just now demonstrated what it can accomplish by approaching through its committee on policy the President of the United States and the Secretary of Agriculture, who then requested its advice on matters of great moment for the scientific work under the government. The question appears to me an open one whether the council of the American Association or the National Academy should be the body to guide the scientific work of the country and its applications to our complicated civilization. Perhaps both might be developed—one corresponding in its methods to the house of representatives, the other to the senate.

The question of the National Academy is a double one;—What should it do as now constituted and what should it do if reorganized on a representative and democratic basis? We have been during my twelve years of membership essentially a pleasant social club. I have greatly enjoyed the meetings, though I doubt whether I should have been justified in attending them if this had not been in the line of my work. I do not think that we can considerably improve on our past performance under existing conditions of membership. We could arrange better social events, such as we have enjoyed this week, and we could arrange a better scientific program, such as the American Philosophical Society has had in recent years. But there are more vital things in the world than dinner and receptions, and a program of papers on all sorts of subjects, even if carefully prepared, would after a while get to be a bore. The bait of membership does not appear to me to be a considerable incentive to good work, either before or after election, and in my opinion nothing has more demoralized science than the payment of scientific work in the fiat currency of reputation and honors. Still under existing conditions the honor of membership and the pleasant associations which follow may be of use, and would be if members were paid a salary as in the continental academies or even had their travelling expenses paid when attending the meetings. I doubt whether much if anything would be gained by enlarging our membership. I am not enthusiastic about a journal to summarize the discoveries of members, or the publication of their researches, though the latter function might be useful if the government will pay the cost. Neither am I hopeful about the likelihood of popularizing science or extending its influence among the public. But let us by all means do the best we can in these and other directions, using the great traditions of the academy, paralleling those of the university, to accomplish the best ends within our reach.

A representative academy based on a true democracy would in my opinion advance science and the welfare of the people to an extent not likely to be exaggerated. I should place its direct and immediate economic value at not less than a hundred million dollars a year. And this would represent the least part of its service. It is a favorite idea of mine that science by its economic applications has made democracy possible, and that when democracy repays its debt to science there will be an advance in the whole conduct of life such as has never before occurred. But it would be futile to consider in detail what might be accomplished by an academy which will not soon exist.

On reading this letter, it sounds somewhat like a public address. I wrote it largely in the train after leaving you last night. This was, however, at a time when the affairs of the academy were vivid in my mind, and if you have no objection I may show it to several of those on whom in my opinion the future of the academy chiefly depends. I enclose several papers which you may not have seen, as I had no reprints of the addresses. I should like to talk over all these questions with you in New York or Baltimore, but best of all here at my house.

Sincerely yours,
[*James McKean Cattell*]

Hale Papers, California Institute of Technology

Samuel James Meltzer[1] to Arthur L. Day,[2] April 13, 1914

Dear Doctor Day:

To my regret I found no time to read the paper of Dr. Hale until a few days ago. I read since two of his other papers in Science. No doubt they are brilliantly written. His paper on the Future of the Academy deals, as you rightly say, with matters vital to the future of the Academy. You say it is the desire of the Council that the members acquaint the Home Secretary with their opinion regarding the more important suggestions which Dr. Hale makes. To discuss the suggestions intelligently it would require more of my time than I can spare just now. Besides, as a young member (not in years) I do not care to force my opinion upon the body of the Academy with too great enthusiasm. But I shall not hesitate to express my opinion on all the points under discussion briefly and concisely.

My title to some opinion is based upon my familiarity with the conditions which prevail in one section of the Sciences in which a marvelous progress was made in this country in the last twenty-five years. I am absolutely sure that the Academy had not a particle of merit in that progress, and I feel quite confident that the Academy will gain no important influence upon the actual progress of sciences, no matter what will be done. The progress is sustained by the larger Universities, by State Universities, by specializing Institutes and above all by the awakened spirit in the intellectual set of the younger men, most of whom have not even heard of the existence of the Academy. Academies in the long past made science; Academies of the present time are intellectual orders conferring membership upon especially meritorious men of science.

An attempt to build up today a creative Royal Society is a most mistaken notion. Two hundred and fifty years ago there were quite a number of savants who were capable of understanding all the knowledge which existed at that time. Today there exist very few men who have a thorough knowledge of the details of their own branch of Science, and there exist not many men in any Academy who can follow intelligently more than 30% of the papers as they are offered now at a single meeting, and the man who can follow more is usually the one who does not contribute great things in his own science. There is a natural mutual exclusiveness between productivity and receptivity. An attempt to make the Academy do things for which it is not fit may terminate disastrously.

I am therefore against the proposition that the Academy should have laboratories of its own for carrying on original research. There will have to be numerous laboratories, as there is no reason why one science should have the preference over another. Where should the money come from for such immense expenditures? Where is the guarantee that the work in these laboratories will be better, the discoveries more numerous and more important than in any laboratory of some University or Institution, and why should it be done—merely to reflect credit upon the Academy—is that a sufficient moral and intellectual reason?

I am against erecting a special Academy building. If there will be no special laboratories, what for should the Academy possess a special building? If there will be a building then, of course, it will be desirable to grace the walls with the portraits of the great men of Science, the great officers of the Academy, etc.; to start developing a library; to have a meeting room, etc. But are these purposes in themselves sufficient reason to make a serious effort to get money for a building?

I am against the idea of using the name of the Academy for instituting courses of popular lectures, that is, lectures for non-scientific and perhaps even uneducated people on scientific subjects. I have no objection to such popular lectures; on the contrary, I believe they are very desirable. But it can not be one of the aims of a *National* Academy of Sciences to provide occasional or regular lectures for a few hundred people in the City of Washington.

I am against the publication of Proceedings, such as of the Royal Society or the Wiener Sitzungsberichte. I had to use them, and I know what a nuisance they are. I am not speaking of the French Comptes Rendus or the Berliner Verhandlungen, because one rarely uses them. The transactions of the Leipziger Akademie gave me trouble because they contained certain articles which were of importance to my line of work—but I could not obtain these publications in New York. In our times of necessary specialization it is folly for one to publish his work in reports of Academies which contain all sorts of things.

But I would suggest that the papers which are read at the meetings as well as the publications of Proceedings ought to popularize their presentation, i.e. that the subject should be presented in such a manner that the underlying principles as well as the progress made could be readily grasped by at least 80% of the members present whose minds are well prepared to comprehend any scientific subject if properly presented. Such presentations would be an education of the members in sciences distant from their own. The same effect could be accomplished upon a much larger circle of readers of such Proceedings.

The aim of the Academy should be to confer honor upon especially meritorious men of science. Men (or women), young or old, who made important discoveries in their sciences, or older persons who for years continued to contribute creditable work in their science, should be eligible to membership. But scientific contributions should be the only criterion of eligibility.

Regarding the number of members we should rather follow the example of the Royal Society than the Prussian Academie. But we ought to be careful in our choice. Above all we ought not to permit the personal influence to be a factor in the admission to membership. By an accidental personal experience I know that the Royal Society is not free from it, and I am afraid that our National Academy also could stand improvement on that score.

I believe that the National Academy should stand especially for upholding the standard of pure science. In recalling examples rather Hertz than Marconi should be cited. We should not erect buildings or laboratories, we should need no money and we should not cater to the good opinions of manufacturers. I am so bold to say that we are harping too much upon the fact that the Government of the United States consults the Academy once in a while on trivial matters. The Academy should be for Science because it is Science and not because an individual who accidentally and temporarily occupies a position of influence in the Government of the party in power pays us, once in a while, a half-hearted compliment.

Universities, special institutions, etc., are studying and advancing the knowledge of pure and applied sciences; it is not a part of their objects to uphold the dignity of pure science as a national asset. The U.S. Government has no room for pure science; it cares only for values measurable by the gold standard. The National Academy of Sciences composed of men advancing the realms of all knowledge for knowledge sake may acclaim for itself the right of being the representative of pure science in this country.

I agree with Dr. Hale that historians, philosophers, economists, philologists, etc., ought to be admitted to membership. It is a National Academy of Sciences and not of Natural Sciences. When they should present their subjects in a manner intelligible to all of us, we all will be the gainers.

I wrote more than I thought I would; pardon now the length.

Very truly yours,
S. J. Meltzer

Hale Papers, California Institute of Technology

1. MELTZER (1851–1920) was a physician born in Russia. While in private practice, Meltzer maintained a private laboratory. From 1904 until his death he was at the Rockefeller Institute. This is by far the most interesting response to Hale's talk.

2. ARTHUR L. DAY (1869–1960) was at this date director of the Geophysical Laboratory, CIW, and Home Secretary of the National Academy at this date.

George Ellery Hale to Andrew Carnegie, May 2, 1914

My dear Mr. Carnegie:

I reached Pasadena yesterday, and now wish to write you further regarding the National Academy, as the statement sent you from Chicago was prepared under the pressure of many engagements, and did not represent the case adequately.

I am perfectly sincere in my desire to see the Academy identified with the Carnegie rather than the Rockefeller interests, and for more than a year I have opposed in the Academy Council any action tending to secure aid from the latter. It is not merely because the name of Rockefeller would severely injur a national organization, perhaps even causing Congress to compel us to refuse a building from this source, and certainly making the grant of a site very unlikely. This is, of course, a serious practical objection, but it is not the one I have had first in mind. My long association with the trustees of the Carnegie Institution and the Carnegie Corporation has convinced me that the Academy, if allied with them, could expect such sympathy and appreciation of underlying purposes as no other body of trustees could give. As director of the Yerkes Observatory and a member of the faculty of the University of Chicago for twelve years, I also had long experience of the Rockefeller methods and some direct dealing with trustees of the Rockefeller Corporation. Without the slightest desire to criticize the Rockefeller trustees, who have done much to advance education and science in the United States, I must nevertheless say that their attitude with a few important exceptions is entirely different from that of the Carnegie trustees. In fact, if you will permit me to say so, the trustees of both corporations reflect in a remarkably accurate manner the personal characteristics of the two founders. This may sound like flattery, but it is simply an honest expression of my opinion, and I could bring independent evidence to support it.

At the present time three members of the executive committee of the Carnegie Institution (Messrs. Welch, Walcott and Woodward), and two heads of its departments (Messrs. Day and Hale), are members of the National Academy Council, which contains no representatives of the Rockefeller Corporation or the Rockefeller Medical Institute.

Feeling as I do, it is only natural that I should wish to see the Carnegie representatives, who have done so much for science by the foundation of research institutions, take the final great step of developing the unlimited possibilities of a national body which stands for the allied interests of American science. Looking toward the future, and realizing the tremendous strides made by Germany in the last quarter of a century, we see that as our natural resources approach exhaustion we must rely more and more on Germany's guiding principle—the fundamental importance of the methods and discoveries of science. The National Academy is in a unique position to demonstrate the value of scientific research to the manufacturer and to counteract the danger, now very apparent, of overlooking the fact that no permanent success can come unless we cultivate pure science as the foundation stone upon which applied science rests. It is easy to improve the dynamo after its fundamental principle is known, but it was Faraday, working in the Royal Institution with no thought of gain, who conceived the first dynamo and made all its subsequent developments possible. In a suitable Academy building the importance of scientific methods could be constantly illustrated and described, thus educating the public, and the manufacturer, great and small, in the procedure which we must follow if we wish to maintain and improve our place in the markets of the world. The great corporations already realize the necessity of establishing research laboratories, but the value of scientific methods must be realized by the minor ones as well. Without a building, the Academy can do little or nothing to aid in this important work.

I might go on to show how the Academy, because of its representative and truly national character, can encourage and help American science in every part of the country as soon as its means permit. The history of the Academy which I have sent you proves that it has already done much in this direction, but the possibilities of the future are enormously greater. In many departments, American science is now on a par with that of the first countries of Europe. The Academy can stimulate the development of the weaker branches, support and elevate the stronger ones, encourage local scientific societies in every state which now lacks appreciation and guidance, preserve memorials of American men of science and illustrate in an historical exhibit the contributions to knowledge made by this country, inform the public of scientific progress in every department, and act, more worthily than is now possible, as the accepted representative of America in the great field of international research.

An Academy building is the prime requisite, and for this reason I have laid before you a request for $750,000. The preliminary design of a building planned to accomplish the above purposes calls for an expenditure of about $900,000, but this can be cut down to $750,000, and still give the necessary space, with an architectural effect as attractive as that of the Pan-American building. As the clearing-house of American science, and its official center in both a national and an international sense, this building would be a contribution of the first importance to our intellectual and material progress.

I have asked the Home Secretary to send you some recent publications of the National Academy, and I can give you much additional information regarding its history and work, our plans for its future, and its place among the Academies of the world, if you care to have it.

Very sincerely yours,
[*George Ellery Hale*]

Hale Papers, California Institute of Technology

10
World War I: Dying Illusions and Vain Hopes

Progress, universal and inevitable, obscured the vision of many in the late nineteenth century and early twentieth century, before World War I. All around was the working out of the course of history—in the spread of industrialization, the eradication of illiteracy, in the elimination of dread diseases, and in the easy triumph of Western arms and ways of life in other parts of the globe. Science, personified and undefined, powered Progress; the two often appeared almost interchangeable.

But not everyone accepted the idea of Progress. There were areas not touched by the Industrial Revolution; remnants of an older order survived even in advanced countries. Radical naysaying was in vogue by the turn of the century. There were those who were outraged at some of the results of the new social order. A few turned Progress and Evolution upside down. Instead of the survival of the fittest yielding the best, the course of evolution was retrogression. History was on a downward slide. That, however, was an extreme position, repudiating a hopeful belief in science. Radical critics and liberal reformers alike looked to accurate knowledge properly applied as the ultimate solution to shortcomings and evils.

Because the sciences had prospered so greatly in the nineteenth century, scientists had little reason to doubt that their research would inevitably advance. Nor was there any serious inclination to regard their achievements as anything but universal in nature; each national community was part of an international movement animated by an unselfish, cooperative idealism. The growth of steam transportation, telegraphy, high-speed printing presses, and the postal service all eased the means of intellectual intercourse. Students flocked to foreign universities. By the second half of the last century international congresses had become part of the ritual of organized science. To an extraordinary degree individuals and organized groups approached the high standards called for by the ideology. Many educated individuals in the Western world felt that science had reached a plateau to which many other parts of civilized society were still struggling.

Noble beliefs sincerely held are not immune from a certain smugness, a tendency to overlook blemishes and to puff up attainments. Conveniently downgraded were ample evidences of bitter personal and national rivalries. Science as an abstract body of concepts and data had universality; scientists—human beings in specific environments—could not evade youthful conditioning nor their own emotions. Like the many radicals who talked of the international proletariat, when the flags were raised at the start of the Great War most scientists stood up and saluted.

Instead of something basic finally uncovered and established by millenia of human striving, the wartime experience demonstrated the fragility of the learned response called "science." One could not always depend on scientists to live up to the ideals of their calling; elementary and baser emotions could prevail. And this even called into question assumptions about the abstract quality of scientific knowledge. If the researchers were not immune to pressures and tensions of the larger world, perhaps the origin and acceptance of the contents of science did not always arise from a chaste internal logic.

The outbreak of a general European conflict in 1914 understandably affected the citizens of the warring states immediately. As the struggle continued and losses mounted, maintaining older routines and beliefs became more difficult. Disruptive, emotional consequences were not limited to the actual combatants. The entire structure of a civilization appeared to crumble. Letters reproduced here from citizens of the contending powers convey only a partial sense of the devastating impact of World War I on both sides. The special anguish of the scientists clearly comes through.

In the United States and in other neutral countries, the outbreak of hostilities produced a concern which quickly changed to horror as the carnage progressed. Openly or tacitly, neutrals began taking sides. Initial affinity for the Allied Powers, the adroitness of British propaganda, and the clumsiness, if not arrogance, of Germany eventually brought the United States into the war. Relatively few scientists in the United States were pro-German. In an effort to maintain neutrality, a small number sincerely tried to find merit and blame on both sides. After the outbreak of hostilities in 1917, evenhandedness largely vanished. A few aspects of this complicated reaction appear in the letters given here. They hardly do justice to the enormous impact of the struggle on contemporaries or to the genuine anguish felt by many scientists. Although German militarism and its embodiment, the Kaiser, were unpopular, much of the German way was admired. Many scientists had studied in German universities. The role of research in the German university was widely emulated in the United States. Almost universally, the Germans were respected for the successful linkage of sciences to practical affairs.

Selections from the correspondence of Jacques Loeb and the Swedish chemist SVANTE ARRHENIUS (1859–1927), a Nobel Laureate, give a wide range of reactions. The men corresponded frequently, and many of the letters survive. Loeb's reactions are particularly interesting. As a product of that Germany so widely admired by American scientists, Loeb became particularly bitter as the war continued. He knew that the liberal, humane, cultured Germany represented only a portion of the reality. As a self-described democrat and socialist, Loeb had no love for the martial spirit so evident in his native country. As a Jew, he was very conscious of how easily nationalism could turn into blatant racism. In his various letters of the period, Loeb was pessimistic about the future of Germany, often in terms strikingly prescient. Somehow, he sensed the possibility of Nazism.

While the war did not end Loeb's beliefs in science as a system of knowledge, it certainly dented his certitude about its civilizing effects. Like other scientists, the war strained and then broke down the complacent belief in the internationalism of science. The process occurred in steps, coming faster in the warring powers than among the neutrals. German war policies produced a strong antipathy. Individual German scientists were associated or identified with these policies, especially as some publicly defended their nation. This produced doubt of scientific findings linked with particular German scientists, doubt existing previously but now emotionally reinforced by images of atrocities in Belgium, mass slaughter before Verdun, or unrestricted submarine warfare. Publicly, the facade of universality began to crack. Among the first indications were the exclusions of Germans from international efforts. One example appears here: the Thorpe letter of June 22, 1918, on atomic weights.

After the United States joined the Allies, there was great solicitude for the views of many scientists in France and Belgium who doubted the possibility of personal relations with scientists of the other side. Given the growth of anti-German feelings among American investigators, the leaders of the scientific community here were inclined to go along with European demands for what amounted to an official international scientific ostracism of Germany. Not everyone agreed; the Pickering letter of August 24, 1918, is an example of dissent. President Wilson clearly differed, as is seen in Hale's memorandum of September 10, 1918. But the policy fitted into ambitions of influential American scientists. Although many individuals and particular organizations immediately resumed friendly relations with German scientists, the official order under the League of Nations and the network of international scientific unions and other bodies barred German participation until 1926 and later in some instances. Greatly resented in German academic and official circles, the international ostracism added to the sense of alienation and humiliation which helped Hitler's rise to power. Forgotten was their own behavior toward Allied scientists during the war.

But World War I did not simply break up an old order nor strip illusions of their credibility. Wars provide opportunities of many kinds simply by shattering many old routines. The accident of an assassin's bullet simply transferred the movement to reform the National Academy of Sciences to a larger stage. To George Ellery Hale, the academy (when reformed by his plan) was the core of the national scientific community. Now he took up the problem of groups and forces outside the core. Hale never doubted where he stood in the struggle between the Allies and the Central Powers. He favored the former. Speaking strongly for American preparedness for war, Hale's action pointed toward an eventual entry into the conflict. In 1915 he unsuccessfully tried to have the academy offer its services to the nation but could not get sufficient support within the organization.

Aspects of this attempt are given in various documents which well represent the flavor of the man and his position during the entire war period.

Two potential rivals troubled Hale in 1915: the Committee of 100 of the American Association for the Advancement of Science, and the Naval Consulting Board. The former was highly visible, thanks in no small part to Cattell's skills as a propagandist, and enlisted the support of an impressive array of academicians and others. But the Committee of 100 had a different purpose; it did not propose to coordinate science but to promote research. What Hale feared was that the war would transform the committee into a kind of administrative or governing body outside the aegis of the National Academy. Although membership and goals overlapped, the committee had leanings for reform activities outside of Hale's vision of the proper scope of the academy. Not that Hale necessarily differed on specific reforms—one might describe him as of the conservative wing of the Progressive movement—but Hale had a different scale of priorities. Research came first and whatever changes were needed for its advancement. Second came a national purpose such as defense preparedness. Other reforms came last and only if they did not disturb the existing order. Despite the seriousness with which the committee was viewed, Hale had little trouble in absorbing it within his wartime program. A postwar revival never amounted to much.

In 1915 the Committee of 100 formally approached the Rockefeller Foundation about the general support of scientific research. By that date the Carnegie Institution of Washington was clearly committed to Woodward's vision of an organization operating research installations. Initially, the Rockefeller interests had limited their support to medicine and related areas as well as more conventional concern in education and social welfare. By World War I, the Rockefeller Foundation was cautiously moving into a broader range of activities. Eventually, this would encompass support of a large number of specific research programs and the development of a key number of institutions. Hale would eventually benefit in his ambitions for an observatory at Mt. Palomar. In 1915–1916 he still thought of himself as a Carnegie man.

The Naval Consulting Board, an official body, also never seriously challenged Hale's program. But its origins made the board appear formidable. Appointed by the Navy Department and headed by Thomas A. Edison, the board was the successor to a similar body in the Civil War. Initially, it was conceived as a body mobilizing inventors and appraising inventions. As such, it provided a seeming solution to the navy's technical problems in accord with popular ideologies. The outcome was significantly different. Most of the board members were professional engineers and scientists. (Robert S. Woodward represented the American Mathematical Society.) They had quite different ideas than Edison. Instead of inventions, the principal contributions of the board were administrative. Here Edison played an important role. Their recommendations led to the founding of the Naval Research Laboratory in Washington, D.C. The board (principally Edison) also became an active proponent of mobilization planning with all its eventual consequences in later decades.

In 1916 Hale wrote, "I am planning a National Service Research Foundation to tie together research in universities, Government Bureaus, manufacturing establishments, medical schools, etc." Formally called into existence by a presidential order, the National Research Council was the operating arm of the National Academy. On the council sat delegates from universities, government bureaus, industry, and professional societies. Hardly the representative academy proposed by Cattell, the council presented a new model for the governance of the scientific community—a corporate model, more particularly a vertical trust. In the cause of efficiency, diverse interests, not merely basic science, met at the council to avoid wasteful competition and to promote production and cooperation. By setting and enforcing standards, marginal research efforts were shunted aside. It was a traditional American voluntarism, powered by moral suasion and braced by the muscle of the state.

Hale busied himself finding ways for the National Research Council to aid the armed services. By 1917, even before the American entry into the war, the council had found a place. Robert A. Millikan became the executive head of the program displaying great talent as an administrator. One of the curiosities of the whole story is that Millikan and many of his associates eventually became uniformed members of the Army's Signal Corps. Millikan had the rank of major but did not let his commission interfere with his successful work with the navy. For many scientists, their wartime experiences were very satisfactory personally and reinforced their faith in the role of theoretical science in human affairs (see, e.g., the comments of Moulton in his letter to Chamberlin of August 8, 1918). Many leaders of government and industry would more or less agree. A presidential executive order perpetuated the council after the war's end.

Perhaps the greatest potential threat to this ideological consensus was the rising community of professional engineers. As a graduate of the Massachusetts Institute of Technology, Hale had both a healthy respect for the power of that profession as well as a concern for tendencies within that professional group to strike off on their own. Questions of influence mingled with questions of deference. Hale took great care to win over prominent engineers to his views. Basically, Hale succeeded because influential segments of the engineering profession desired an applied science status rather than a role as inventor or routine practitioner. Despite all the rhetoric about basic research, many of the leading scientists had a rather relaxed and cordial attitude to applied research. Few had the hostile class or ideological attitudes of many of their opposite numbers in Europe.

The wartime success of the National Research Council further stimulated Hale's thinking about the future. Enamored with the council as a model, Hale wanted it adopted in other countries and in international ventures. That particular organizational heritage still survives in a number of nations. To further that end, Hale went along with the exclusion of the Germans from international organizations, which was according to his own beliefs. Of course, the academy would act as the national body for the United States.

Early in 1918, the Rockefeller Foundation entered into the postwar planning discussions in a crucial manner. Under the prodding of Simon Flexner, the foundation raised the possibility of founding a great research institute for physics and chemistry. In view of the success of the Rockefeller Institute, an analogue for the physical sciences was not an unreasonable proposal. That offer precipitated a crisis of soul-searching. Hale was willing initially; after all, he headed an astronomical research group of the Carnegie Institution. In fact, Hale briefly flirted with the idea of putting the proposed institute under the academy. Others in the council sided with Hale, but Millikan prevailed by a narrow margin. He argued for placing research bodies at universities; almost as an afterthought Millikan asked for fellowships. Hale eventually came around to that view, perhaps influenced by his concern for the emerging California Institute of Technology. Since the foundation was not ready to support research in universities on the scale proposed by Millikan, they compromised by starting the National Research Council Fellowships in 1919. Initially limited to physics and chemistry, the program was subsequently extended to other disciplines. The establishment of the fellowship program was a tacit recognition of the primacy of the universities in basic research.

As the leading scientists departed Washington in 1919, they left behind two tangible results of Hale's program: the National Research Council, and the fellowship program. From the standpoint of the post–World War II era, their return to the university appears odd. Subsequent events produce almost a sense of unfulfillment. We expect a national policy and an array of supporting programs. Instead, the council had a role of modest usefulness.

If one looks at the events from the perspective of the leading participants, a quite different interpretation emerges. The leaders involved did not conceive of the federal government having any major role in basic science. In a few traditional areas, federal bureaus did basic research related to their missions. Many scientists did not see any way of avoiding supposed constitutional limitations or perceived popular attitudes. Nor did this bother men like Hale and Millikan. Conservative in their political orientation, they favored limited government and reserved fundamental science for the private sector. Only one exception was admitted, and that was national defense, a key exception in the subsequent history of the sciences.

They did have a program, if only tacitly. Starting from the belief in abstruse theory, the program stressed the promotion of conditions nurturing the development and discovery of great men. The fellowships provided an obvious important step in the process. But, as in the early days of the Carnegie Institution, there remained the question of how to aid the exceptional men and women who proved themselves in their early work. In the twenties the academy tried to solve that by raising an endowment—the National Research Fund.[1] As the depression put an end to that effort, it is not certain exactly what was contemplated. Most likely the aim was something like the research professorships of Richards and the research associateships of Woodward. Those not worthy of such support would man the lesser posts in universities, corporate laboratories, and government. At the same time, private philanthropy—for example, the Rockefeller Foundation—would fund needed growth in research. A sincerely conceived program, it yielded much of value.

Yet the basic conception was flawed. In retrospect, the possibilities of support from the private sector were overestimated. More important, there was a serious misreading of the conditions prevailing in the United States. Just as he had understated the status of science in 1913–1914, Hale (and Millikan) acted out of a sense of scarcity. Problems and shortcomings existed. But the period 1920–1940 was one of great growth and intellectual accomplishment in many scientific fields in the United States. A half-century or more of preparation produced a very rich harvest. The size and diversity of the research community simply defied the kind of simplified coordination provided by the National Research Council during the war years. Concentrating on a small number of scientific geniuses became a worthy but anachronistic goal.

During the years between the two World Wars science had two general problems. These will appear in various guises in a number of the documents in the concluding chapters. Usually, scientists in public discourse evaded these matters. First, despite all their achievements as a group, the scientists were uneasy about their status in their society. It was a matter of seeking deference for their role as producers and transmitters of knowledge, not simply as high-brow tinkerers. The second problem was closely related to the first. The United States contained the first mass scientific community; how could it provide continuity of support so as to satisfy both the needs of a democratic society and the traditions and folkways of a rather peculiar and strongly motivated group?

1. Daniel Kevles's *The Physicists* (New York, 1978) came out too late for use in this book, but it is a very good source for this period. Also very useful is Ronald C. Tobey, *The American Ideology of National Science, 1919–1930* (Pittsburgh, Pa., 1971).

Bertrand Russell to Norbert Wiener, August 29, 1914

Dear Wiener

There is no reason why you should not come back to Cambridge because of the war—you will be just as safe here as in America. It is true that Cambridge presents an unusual appearance—half the undergraduates have gone into the army, there are 20,000 troops in the town, & Heirle's Court has been turned into a hospital (so that I have had to move into other rooms). I dare say if you come you will be my only pupil—English people are not taking usual interest in mathematical philosophy just now. But if you feel disposed to come, there is no reason why you should stay away.

Of course if you decide not to come the only alternative is to stay in America. And I imagine you already know all that America has to offer. The two men who most impressed me were Moore (E. H.) & Dewey. I had not been impressed by Dewey's writings, but when I came to argue with him he seemed to me very good.

This time makes me wish I were an American. This universal barbaric madness is too horrible.

Please let me know what you decide.

Yours very sincerely
Bertrand Russell

Wiener Papers, Massachusetts Institute of Technology

Jacques Loeb to Wilhelm Ostwald, October 10, 1914

My dear Dr. Ostwald:—

I naturally thought very often of you during this war and the fact that I have not written to you was due to the general impression here that letters to Germany will not reach their destination. I have not received a single reply thus far to the letters I have sent to Germany.

I hardly need to tell you that I am relieved that the war thus far has been essentially outside of the boundaries of Germany, and if it is to continue, which seems almost certain, I hope that Germany will be spared the horror in its own territory. You know that the sympathies of the Americans are to a large extent with the French, especially the Belgians. The more cultivated people here feel that first of all it was barbarous on the part of Austria to declare war on the whole people of Servia, who are surely not all murderers. There is a strong feeling here that the case of Servia called for the Hague Tribunal, but not for war.

The second reason that has hurt the cause of Germany is the violation of the neutrality of Belgium. I have heard people say that there are cases in which it is more moral to perish than to preserve yourself; but I give you only what you hear expressed. The devastation of Belgium is a nightmare. There are very few people here who are willing to agree that the resistance of the Belgians to the demand for the passage of the German army was unreasonable.

The third cause which animates the Americans against the Germans lies perhaps in the former speeches of the Emperor William, when he glorified the army on every suitable and unsuitable occasion. While the Americans sympathize with the German scientists and the German people, they do not sympathise with the military caste in Germany, and they have a strong desire that militarism disappear at the end of this war. The general feeling also here is that in the long run Germany cannot win. In the last point they are surely mistaken, because I am just as convinced that Germany is going to win, though the victory may perhaps not be as striking as in the war of 1870.

As far as the results are concerned, of course, nobody can predict, but I vividly recollect the article you wrote about Russia some time ago, before the war. You pointed out that each war led to a strengthening of the reactionary elements, and I am pessimistic enough to expect it of this war too. The Russian government has given great promises of future freedom which nobody expects to see kept. England seems to be the only country in which people dare to protest against war.

I do not know whether you have time or are in the mood for writing any letters, but if you could let me know in a few words how you and your family are I would be greatly obliged to you. I am afraid that your sons have gone to the front, but I hope that they are well.

I remain with kindest regards,

Yours most sincerely,
[*Jacques Loeb*]

Loeb Papers, Library of Congress

Jacques Loeb to Svante Arrhenius, October 23, 1914

My dear Professor Arrhenius:—

Heartiest congratulations upon the arrival of Anna Lisa. I like the name because I have two Annes in my family too, as you know. They are both well and send their best love and wishes to the new namesake. I hardly need to tell you how much your letter interested me. I was amused to learn that you too are bombarded with statements from the English and the German scientists, both defending the justness of their own cause and the wickedness of the cause of the others. I begin to admire the dignity of the French, who keep still in the matter.

I fully share your sympathy for the poor Belgians; it is beyond my belief how any German can stand up and defend the conduct of his government toward these poor people. The demand of a contribution of war from these ruined cities and these miserable people seems like an uncreditable piece of barbarism. I have met Van der Velde[1] and his wife here and it is perfectly pitiful to hear their account. Of course, in a way the English and French are also not without guilt, inasmuch as they induced the Belgians to fight.

That Ostwald and Richet[2] now should try to get the Italians and Swedes involved is almost unbelievable. The madness of the people in this war is so great that we can only compare them with the crusaders, and I really, in order to save myself from further worry and depression over the situation, have begun to look at it from a humorous standpoint, though that may be at times rather difficult. But I certainly refuse to take seriously the Emperor and his military caste, or the military caste of any other country, and I am rebellious against being confronted every day with these deeds of barbarity to the extent that they be accepted as history on a grand scale. To amuse myself, I have written a little article on the "Freedom of Will and War"[3] which I shall send you when it comes out, though I am afraid no printed matter will reach you. The worst of it is that one cannot be very optimistic in regard to an improvement in the situation when the war is over. Militarism will not cease after the war; on the contrary, I am afraid it will increase, inasmuch as the English are going to have the same general duty to serve as the Germans.

In America everybody, with the exception of a few Germans, is opposed to Germany and it is perfectly surprising how strong the feeling runs, although to America the English are more dangerous in the future than the Germans. The reason for this peculiar phenomenon is based, as far as I can find out, on two facts; one is the violation of the neutrality of Belgium and the cruel treatment of the Belgians, and the second fact is the extreme racial conceit of the Germans, who, it seems, consider their civilization superior to that of the rest of the world. I have heard it said from more than one scientist here that if the Germans win in this war there will be no more living with them on account of their arrogance.

It seems to me the root of this latter evil lies in that form of insanity, which originated with the philologists and literateurs, that intellectual and moral superiority was a matter of race, while in reality it is a matter of strain and occurs in any race. Though the Slavs believe their race is the chosen one and the Germans believe that the Teutonic race represents the superman. We in America are beginning to be troubled with that racial insanity too and it is exploited by men of the type of Roosevelt and others to prepare the possibility of a war with Japan. The fear of the yellow peril is beginning to inflame the minds here and we really have to be grateful that we have as calm a man as Wilson at the White House since otherwise, if Roosevelt were still at the helm, we should have already war with Mexico and should have further war with Japan.

In a way I am thankful that the war has at least had one effect: to call attention to the viciousness of the Teutonic conceit. Men like

Treitschke,[4] Bernhardi,[5] Houston Chamberlin,[6] and the other literateurs who have stirred up racial antipathy are now openly ridiculed in the American and of course also in the English press. I only wish that that slight gain would not be lost after peace is made and that we would not hear any more of the superlative value of Teutonic civilization or of Slavic civilization. The French and the English on the whole have been much better in that respect.

I hear absolutely nothing from Germany except that I receive the printed campaign literature, and I do not know how my friends have fared. Mrs. Loeb's sister in Zürich, Mrs. Gaule, has a son in the German army who has already been wounded twice and probably will be sent to the front again if he survives the second wound; I suppose they will keep on that way until he is killed off. And with all that, the Germans are as enthusiastically ready to sacrifice their life as the crusaders were; it is a pity that we cannot construct insane asylums to intern the warring nations, it might be cheaper and more satisfactory in the long run.

I forgot to mention that the people in America feel that the German scientists hurt the cause of Germany by sending out such emotional protests rather than calm statements which consider the other side too. I really do not understand how Ostwald can seriously recommend, as he has done in his last Sonntagspredigt that the German Emperor should be made President of the United States of Europe after the war. I am afraid that Germany will not dictate the terms. In America the feeling prevails absolutely that whatever may happen Germany will be compelled to make good the war indemnity and as much as possible in any other way the damage she has done in Belgium. It is also apparent that any other form of peace will be considered an injustice by the Americans.

I have been back in New York for several weeks and am beginning to get my laboratory into running order again. I am continuing the work on the effects of light and am playing with a number of other problems, but thus far have had nothing to report.

I remain with kindest regards to yourself, Mrs. Arrhenius, and the children, and with the hope that you may be spared any real war experience,

Yours very sincerely,
[*Jacques Loeb*]

Loeb Papers, Library of Congress

1. ALBERT J. J. VANDEVELDE (1871–[?]) was a chemist, the director of a brewing school at Ghent.

2. That is, Loeb was equally condemning the efforts of the German scientist Ostwald and the French scientist Richet.

3. *New Review* 2 (1914): 631–636.

4. HEINRICH GOTTHARD VON TREITSCHKE (1836–1896) was a historian who saw in the German Empire the culmination of history.

5. FRIEDERICH ADAM JULIUS VON BERNHARDI (1849–1930) was a solid German nationalist whose writings on the coming war provided fodder for British propaganda.

6. British-born HOUSTON STEWART CHAMBERLIN (1855–1927) spent his life in Germany. Richard Wagner's son-in-law, Chamberlin was a racialist theorist applying his vision to the history of civilization, idealizing the Teutons and excoriating the Jews. Like many others in Germany, he was unconsciously working to make the Vernichtungslager the culmination of German Kultur.

Wilhelm Ostwald[1] to Jacques Loeb, November 6, 1914

Dear friend Loeb:

Thank you very much for your letter of October 10, which I received on November 6, that is about double the normal time. Other letters from America reached me substantially faster. The fact that news communications between America and Germany are so slow and irregular naturally is caused by England, which still considers itself ruler of all oceans and its removal will be one of the main purposes of the present war.

I thank you very much for your detailed news regarding the American feelings. Since the public opinion is based over there a majority of the time on the not only onesided but often also completely false news from the REUTER-Bureau, I am not surprised that the public opinion is so wrong and unjust. I will discuss your points individually, not because I believe that I have to convince you personally, but so that you could explain to our American friends the German viewpoint.

1. When Austria declared war against Serbia the territory had been guaranteed, and it would have been time enough later on to get involved, in case the original promises would not have been kept. Most important this was no sufficient reason for Russia to declare war on Austria, since Austria did not attack any Russian rights.

2. *Belgium*

At the present there is completely clear evidence available, through the occupation of Brussels and the perusal of the local Archives, that Belgium, France and England have had extensive preparatory negotiations for joint war operations. French artillery officers were fully familiar with the Belgium fortresses while this was kept completely secret from the Germans. Packages have been found in fortresses of northern France which had been labeled, "To be opened only in case of mobilization," and which contained maps of Belgium and Holland. The breach of neutrality was therefore not executed by us but long before by Belgium with the other two countries together, and we acted like the merchant who does not accept a forged check but undertakes steps himself to render the forger harmless. Luxemburg, which was in exactly the same situation as Belgium, did not oppose, because it retained actual neutrality. And its citizens have no complaints. The same would have happened in Belgium despite the prior breach of neutrality, if the people of Belgium in their trust for British and French help would not have resisted with arms, not only the military but the civilians also. The events of 1870/71 already confirmed that the "Franktireur-War" is as mean as it is useless, but the inability to learn from history is a special peculiarity of the French and related nations. The lamentation over destroyed art objects is for the most part hypocrisy. How many people are there in the world who ever saw the Cathedral of Reims, and how many of them were really impressed? Besides, the Germans had occupied Reims for a week without touching a single stone, and only after the French used the Cathedral for war purposes did it suffer through war actions.

It is also my wish that with the end of this war the entire military activity may disappear, but I do not think it is realistic, as long as the barbaric Russian nation exists.

I personally see in the German military a type of organization, thus a higher civilization, which, as long as the military is not used for war, would be advantageous for the nation. In order to clarify my view on this and several closely related topics, I am sending you simultaneously some Sunday-Sermons, which I wrote during these war days.

Recently I have had the opportunity to become acquainted with leading personalities of the Foreign Office. And what I saw and learned there gives me hope that the "reactionary wave" after the war should not be as bad this time here. First of all the new attitude of the Government towards the social democrats, which is now recognized as an equal civil party, has removed one of the most serious impediments for a healthy political development in Germany. And it will be up to us, the progressive thinkers, that this gain will also be preserved after the war.

As you can see, generally speaking I am in an optimistic mood, although I am not a stranger to personal worries. Of my three sons, the oldest and youngest are soldiers, although fortunately neither close to the battlefield. Especially Wolfgang writes to me frequently. In general he is satisfied, since he successfully tries to keep up with the skills of peace in the active service. For example he has been actively involved in publishing a newspaper, which was started with the help of his fellow soldiers, a number of them printers, and with the material found in a deserted printing office in northern France. Scientifically, however, he cannot do anything at all.

With kind regards to you and yours.

Yours very truly,
W. Ostwald

Loeb Papers, Library of Congress

1. This is a translation from the original German.

Emil Fischer[1] to Theodore W. Richards, November 16, 1914

Dear friend:

Your kind letter of October 10 reached me just a few days ago, it took approximately 5 weeks for its journey. This is the best evidence how much the horrible war under which half of the world is suffering interferes with transportation.

You know that I foresaw the disaster. But the reality by far surpasses all we imagined then. Although Germany has been fortunate so far to see foreign troops only in a very small area, that is, in one part of the province of East Prussia. Therefore, everyday life has not changed as much as one would think. For example, you hardly would notice here in Berlin any difference if one excludes the serious mood of the people. But some families already are experiencing great grief through heavy losses in the battlefields, including our academic circles. Nernst, Sachau[2] and the chemist Professor G. Kraemer[3] each lost one son. The son of von Planck has been wounded and is a prisoner in France. Some of my associates have died in combat and we are anticipating daily new losses.

So far my 3 sons are fine. The oldest is at Verdun as a Reserve Officer. The second is working in a field hospital as resident physician and the youngest, also a medical student who has had 4 semesters, works on a hospital train between France and Halle. I have had great problems with him, since he wanted to serve with a gun—in line with the great enthusiasm of our youth. In my opinion, however, because of his youth, he does not yet have the necessary physical strength. In fact, several regiments rejected him as presently unfit. He was so unhappy about it that I finally sent him to my friend Abderhalden in Halle, who luckily assigned him to the care of the wounded within his own organization. My son is satisfied since then. This typifies the general attitude of our youth, which far surpassed our highest expectations of their willingness to die. One who is not now in the armed forces or at least involved in the care of the wounded or ill is ashamed of his existence.

Even us old ones are trying to be of some use and you should not be too surprised when I tell you that I am now much more frequently seen in the War Ministry than in the Cultural Ministry. Since you know that I am a peace-loving man, you should be able to judge for yourself the present psychological condition of the German public.

We do not know what the final effect of the war will be, but we are all convinced that we will have to fight to the utmost tolerable, after all our own existence is at stake.

The war opened a huge rift between European Nations and the understanding of a cultural unity, about which we have talked often and which I have always clearly supported, seemed to have been completely lost. Even the scientists do not understand each other any more. Each party accuses the other of fanaticism and dishonesty. It appears to me best that all further discussions of the cause of the war be postponed until peace is reestablished and all minds are calm again. I fear, however, that this time is far away. Any sign of true human understanding is especially gratifying in these times of intense political charge and because of this I thank you very much for your letter filled with personal sympathy. I can insure you that I too will never forget the few months I corresponded with you and which I enjoyed so much and that your friendship is a most valuable gift. I have to admit frankly that despite the animosities which exist in Germany against the political England, my personal feelings towards the British colleagues has not been influenced by all of this.[4]

Good bye and kind regards
Yours truly
Emil Fischer

Richards Papers, Harvard University

1. This is a translation from the original German.
2. "Sachau" is what the typed letter says, but no one of that name is known to us.
3. GUSTAV WILHELM KRAEMER (1842–1915) was in industrial research.
4. See Gerald D. Feldman, "A German Scientist between Illusion and Reality: Emil Fischer, 1909–1919," in *Deutschland in der Weltpolitik des 19. und 20. Jahrhunderts: Fritz Fischer zum 65 Geburtstage,* ed. I. Geiss and B. J. Wendt (Dusseldorf, 1973), pp. 341–362.

Jacques Loeb to Svante Arrhenius, December 14, 1914

My dear Professor Arrhenius:

I have just received your letter of November 24th, which has interested me very much indeed, especially what you tell me of the conditions in Russia and of Ostwald's campaign.[1] I agree with you that the whole war is against liberty and humanity and that the aftereffect will last for many years. The reactionary spirit caused by the war is even noticeable here where men like Roosevelt and the military party try to use the war to promote their own personal interests. Roosevelt hopes, by arousing the war spirit, to gain the favor of the army people and thereby become a candidate again for the presidency. The stuff he writes is perfectly scandalous and unscrupulous.

As far as Ostwald is concerned, I think his exaggerations on the German side are fully equaled by those of his friend Sir William Ramsey[2] on the English side. I suppose you have seen Ramsey's article on German Science in NATURE sometime in November and now he has another article, it seems, on "How to Destroy German Commerce."[3] The English seem to be now in a condition of the same barbarian excitement in which the others are. My friend August D. Waller,[4] from London, called on me on his way home from the British Association Meeting and informed me that all the Germans were thieves, robbers and vermin; not in a joke but in all seriousness. I did not quite know what kind of a face to make. He also informed me that he had asked Prof. Pank,[5] the Geographer of Berlin, who was an official guest of the English at the British Association Meeting, to leave the table because he could not stand seeing a German eating himself full at the expense of the English. Have you ever heard of such madness before? The joke of it all is that Waller has a great many friends in Germany and that his daughter has been on very intimate terms with the daughter of Emperor William. Last year he told me that the two girls corresponded regularly.

I had a long letter from Ostwald, very optimistic in character, in which he tries to convince me that the violation of Belgium's neutrality was amply justified. I am very fond of Ostwald but I do not know any more what to say to him because his judgment, as that of most Germans, is completely obscure. They literally seem to be lunatics.

The whole trouble comes from their identifying themselves with their governments and their diplomatists. I think, in all seriousness, that as soon as this war is over we shall have to begin a campaign against the racial conceit which has been fostered systematically in Germany, Russia, and possibly in other countries, by irresponsible agitators who were tolerated if not supported by their governments. Unless this idea of "racial superiority" is abandoned, the hatred among the various nations will continue. Here in America too we are having no end of difficulties caused by the fanatical maltreatment of the Negro and the Japanese. If this agitation does not stop, it will land us in a war with Japan—a situation which would probably be welcomed by Mr. Roosevelt and the militarists. Is it not incredible that such childish silliness can exist in our age of science?

We send you and your family our heartiest wishes for Christmas and the New Year. We cannot send any Christmas cards to our German and English friends this year; it would almost sound like satire.

With kindest regards to all of you,

Yours very sincerely,
[*Jacques Loeb*]

Loeb Papers, Library of Congress

1. For some aspects of Ostwald's wartime experiences, see Niles R. Holt, "Wilhelm Ostwald's 'The Bridge,' " *British Journal for the History of Science* 10 (1977): 146–150.

2. RAMSEY (1852–1916), the Nobel Prize–winning chemist, was at this date emeritus at University College, London.

3. Analogous articles were coming from the pens of German scientists in addition to Ostwald.

4. AUGUSTUS DÉSIRÉ WALLER (1856–1922) was a physiologist at the University of London.

5. ALBRECHT PANCK (1858–1945) was known for his work in geomorphology.

Ernest Rutherford to George Ellery Hale, April 17, 1915

Dear Hale

You will be glad to hear that I am sending off to-day the manuscript of my lectures, and the diagrams to go with them. I always find it hard work to overtake past obligations; but I hope the lectures will read well enough to suit you. I am sending the MSS and figures direct today.

I have been exceedingly busy at research, and am trying to investigate the conditions of excitation of X rays over a very wide range of speed. This has entailed a good deal of preparation and hard work, but I hope the conclusions will be worth it. It looks to me that for slow speed electrons the greater part of the energy E of the electron goes into the wave form given by Planck's formula $E = h\nu$ but as one might anticipate this equation no longer holds when one penetrates deep into the atom. I am hoping with luck to test the relation up to 200,000 volts using a Coolidge tube,[1] which is a *great* American production.

By the way, we went [on] a small motor expedition to bring our daughter back from school at Easter time, and we stopped for the night with the Schusters. Their son had just been married the week before to Edyth Goodall the actress, and he has since departed with his squadron to Egypt.

Evans and Bohr in my laboratory are very keenly interested in spectrum matters, and the former is experimentally testing a number of Bohr's views in regard to the spectrum of hydrogen and helium. He seems to think he has got a number of interesting results in regard to the change of width of spectrum lines. You will have seen Evans' paper on the helium spectrum in the Phil. Mag.[2] I think it important to concentrate attention at the moment on the spectra of hydrogen and helium, for if we cannot explain them we cannot hope to make much progress with heavier elements.

You know as much about the war as I do, but we are anticipating during the next week a visit of Zeppelins to Manchester, and everyone is pleasantly excited at the prospect. The psychology of the German mind is so different from ours, and is really an interesting study. I have come to the conclusion that their mental attitude is exactly that of an overgrown bullying schoolboy. I think this is borne out by their threats of reprisals and frightfulness generally. I see by the papers that they are now bullying your inoffensive President. I imagine that that sort of thing will go on until Wilson's patience is exhausted. The Government here seem to be making headway with the liquor question, and the organisation for turning out great supplies of munitions of war. Everybody seems very confident and has given up worrying about submarines. It is generally reported that a number of the German submarines have been caught by methods not to be mentioned, and are now being used with English crews. Apparently a very large land force is being collected for the Dardanelles and after. I presume this is the reason for the delay in the operations, for I know that troops are going there from all parts of the Empire, as well as a division of French and possibly other powers.

Spring is now just coming on with us, and the weather is far more pleasant and bracing.

With kind regards.

Yours very sincerely,
E. Rutherford

I have Barnes[3] of Bryn Mawr working with me. He is a good fellow.

Hale Papers, California Institute of Technology

1. WILLIAM DAVID COOLIDGE (1873–1973) was a physicist specializing in the study of X-rays. An 1899 Leipzig Ph.D., he was director of the General Electric Laboratory from 1932 to 1940, and vice-president and director of Research at General Electric from 1940 to 1944.

2. E. J. Evans, "The Spectra of Helium and Hydrogen," *Philosophical Magazine*, 6th ser. 29 (1915): 284–297.

3. JAMES BARNES (1878–[?]) was a Canadian physicist then at Bryn Mawr.

Bertrand Russell to Norbert Wiener, May 15, 1915

Dear Wiener

I was glad to hear you had got safe home, & I meant to write, but one thing after another kept me busy. It is just as well you did go home, seeing how the Germans have been going on since. Will you enlist when America goes to war, & reappear in Khaki?!

Feeling here is enormously more fierce than it was. We expect the war to go on for years longer, & only to stop when all the adult males on one side or the other are dead. The Asiatics look on with glee, & realize that the day of the white man is at an end. European civilization, as it has existed since the Renaissance, is, I think a thing of the past.

Yours sincerely
B. Russell

Wiener Papers, Massachusetts Institute of Technology

George Ellery Hale to William Henry Welch, June 10, 1915 (first draft)

Dear Doctor Welch:

In the act of Congress incorporating the National Academy of Sciences it is provided "that the Academy shall, whenever called upon by any department of the Government, investigate, examine, experiment, and report upon any subject of science or art." In accordance with this provision, the Academy has assisted the Government in the solution of a great number of problems. Many of these were connected with the Civil War, when reports were made on the protection of the bottoms of iron vessels, on magnetic deviation in iron ships, on wind and current charts and sailing directions, on the explosion on the United States steamer Chenango, and other subjects.

We are now facing the possibility of another war, in which the methods of science will be utilized in the most effective manner by Germany, a nation which has reached its present position in the world mainly through the equally intelligent application of scientific principles in its industries. No nation, however rich in money, men and natural resources, can successfully compete with Germany in peace or war without adopting similar means.

Here is the great opportunity of the United States. Our men of science and our inventors are more original and resourceful than the Germans, who have achieved their ends largely through the intelligent and painstaking elaboration of ideas borrowed from other countries. The well-known cases of the aniline dye industry, which sprang from the discovery of mauve by Sir William Perkin, and the manufacture of machine tools, largely derived from American models, sufficiently illustrate this point.

Under these circumstances I believe that the National Academy should at once offer its services to the President, and throw its best energies into a task which will call for great sacrifices by the American people. No injustice could be greater than that which sends to the front men inadequately trained and insufficiently supplied with every possible appliance for offence and defence. We will doubtless avoid the use of poisonous gases, but we must equip our soldiers with suitable protection against them. The detection of submerged submarines at a distance, and the protection of ships against torpedo attack, are vital questions to which every attention should be given. I have heard of a device which appears to me very promising for the former purpose, and I am strongly of the opinion that means of causing torpedoes to explode before they strike a vessel can be worked out.

These are merely typical of the scores of questions which a war would raise. To solve a single one of them might be the means of saving thousands of lives and materially shortening the war. No sane and well-informed man will make the mistake of underestimating the power of Germany, now so evident in her successful attacks on Russia. It is easier to organize the best brains of the country than to raise enormous armies, and the effectiveness of our fighting force can be greatly multiplied by the work of men who may be physically unfit for the trenches.

When we think of such members of the National Academy as Flexner, Theobald Smith and others of your own profession, who could do so much in the fields of medicine and surgery; Elihu Thomson, Graham Bell and Pupin, whose inventions have revolutionized electrical engineering and made possible long-distance telephony; and our long list of physicists, chemists, physiologists and men of ideas and of action in other departments, it is evident that the Academy could do much of itself. But in my opinion the ablest men outside its ranks should be invited to join its committees, thus concentrating in a single focus the best thought of the country.

The Government has many agencies, both civil and military, from which to obtain advice. But the Academy could supplement these in a very effective manner by organizing the best thinkers outside the Government service, directing their attention to the problems raised by the war, and communicating their suggestions from time to time to the President.

I would therefore recommend that if the Council approves, you will see the President and offer our services. I have no doubt that many would volunteer to devote a part or in some cases the whole of their time to the work. For myself, I may say that I should be glad to act in any desired capacity, and to give as much time as my obligations to the Carnegie Institution will permit.

I would suggest that immediate action on the part of the Council would be advisable the moment war becomes inevitable. Various other organizations may plan to make similar offers to the President, but the terms of our charter and the possibility of securing the cooperation of the best men not already committed, should cause the National Academy to be first in the field.

Very sincerely yours.

Note for E. G. C.[1]

Personally I should prefer to have our chief object the perfection of means of defence, as killing people does not appeal to me. But I suppose the nation must fight if war is declared.

G. E. H.

Conklin Papers, Princeton University

1. Edwin G. Conklin to whom a copy of this draft was sent.

Edwin G. Conklin to George Ellery Hale, June 23, 1915

My dear Hale,

I have just received your letter of June 18th enclosing a check to pay for my telegram to you which of course you ought not to have sent me.

I like the second draft of your letter better than the first and see no objection to your sending it at once to Welch[1] in view of the fact that he is going to China soon according to the newspaper clipping which I sent you in my last letter and which I saw after I had written you. I am sure that Welch will not favor any action at this time but it seems to me that it would be appropriate for the National Academy to offer its services to the government as soon as Congress begins to consider the national defense. The only difficulty which I can see arising from such an action would be the possible resentment of certain army and navy officers. You know how sensitive some of this genus are, some of whom would rather fail following their own devices than to receive aid from any other source. It is probable also, I think, that the President would feel obliged to consult the General Staff or the Naval Board on all matters concerning the national defense. It might create an awkward situation for the Academy if it were to volunteer its services only to be told by such officials that they were quite able to take care of such things themselves. But you have doubtless considered this aspect of the matter and feel with me that the Academy ought to volunteer its services when the proper time comes in such a way as to provoke as little opposition as possible.

I think your letter is admirable and do not see how it can be improved though I would advise delay in action until Congress considers the national defense unless the approaching absence of Dr. Welch should make it advisable to communicate with him at once.

With best regards,

Sincerely yours,
E. G. Conklin

Hale Papers, California Institute of Technology

1. A revised version went out on July 3.

George Ellery Hale to William Henry Welch, July 13, 1915 (Western Union day letter)

Further reflection and conference with T. C. Chamberlin convinces me that Academy is under strong obligations to offer services to President in event of war with Mexico or Germany. I had already concluded that to disarm possible criticism and insure success we should concentrate attention on medicine and surgery but include representatives of other subjects on committee. Daniels[1] announcement today regarding Edison confirms this. Everything depends on chairman who must represent medicine or surgery. Flexner is the ideal man. Could you possibly induce him to serve and perhaps secure use of Rockefeller Institute for research problems. Council is widely scattered but I hope you will at once secure their opinion by telegraph. Day's address is Drakesbad, Westwood, California. We believe that all preliminary steps should be completed before you sail so that Conklin an intimate friend of President Wilson could at once offer services in case of war. If you disagree and desire conference please wire me.

George E. Hale

Hale Papers, California Institute of Technology

1. The North Carolina publisher JOSEPHUS DANIELS (1862–1949) was secretary of the navy in the Wilson administration.

Ejnar Hertzsprung[1] to George Ellery Hale, July 29–31, 1915

Dear Professor Hale,

Before I return to Potsdam from Copenhagen after a short stay I take occasion to write you a few lines about my war-impressions. It is of course difficult to talk about this matter at such a distance and without reply. Perhaps I cannot tell you anything new. My horizon is of course very limited as most of the people, with which I come into contact, cannot be taken as fair representants of "the people."

It is peculiar for us to live in surroundings—"the darkest Germany" as somebody said—without any consent with the manner of thinking there prevailing. I am as strange to that, as I would be e.g. to the infallibility of the pope, if I were observing the stars at the Vatican observatory.

The first and most pronounced thing to meet with, as soon as the war broke out, was the troubling of the mind seizing even the men, who in time of peace are considered as the guarders of wisdom. I suppose, it will be about the same thing on both sides especially in France and Germany. Perhaps England was at first the least affected, because this country has—with a slight exaggeration—so far taken no active part into the war. Next, the bad german diplomates, perhaps nobody has done so much harm to the judgement of Germany's position as the professors with their manifests. There is an entire lack of understanding of what to say to excite the impression desired in neutral minds. Rather the contrary is reached. It has therefore been an interesting observation for me to see, that the higher degrees in german administration are relatively cool-minded. The men, who make and regulate the "Stimmung" in the country, are not themselves below but above that "Stimmung." In this respect the following question to a neutral man is characteristic: "Is it really clever, that we have denied every atrocity in Belgium? Ought we not to have admitted a few?"—in order to make the denying of others more easy to believe. The same "neutral" of our acquaintance has visited the german fronts in east and west. His impression is, that the Germans have burnt more in Belgium, than the Russians in eastern Prussia. But several objections may—as he says himself—be made to general conclusions drawn from this fact. First there are more houses in Belgium per square-kilometer and therefore more occasion to burn, secondly the gabels are often left in Belgium making a sad impression, whereas in eastern Prussia the farms have often been made even with the earth, and therefore are not to be seen. Finally there has probably been shooten more on the german soldiers by Belgians than on the russian ones by inhabitants of eastern Prussia. Talking about this last matter, the "Oberpräsident" of the province Eastern Prussia said: No, franctireurs there have not been here. I can guarant you that. "Dazù sind meine Ostpreùssen zù feige und zù dùmm."[2] This consideration is considerably cooler than what you are used to meeting with among german professors.

It is remarkable, that in Germany the official reports of the enemy are not suppressed as it seems to be the case in France and England. The neutral papers are easy to get. Even in the german press itself the reports from the other side are reprinted—at least at the present time. At the beginning of the war the defeat at the Marne was kept secret for weeks. On the 19th of July I saw in a bookshop in Potsdam a french newspaper only two days old for public sale! Another question is, what this fact shows. The Germans are so well educated, that such papers can be given to them without danger. They know, those things are only to be read for fun.

The average german has a great want to think his own country to be right. Perhaps this may be taken as a good sign, but the resulting obtruding selfgoodness acts repulsive on neutral minds. I fear, the switz writer Spittler is nearest to the truth, when he maintains, that a state is no system of morality and includes the history of the world in one single sentence: "Jeder Staat raūbt so viel, er kann, Punktum"—die Pausen dazwischen von Verdauungsbeschwerden und Ohnmachtsanfälle nennt man Frieden.[3]

The realisation of "The united states of Europe" seems farther away than ever.

With best regards to yourself and all at Mount Wilson.

Yours sincerely,
E. Hertzsprung

Hale Papers, California Institute of Technology

1. HERTZSPRUNG (1873–1969) was a Danish astronomer at the Potsdam Observatory from 1909 to 1919. He had been at Mt. Wilson in 1912.

2. "My East Prussians are too cowardly and dumb for that."

3. "Every State robs as much as it can, period." Peace is the pause between the pains of indigestion and the fainting fit.

Jacques Loeb to Wilhelm Roux,[1] October 5, 1915

My dear Professor Roux:

I was delighted to hear once more from you and to see that all is well with you personally. I have translated your letter into English to the best of my ability and sent it off to the NEW YORK TIMES, Current History Department. (I enclose a copy of the translation.) I shall be very much surprised if they will publish it since the NEW YORK TIMES is the worst representative of that group of millionaires who exploit the war for money.

The situation which has been brought about by the war in this country is most deplorable. New war industries have arisen everywhere and the profits which they are making are said to be fabulous. The outcome of it is that we are already threatened with a terrific development of militarism for the simple reason that the builders of these factories are not willing to stop when the war is over. Moreover, knowing them to be utterly unscrupulous, I am afraid that these same business interests will try to create wars wherever this is possible, simply to have a chance to keep their factories and profits going. It is thus obvious that this war will set back civilization not only in the countries which are actually engaged in the war but also in the neutral countries like the United States.

The situation is so bad that I have gradually given up reading the papers, because it is a deplorable spectacle to notice such a general condition of prostitution. Wilson, who originally was for peace, now has to yield to the pressure of those who want armaments. Only one man has stood by his ideals, and that is W. J. Bryan,[2] who resigned his position as Secretary of the Exterior under Wilson. However, it is at least a blessing that we have Wilson instead of Roosevelt at the White House. With Roosevelt in the White House we should be mixed up in the war directly.

I think it is a wise thing to postpone the work on the Handbuch until after the war is over. As matters stand at present nobody can foretell how long it will last. People here, who profess to know, are inclined to believe that we may count upon two years more of warfare. The only salvation is to rush into scientific work and forget as much as possible the misery of the situation. It is a great pity that the war has also interrupted all scientific relations between the various countries. It is too risky to send any manuscripts abroad now, but after the war we must all work together to see that international relations are restored and the enmities forgotten.

When you see Bernstein, kindly remember me to him.

I remain with kindest regards to yourself and your family,

Yours most sincerely,
[*Jacques Loeb*]

P.S. You do not mention your sons. I am afraid that they are also in the war, but I hope that they have escaped injury.

1. ROUX (1850–1924), the embryologist.

2. WILLIAM JENNINGS BRYAN (1860–1925) resigned because he felt Wilson was not being properly neutral. Three times defeated in his bid for the presidency, Bryan after the war gained a degree of notoriety in the history of the sciences in the United States during the Scopes Trial (1925) when he opposed the teaching of evolution.

[Loeb translation of Roux's statement of Sept. 14, 1915]

In reply to your request, which I received with much delay, I take the liberty of submitting the following statement since I have not published anything on the subject.

I will confine myself to the discussion of one point which seems to me to be the most essential for the termination of the war. This point is the fact, which, as far as I am aware, has never been denied, that, without the furnishing of war material through the United States this unparalleled world war would, if not be at an end, at least approach its end. It is also certain that the billions of dollars which a number of well-to-do men make out of this business cause the loss of life, health, happiness, and economic existence of many millions of human beings of seven different nations, and also destroy immeasurable values of civilization.

I, therefore, dare to raise the following question: Which state is morally higher, the one which begins the war in the justified or erroneous belief that her existence is menaced and which continues it in this belief or that state which, without her existence being threatened, continues this endless war misery in the world or tolerates it merely to make money out of it? Perhaps the raising of this question might induce the one or the other to think and express his opinion about it.

PROF. WILHELM ROUX

Loeb Papers, Library of Congress

George Ellery Hale to Edwin G. Conklin, October 12, 1915

My dear Conklin:

It is interesting and a bit depressing to recognize that the Advisory Naval Board, as now constituted, is probably a stronger and decidedly more representative body than it would have been if organized by the National Academy. I of course think that the chairman should have been selected by the Board, and that some other changes might have been advantageous, but these are matters of detail. As it stands, the Board should be able to give excellent service, especially if the experimental laboratory is built.

Such unpalatable truths should be useful in emphasizing the necessity of broadening the scope of the Academy, rendering it more representative, and bringing it into closer touch with outside affairs. Incidentally I hope we may ultimately raise its reputation to a point where it may penetrate to a sanctum of the Secretary of the Navy.

I am endeavoring to apply this experience in a practical way, and hope to have a chance to talk it over with you before long, though I may not be able to attend the November meeting.

Very sincerely yours,
George E. Hale

Conklin Papers, Princeton University

Svante Arrhenius to Jacques Loeb, December 10, 1915

My dear Professor Loeb:

It is a funny date now today, it is the day of Alfred Nobel, who wished to use his fortune for advancing peace amongst the nations. It is very cold to-day, −15.5° C. at 7:30 A.M., which makes it still more bitter for the poor fighting soldiers. It is as if Nature mourned on cause of the wickedness of mankind. The whole summer was too cold, about 1.5° C. Perhaps therefore the seas round our coasts were not warmed up as much as in common years and the early winter has causes. Mrs. Arrhenius has a niece, who was in Germany, and came back a week ago. She said that everyone there is longing for peace most eagerly. The soldiers say that the present underground war is so dreadful, that hell must be much better. During the fighting the soldiers work themselves up into a kind of furious madness, they only kill and destroy everything. They have only the feeling that otherwise they will have a greater chance to become killed or spoiled themselves, and further they have no responsibility but to their officers, who command them to give up all considerations of humanity. Of course my niece is a pro-German and does not express herself in such words. Some days ago there was a little riot in Berlin, the poor people cried in the street for peace and for bread. It cannot go on in this manner for very long time. I think it is still worse in Austria, where the "organization" is not so good. How shall it then be in Belgium and other provinces invaded and devastated by inimical armies.

A fortnight ago Charles Richet came through here. He went to Roumania for convincing the people there that they ought to go with the "entente" or at least to be neutral. He was formerly a fervent pacifist, but now he was the incorporated hated against the Germans. He has five sons in the war, one is very badly wounded and prisoner and he hears very little from him, probably also from the others. My niece spoke to an old German soldier of about fifty years. He had four sons fighting and did not know a bit of them, where they were or if they lived. It is clear that this will give a holy fury against the enemy. Better were if people looked on the merchants who have found the war desirable and the generals who preached the necessity and moral value of the war and provoked it. Ostwald, another pacifist, has now given up his position as president of the "Monistenbund" and was elected honorary president. His periodical "Monistische Jahrhundert" shall not appear during the war. He has given up Ido and is now working for "Weltdeutschen."

For me everything is proceeding all right. I am very content with the development of the children. My eldest son has made his military service, except for three months in next summer. He now is a student of botany and chemistry. He works in my laboratory with the viscosity of salt-solutions, preparing figures, which I need for my theoretical work. Of course I am very glad for that. Mrs. Arrhenius and I unite in the warmest wishes for yourself, Mrs. Loeb and your children. We hope that you are all well and that you will have a nice Christmas and a happy new year.

Yours very truly
Svante Arrhenius

P.S. I say you my best thanks for your many memoirs, which I have read with the greatest interest.

Loeb Papers, Library of Congress

Theodore W. Richards to Woodrow Wilson, January 13, 1916

My dear Sir:

On the basis of our brief but pleasant meeting at Haverford in 1908, I venture to address you on a subject which is greatly on my mind.

In the first place, I wish to assure you of my profound sympathy for you in the very difficult task of solving, with honor to the country and yet without useless participation in this terrible war, the international problems which confront us, and I congratulate you on the present outcome of the controversies with Germany.

In view of the conflicting opinions which seem to be held by various persons calling themselves Americans, I wish earnestly to tell you that my sympathies and judgment are entirely with the Entente Powers and entirely against Prussia. I have come to this state of mind from a careful study of the situation and in spite of my long experience in Germany, my intellectual debt to that country, and the fact that I formerly had many friends there. The German treatment of Belgium and of Serbia, the Lusitania outrage, and the unwarrantable murder of civilians in unprotected English towns by air raids, together with the abominable German propaganda in the United States and Mexico, make a case so strong against the Teutons as to leave no room in my mind for sympathy with them. Moreover, I believe that the war was primarily started by Prussia with aggressive intentions.

I should like to add that I know these feelings to be shared by nearly all of the intelligent people with whom I come in contact in this part of the country, even including some of those of German blood. I have been strongly urged in the last few weeks to sign proposed public manifestos expressing our abhorrence of Prussian methods; but I prefer to write to you personally.

These statements are simply to express my feelings and judgment. They do not pretend to offer suggestions as to possible action. Indeed I do not know that any other action beyond that which you are taking is advisable. Although some intelligent people here would like to have us break with Germany, I should greatly regret our entrance into the war, if for no other reason because I think we can serve the cause of the Entente Allies better by remaining out of the maelstrom. I hope, nevertheless, that we shall do as little as possible to hamper the Allies in their noble fight for the freedom of the world.

With assurances of high esteem, I beg to remain,

Respectfully yours,
[*Theodore W. Richards*]

Richards Papers, Harvard University

Jerome D. Greene[1] to Simon Flexner, March 13, 1916

Dear Dr. Flexner:

I shall be glad to see Professor Cattell and Professor Pickering on the subject which they have brought to your attention, for I suppose you think it wise that I should do so. Of course they are coming on behalf of the Committee of One Hundred.

I have often thought that the promotion of research through subsidy had not been very satisfactorily worked out in this country. You have had a rather discouraging experience with it at the Rockefeller Institute, and the Carnegie Institution at Washington has come more and more to prefer working through its own laboratories. Whether the very general discontent with the methods of the Carnegie Institution is due to the disappointment of those who sought subsidies in vain, or whether it is due to the somewhat autocratic control of the Institution, you will know much better than I. I cannot help thinking that a benevolent plutocrat who didn't have to bother with precedents and policies, as you and I have to do as Trustees, could do a lot of good for science by arbitrarily selecting now and then a man or a group of men to be aided by the subsidy method. Half or three-fourths of the resulting investigations might be sterile, but the remaining quarter would justify the whole expenditure. Yet Trustees do not like to face any considerable proportion of futile or sterile investigations.

Of course investigations may be unsuccessful in the sense that they do not result in well defined discoveries and still be thoroughly worth while as providing the scaffolding of detailed knowledge on which future advances will be made. I take it that every department of the Rockefeller Institute is successful in this sense whether it yields a definite discovery or not.

I am passing these reflections on to you in order that you may see the background on to which Messrs. Cattell and Pickering will project their image of the enlightened money bag.

Sincerely yours,
Jerome D. Greene

Rockefeller Foundation Archives, Rockefeller Archive Center

1. GREENE (1874–1959), an attorney and banker, was very active in educational and philanthropic organizations. After working with Eliot of Harvard and the Harvard Corporation, Greene was general manager of the Rockefeller Institute (1910–1912), an advisor to John D. Rockefeller (1912–1914), and secretary of the Rockefeller Foundation (1913–1916).

Jerome D. Greene to [?],[1] April 18, 1916

Dear Sir:

In any fundamental consideration of the field of service of an institution like the Rockefeller Foundation, it is impossible to leave out of account the importance of scientific research. Nearly every other branch of philanthropic effort, even including the promotion of education in its narrower sense of school and college, teaching as distinguished from research, the doubt is bound to rise now and then as to whether such artificial nourishment as may be given by a rich Foundation is likely to be of permanent and unqualified good. It certainly will not be so if it fails to draw forth from the community its own effort and its own financial resources to an extent immeasurably far in excess of the money received from the bounty of a single institution or individual. It has been the characteristic of the educational and public health work already done by the Rockefeller boards that its promoters have been constantly alive to this fact. The danger of diverting a great Foundation from its true function of experiment, discovery, initiative, and demonstration, to being a mere bag of money, relieving the community from the necessity of wholesome effort and self reliance, has been guarded against with unremitting vigilance, a vigilance, the necessity for which has been proved by daily experience.

There is one kind of philanthropy which is not open to this particular risk, although it has difficulties and dangers of its own, namely, the extension of the bounds of human knowledge. The discovery of new facts and new laws covering the world in which we live, biological, physical, economic, is in itself a service which may be called thoroughly good without drawback or qualification of any kind; and experience has shown that all branches of scientific knowledge are worthy of equal respect, so numerous and unpredictable are the interrelations of knowledge in the various branches. A good example of this truth is found in the enormous usefulness of the science of entomology to medicine and agriculture, a usefulness which could never have been attained had not the characteristics and habits of various bugs interested for generations men whose sole incentive to study was their love of nature and truth without any regard to its practical utilization.

Thus far the funds of the Rockefeller Foundation other than those expended for war relief, have been applied very largely to propaganda of existing knowledge through the hookworm work in the Southern States and foreign countries and through contributions to the National Committee for Mental Hygiene and the National Committee for the Prevention of Blindness. These are all admirable undertakings, calculated to make effective the results of research, and they are free from any of the dangers or drawbacks referred to above.

In addition, and counting funds designated by Mr. Rockefeller from that part of the Foundation's income which remains at his disposal, $2,000,000 has been pledged to the Rockefeller Institute for Medical Research. . . .

Rockefeller Foundation Archives, Rockefeller Archives Center

1. This is a draft surviving in the foundation records with no indication of the recipient. A hunt among obvious possibilities yielded nothing.

Ross G. Harrison to Edwin G. Conklin, April 29, 1916

Dear Conklin:

I would ask your attention to an important matter, which has only casually come to light, but which deserves the widest publicity. It came to my knowledge through an official announcement of the American

Red Cross which was received by persons who have sent hospital supplies to the Red Cross depot at Bush Terminal for shipment to Germany.

The announcement, of which a copy is enclosed, states that the British Government refuses hereafter to permit Red Cross supplies from America to the Central Powers to pass the blockade, and it indicates that the American Government has assented to this decision without having made it public.

Two matters of grave importance are involved. One is the far reaching effect of the decision itself, which virtually forbids the international activities of the Red Cross, in flagrant violation of the Geneva Convention. The other is the peculiar inaction of our Government in submitting to the nullification of the principles of humanity and neutrality in the relief of war suffering, when at the same time it proclaims our mission as "spokesmen for the rights of humanity," and threatens to break friendly relations with Germany because of alleged inhumane acts.

If the inconsistency and injustice of our Government's position can be clearly brought before the people, it surely must influence public opinion against our entering the war as the ally of one party, when both have offended against our rights.

Participation in the war would mean a country divided against itself. It would mean the outraging of the deepest convictions and affections of millions of our people. It would rend asunder our national unity for a generation.

Will you not use your influence to avert such a disaster, and help in every way you can to give publicity to this affair?[1]

Very truly yours,
Ross G. Harrison
Professor of Comparative Anatomy
Yale University

Conklin Papers, Princeton University

1. Conklin's reply of May 16, 1916, diplomatically assured Harrison of support in getting Red Cross supplies to Germany.

Willis R. Whitney to Edwin G. Conklin, April 29, 1916

Dear Sir:

The enclosed galley proof is sent you because it contains the essential points of the Newlands Bill,[1] which will probably soon be discussed in Congress. I really want your favorable opinion of the plan, if you will give it, and as soon as possible.

I am taking interest in this Bill because I believe it is the most promising effort actually well under way to bring about conditions of research in this country at all approaching what we ought to have. I believe that it can be demonstrated by this experiment that there are and have long been in many colleges groups of men who could and would do good research work if they were permitted or encouraged. They might, and probably would collect about them in a few years post-graduates and men doing advanced work, which would then warrant greater expenditures. I think the scheme of distributing this small amount to the different land grant colleges is one of the best and quickest ways of discovering the possibilities, and if this scheme were organized under the Department of the Interior so as to cooperate with other departments, as is planned, no scheme I can think of would be better.

The Massachusetts Institute of Technology, Purdue, Wisconsin, Illinois, Cornell, and other such colleges come under this Newlands Bill. I know from my own experience at the Institute of Technology, where Noyes and others chipped in their own money to start research in physical chemistry, that such a scheme would be well worth trying. I think research in physics, for example, has never had opportunity there, because of lack of funds. When I think of the work accomplished by the groups of research men under the professors at such German universities as Bonn, Freiburg, Heidelberg, Leipzig, Göttingen, Berlin, etc., I see no reason why we ought not to force ourselves to make at least a beginning. Most of our American teachers and research men had to go abroad for their first experience in research.

If the amount of money represented by the Newlands Bill proposal were centered in a single organization, it would not, to my mind, serve the purpose half as well. In the first place, it would not discover the men in our colleges who are now in position to do advanced scientific work. It would cut out local interests and possible state appropriations, and would not take advantage of the enormous amount of available apparatus in the colleges.

This experiment station scheme should constitute a good foundation for scientific cooperation in our country, and would serve well in conjunction with the naval research laboratories when these are built.

The fact that the Newlands Bill has been prepared and presented is, to my mind, a very important step in itself, and the impracticability of doing as much for the desired end by any other measure within any reasonable time would insure my own support for such a measure and would prevent my making at this time suggestions which might delay it, even if I thought such suggestions constituted improvements.[2]

Yours very truly,
W. R. Whitney
Chairman of Chemistry and
Physics Committee
U.S. Naval Consulting Board

Conklin Papers, Princeton University

1. Daniel Kevles, "Federal Legislation for Engineering Experiment Stations: The Episode of World War I," *Technology and Culture* 12 (1971): 182–189.
2. In his reply of May 11, Conklin expressed reservations about the experiment station precedent but would agree if that were the only way to get more research funds.

Excerpt, Minutes, Board of Trustees, May 24, 1916

APPLICATION FROM THE COMMITTEE OF ONE HUNDRED ON SCIENTIFIC RESEARCH OF THE AMERICAN ASSOCIATION FOR THE ADVANCEMENT OF SCIENCE

The American Association for the Advancement of Science appointed at its annual meeting in 1913 a Committee of One Hundred on Scientific Research, with President Charles W. Eliot as Chairman and the following Executive Committee: E. C. Pickering, Chairman; Charles D. Walcott; William H. Welch; Edmund B. Wilson; J. McKean Cattell, Secretary; Dr. Simon Flexner is a member of the sub-committee on Research Funds. Dr. Jacques Loeb and Theobald Smith, of the Rockefeller Institute for Medical Research, are also members. The membership of the Committee unquestionably includes the most productive scientific investigators in the country, such as Professor E. G. Conklin, of Princeton, Dr. C. B. Davenport, of the Carnegie Institution, Professors J. R. Angell,[1] T. C. Chamberlain, A. A. Michelson, R. A. Millikan, F. R. Moulton of the U. of Chicago, Professors W. M. Davis,[2] W. G. Farlow,[3] T. W. Richards, W. M. Wheeler[4] and Reid Hunt,[5] of Harvard, Professors W. H. Howell and F. P. Mall, of Johns Hopkins, Professors T. H. Morgan and E. B. Wilson, of Columbia, Professor A. A. Noyes, of the Massachusetts Institute of Technology, etc.

The communication has recently been received from Professor Edward C. Pickering, who has succeeded President Eliot as Chairman of the Committee of One Hundred, asking the Rockefeller Foundation to contribute the sum of $50,000 for immediate use in aiding institutions and individuals in researches for which provision cannot otherwise be obtained. The arguments presented in favor of the proposed appropriation may be summarized as follows:

The present is an especially favorable time for aiding scientific research. Its value is now appreciated as never before. The advantages gained by countries where it has been fostered are admittedly very great.

The six hundred universities and colleges of the United States spend one hundred millions annually in teaching and comparatively little for research in pure science. Consequently the scientific output of the United States compares unfavorably with that of other civilized nations. A selected list of living eminent men of science shows that the United States has produced no greater number than Saxony, although its population is about twenty times as large.

The Carnegie Institution has practically abandoned the field of aiding individual investigators, only a small part of its income being available for this purpose.

Present conditions in Europe are likely to leave scientific men of the greatest ability with very limited resources. An astonishingly large number of promising young men of science have been killed in the war.

At a meeting of the Executive Committee held April 11, 1916, the Secretary was instructed to present a memorandum on some of the general considerations suggested by this application from the American Association for the Advancement of Science, and the following considerations are therefore presented to the Trustees:

The useful services of the Foundation in the future, as the useful services of the related boards have been in the past, are likely to be divisible into two main divisions, the discovery of new knowledge bearing on human welfare, and the dissemination of this knowledge by various educational methods. The best examples of these two kinds of work are afforded by the Rockefeller Institute for Medical Research and the hookworm work in the southern states. In the development of the Rockefeller Institute, the Trustees and Scientific Directors have wisely determined that the work of the Institute should not be confined merely to investigations having the most obvious, direct bearing on the treatment and prevention of disease, but should also include research into chemical, physical, and biological problems that might be assigned to the realm of pure science rather than applied science, were it not for the repeatedly demonstrated importance of maintaining this distinction. It may fairly be maintained, therefore, that scientific truth is not only worthy of search for its own sake, but is almost certain to have sooner or later practical applications to the use and enjoyment of man.

Rightly or wrongly, the opinion prevails among men of science of the highest standing in the United States that the existing resources for the promotion of research are lamentably short of the amount they should be to enable the United States to do its part in this branch of human activity. It was thought that the Carnegie Institution, of Washington, with its endowment of ten million dollars, subsequently increased to twenty-two million dollars, would go far toward meeting the need, but this hope has been disappointed, for the Carnegie Institution is now devoting its resources very largely to the upkeep of its own research establishments, such as the Nutrition Laboratory in Boston, the Geophysical Laboratory in Washington, the Solar Observatory in California, and the non-magnetic ship. Mr. Carnegie's last gift of ten million dollars to the Carnegie Institution was prompted not by any general apprehension of the needs of scientific research, but, as he has himself avowed, by the enthusiasm evoked by some of Professor Wilson's[6] astronomical discoveries in California. Moreover, the administration of the Carnegie Institution has been the subject of very general complaint because of the exaggerated control of the executive officer in determining the grants and appropriations nominally made by the Trustees. Meanwhile, in many fields of scientific work men of absolutely first-rate ability are cramped and thwarted in their work for lack of very moderate sums of money, which, added to the considerable investment they already have in experience and working equipment, would suffice to bring highly valuable researches to completion.

There are two ways in which the Rockefeller Foundation might look upon the proposition to make possible numerous grants in aid of research. It may be regarded like any other proposition which would involve Mr. Rockefeller or any of his Boards in carrying part of the ordinary load of education and charity which each community ought to carry for itself. There is much to be said for this point of view, for unquestionably the community must realize its responsibility for research as it recognizes its responsibility for hospitals and schools, and it is conceivable that grants in aid of research might be so lavishly and indiscriminately scattered abroad as to deaden rather than to enliven the responsibility of the community in this regard. On the other hand, the proposition may be regarded as offering the opportunity to do a service to science and to human welfare which, to a large extent, at least, will be seriously deferred, if it does not actually fail of accomplishment, but for such outside aid. The great argument for aiding research is that knowledge breeds knowledge, it might almost be said in geometrical proportion, and the reward of prompt aid where it is really needed is to be found in the enormous and far-reaching fecundity of the ensuing benefit.

It would be premature to present at this time a definite proposition for the aiding of research by the Rockefeller Foundation. The suggestion of the American Association for the Advancement of Science presents some serious difficulties of administration which would require further attention before any recommendation could be made. The importance of the whole subject or research, however, would seem to need no argument at this time. In all that is said about the importance of military preparedness no single measure advocated begins to compare in importance with the proper mobilization of the resources of this country for research, and the fact of greatest significance in this connection is that without exception all the benefits accruing from research will be no less valuable for peace than for war.

As the basis of discussion and possible action on this subject the following resolution is proposed:

> RESOLVED that the President be requested to appoint a special committee of this board to inquire into the needs of scientific research in this country and to report at a subsequent meeting of this board.[7]

Rockefeller Foundation Archives, Rockefeller Archive Center

1. JAMES ROWLAND ANGELL (1869–1949) was then at the University of Chicago. In 1919 he was chairman of the National Research Council, president of the Carnegie Corporation in 1920, and from 1921 to 1937 president of Yale University.
2. WILLIAM MORRIS DAVIS (1850–1934) was a geologist and geographer at Harvard, particularly known for work in landforms.
3. W. G. FARLOW (1844–1919) was a botanist at Harvard.
4. WILLIAM MORTON WHEELER (1865–1937) of Harvard was a zoologist interested in insect behavior.
5. HUNT (1870–1948) was a pharmacologist at Harvard.
6. A slip of the pen. He had to mean Professor Hale at Mount Wilson.
7. After Greene's resignation on September 1, 1916, the appointment of the committee was indefinitely postponed.

Henry N. Russell to Frank W. Dyson,[1] June 27, 1916

My dear Sir Frank:

When I realize that your good letter has gone for half a year without an answer, I begin to wonder whether I deserve to remain in respectable scientific society. But, without further apology, except to cry *Vae, mea culpa,* let me respond to your good wishes in the spirit of cordial friendship in which they were offered.

I am glad indeed to hear that the War has so far brought you no more personal problem than you speak of. It is hard here to realize what it really means, though this, I hope, is not for lack of interest and intense sympathy. As you may have heard, Princeton is decidedly on the right side. We have done a bit more for the War Relief than most other towns, and, out of about three hundred boxes and cases of things, I believe *three* have gone to Teutonic recipients,—these being sops to the neutrality idea in the earlier days, and contributed by the scattering Germans in town. Recently, we have all been busy with the Allied Bazaar in New York, Princeton being the headquarters of arrangements for New Jersey. It may amuse you to hear of my share. Having nothing salable of my own, I wrote to Metcalf,[2] and begged of him six unnamed asteroids. Though I found that interest in celestial real estate was not so keen as in some of the other attractions, I managed to dispose of the right to name four of the little fellows, and netted two hundred dollars for the good cause.

A cousin of mine is drilling his battalion at Toronto, and will be at the front before long, so I am not wholly out of personal touch with things. The nearest it has come to me yet was when a dear friend of mine—the daughter of Professor Baldwin[3]—whom I have known since she was a baby, was terribly injured on the Sussex. But what hits me hardest is to have to get used to being ashamed of my country—or at least of the people who are running it. I have really hoped once or twice that we would do something intelligent, if not useful; but I am still waiting. The political campaign just beginning being about my last hope. It is hard to feel any interest in this Mexican mess that we appear to be getting into, though I fear that it may be a rather serious business, considering our utter lack of preparation, but, even at this distance, the real war news takes so much of my time and thought that work is in danger of coming second.

Forgive my boring you with these details of personal opinions; but I cannot resist the temptation of expressing the opinions of one of the Americans—and their name is Legion—who emphatically do not agree with a certain prominent gentleman that sanity is the peculiar and exclusive attribute of the neutral peoples.

Our work here goes well. We have about 160 more of the lunar plates finished, with results of about the same precision as before. We are still at work on eclipsing variables. One of Dugan's[4] monographs has just come out, and another is ready—on RZ Cassiopeiae. We are reducing Harvard photographic observations of eclipsing variables, too, and are getting excellent light curves, and finding that it is a general rule that the faint companions—which in the cases we have selected are larger than their brighter primaries, and totally eclipse them—are much redder than the primaries, the mean difference in color index for the first five stars being about 0.7 mag., as against a difference of visual surface brightness averaging about five times as great. As these red stars are undoubtedly considerably less dense than their white companions, it all fits in very nicely with my notions of stellar evolution.

Most of my own spare time has gone into the revision of Young's[5] text-books on Astronomy, which I find an interesting, but lengthy job. Perhaps this is because I am doing it from the ground upwards, but I find that I am learning a lot of astronomy out of it. The two papers on albedo, &c, in the Astrophysical this spring grew out of the necessity of getting values of albedo that I could trust to go into the book. There was nothing for it but to derive new ones.

While I am speaking of this, I am going to beg from you the privilege of reproducing some Greenwich photographs for the new book. There is a beautiful series of Mercury in transit in 1914, which I would like, and a slide, which I saw at Harvard, showing Jupiter and satellites vi, vii and viii, and on the same plate, which is extraordinarily interesting.

We are just off for our summer vacation in three days, and are all very well. My little daughter Lillie, one of the twins, who has given us a good deal of anxiety, is coming on nicely. She has some congenital trouble with the coordinative mechanism in the cerebellum, but by proper training the higher centers can be got to take on the work of the deficient regions, and the outlook is now very encouraging. Her mentality is not at all affected, thank God.

With kindest regards to Lady Dyson and all your family.

Very sincerely yours,
[*Henry N. Russell*]

Russell Papers, Princeton University

1. DYSON (1868–1939), Astronomer Royal, 1910–1933.
2. JOEL HASTINGS METCALF (1866–1925) was a discoverer of asteroids.
3. The psychologist James Mark Baldwin reacted to his daughter's injury by sending an open telegram to Wilson. The sinking of the *Sussex* is not well remembered in an era of many acts of violence at sea. Hale, however, used the emotions of this event to get a Presidential Executive Order to establish the National Research Council in 1916.
4. RAYMOND SMITH DUGAN (1878–1940) was an astronomer at Princeton.
5. CHARLES AUGUSTUS YOUNG (1834–1908) came to Princeton in 1877 from Dartmouth.

Thomas Hunt Morgan to Albert Francis Blakeslee,[1] October 19, 1916

Dear Dr. Blakeslee:

I think I can promise a paper for the Christmas meeting of the Naturalists, but whether it will be on the effects of castration in fowls or whether on some fly work, I can not say just at present, as the course of experiments in the next two or three weeks will decide which is the better material to report. Both Sturtevant and Bridges have some interesting as well as important things that they could give if there were room for them on the program—or they might replace my own report if that seemed preferable. I am glad you asked Goldschmidt,[2] as he will give an interesting talk.

In regard to Punnett,[3] my opinion, after talking to Cattell and Wilson, is that it would be difficult to raise the money; that the amount is really not large enough to extend such an invitation to him to come so far at that time of year; and that in all probability Punnett has nothing in particular to say that he has not already said. I have attended so many meetings of committees at which large hopes were held out for obtaining small sums that I have become rather cynical in regard to what is likely to take place.

I am inclined to think that no one is likely to notice whether our special program of papers has Germans, Russians, Turks, Austrians, Italians or even Roumanians on it. It is safer, on the whole, to ignore the question than to make any pretence of striking a balance between the warring countries.

Pardon my delay in answering your letter. It came just as I was leaving for the West and this is the first opportunity I have had to take up the matter.

I hope when you are in the city you will drop in and see us some day. I shall be glad to have you take lunch with me any day at one o'clock.

Yours sincerely,
T. H. Morgan

Blakeslee Papers, American Philosophical Society

1. BLAKESLEE (1874–1954) was a botanist who discovered sexual reproduction in bread molds. Most of his career was at CIW's Cold Spring Harbor Laboratory of which he was director from 1936 to 1941.

2. The German biologist RICHARD B. GOLDSCHMIDT (1878–1958) was stranded in America by the war. From 1936 to 1948 he was at the University of California, Berkeley. He was an opponent of Morgan's. See Garland Allen, "Richard Goldschmidt's Opposition to the Mendelian-Chromosome Theory," *Folia Mendeliana* 6 (1970): 299–303.

3. REGINALD G. PUNNETT (1875–1967) was professor of genetics at Cambridge University (1912–1940).

Willis R. Whitney to George Ellery Hale, January 9, 1917

Dear Dr. Hale:

I want to get you to think a moment on the Newlands Bill. President Pearson,[1] of Iowa State College, has proposed a modification as follows: "That the Federal Government should appropriate for engineering research work $15,000. to go to each land-grant college, *provided* that the state appropriate at least as much to be expended for a similar purpose and thru such agency as the state may select."

I honestly believe that this is a better plan than anything that is likely to be proposed and carried out in the near future, and I would personally favor it because it would give a means in the present Congress, of seeing if such appropriations could be made without oppositions. The only alternative I see is some long drawn out work on the part of committees of people who are not intensely interested. After attending the recent meeting in New York of the Committee of One Hundred on Scientific Research, I made up my mind that when the number of scientists gathered together is more than two or three, the Lord is certainly in some other place. Twelve year old children can do more constructive work in an hour than a committee of one hundred scientists could do in a year, as our American scientists are actually constructed.

I believe it is my duty to help those who want the help along the lines of constructive experimental work and that, while the land grant colleges themselves would devote much of their attention to practical engineering problems quite remote from pure science, the state universities would, on the other hand, be almost forced to devote equivalent efforts to scientific research if the Newlands Bill, amended as proposed, became a law.

I was so sorry that Mr. Carty[2] did nothing on the oil matter that I have made up my mind to do what little I can do without reference to the aid of others. On the other hand, I should be very much pleased if you could see your way to advise on this matter and offer me at least your personal opinions. If there were good criticisms against the proposal of President Pearson and, in addition to the criticisms, suggestions were made which looked to be better, I should of course want to aid the most promising outlook.

If you can talk with Noyes[3] about this, I should also be interested in his views, for I have not written him directly. I assume he is in the West.

Yours very truly,
W. R. Whitney

Hale Papers, California Institute of Technology

1. RAYMOND ALLEN PEARSON (1873–1939) was president of the University of Maryland 1926–1935.

2. JOHN J. CARTY (1861–1932), chief engineer of the American Telephone and Telegraph Co.

3. A. A. Noyes.

George Ellery Hale to Robert A. Millikan, March 1, 1917

My dear Dr. Millikan:

In the present crisis, and in view of the primary purpose of the National Research Council, you will need no suggestion from me to indicate that the work of your Committee should be concentrated for the time being on the research problems of national defense. In some fields of science the possibility of important contributions may seem rather remote. But by referring to the files of Nature, the Comptes Rendus, the Scientific American, and numerous other scientific and technical journals, you will be interested to see how many aspects of science are actually involved. Thus I am acquainted with astronomers who have contributed toward the solution of war problems by designing a range-finder for use against Zeppelins (Newall); devising an instrument for detecting the exact position of shrapnel in the body (de la Baume Pluvinel);[1] making mathematical calculations needed in the development of airplanes (Chrétien);[2] devising and applying a collimating instrument for adjusting binoculars (Dyson); and otherwise utilizing their knowledge wherever they can be of most service. Sometimes there is ample opportunity for service in one's own field; this is notably true in chemistry, physics, and medicine. In other cases, such as those just mentioned, an investigator used the experience or the technique gained in his own department to solve problems lying wholly outside of his customary experience. Professor Starling,[3] the physiologist, is in charge of the asphyxiating gas work in England; Count de Noüys,[4] the French physicist, has developed for Dr. Carrel a formula by which the time of cicatrization of wounds can be computed with great precision. Sir Joseph Thomson has worked on a variety of problems, including a device for detecting submarines. Lord Rayleigh, who is equally busy, has given special attention to researches on the development of airplanes. Dr. Harker[5] is studying processes for the fixation of nitrogen. M. Fabry and M. Cotton[6] have devised a new type of range-finder (for locating field artillery) and have done other work of importance. The Duc de Broglie[7] is busy with naval problems. Many other illustrations might be cited, but these will suffice to indicate that almost any capable investigator should be able to contribute in a useful way.

The first work should be the formulation of a large number of problems, which can easily be done with the aid of the journals already mentioned. These will then be transmitted to our Military Committee, which will indicate those that are most important to investigate. Research will then be undertaken in the laboratories of universities and other institutions, where investigators will certainly be afforded time for work of this nature, without reduction of salary, under the existing war conditions.

The Military Committee is already formulating problems, some of which have been referred to other committees to study. But its members are extremely busy with military duties, and all possible assistance should be given every committee. The very essence of preparedness is not to wait for a need to develop, when it is too late to meet it, but to *foresee* the need, and institute the researches required for its solution.

In the face of war, every loyal man of science should be willing to drop his present work, wholly or in part, and devote his time and attention to researches on military problems. No one should hesitate because he faces new conditions. His experience as an investigator in any field will serve him well. It should not be forgotten that many of the greatest discoveries have been made by men of science who have come with fresh vision into a new department, where freedom from the hampering effect of habit and tradition has more than compensated for deficiency in special experience.

A meeting of the Chairman of Committees of the National Research Council will be held at the Cosmos Club, Washington, on Sunday, April 15th at 3 P.M., to receive reports from its members on the steps that have been taken to formulate and investigate national defense research problems and other subjects. Please be prepared to report at this time, and also at the meeting of the Council, which will be held on the following Wednesday (or Thursday morning, if the Local Committee cannot allow sufficient time on Wednesday afternoon).

Very sincerely yours,
G. E. Hale,
Chairman

Hale Papers, California Institute of Technology

1. AYMAR, COMTE DE LA BAUME PLUVINEL (1860–1935), had worked in astronomy.
2. HÉNRI CHRÉTIEN (1870–1956), a French mathematician.
3. ERNEST HENRY STARLING (1866–1927), professor of physiology, University College, London.
4. PIERRE, COMTE DE NOÜY (1883–1947), was a phsyicist who worked on applying molecular physics to biology.
5. JOHN ALLEN HARKER (1870–1925), director of research, Ministry of Munitions, 1916–1921.
6. CHARLES FABRY (1867–1945), University of Marseilles. AIMÉ AUGUSTE COTTON (1869–1951) was then at the École Normale Supérieure.
7. LOUIS-CÉSAR-VICTOR-MAURICE, DUC DE BROGLIE (1875–1960), was a French physicist who had served in the navy before turning to science.

Thomas Hunt Morgan to Charles B. Davenport, May 9, 1917

Dear Davenport:

I received your letter some time ago in regard to the War-Eugenics question. It is a tremendous program you have planned and would require a regiment of scholars to handle it. I think it possible that the Peace Foundation here might be interested in it. If I can get a word with Keppel,[1] who is now in the War Department as one of the assistant secretaries, I will find out if they intend to look into such matters, although I doubt if they have any money to spare.

I had a less ambitious plan which might possibly be put through with the help of Sturtevant and Gowen,[2] namely, to calculate on the basis of Mendelism and probability what effect the draft of a given percentage of the population would be expected to have on the subsequent generations. We have sketched briefly the results and I think it is surprising to find how small is the actual result. Perhaps I can get from the historians some definite data relating to the number called in some of the great wars in relation to the population which will in a rough sort of way give an actual basis for applying the theoretical conclusions. It seems to me that it will be more advisable to cover a small part of the field thoroughly than to splurge over a far-reaching topic. Let me have your opinion some time.

Sincerely yours,
T. H. Morgan

Davenport Papers, American Philosophical Society

1. FREDERICK P. KEPPEL (1875–1943) was president of the Carnegie Corporation from 1923 to 1941.
2. John Whittemore Gowen (1893–1967) was a geneticist then at the Maine Agricultural Experiment Station. From 1926 to 1937 he was at the Rockefeller Institute's laboratory at Princeton, N.J.

Jacques Loeb to Svante Arrhenius, July 9, 1917

My dear Professor Arrhenius:

I have been in Woods Hole and I have seen Osterhout, who is very flourishing. I have been very grateful to you for your kind words about my book. Lillie, who reviewed it in the Journal of the American Chemical Society, is a good deal of a mystic, as are the majority of zoologists; they do not realize that all life phenomena are determined by rigid laws and that these laws can have a mathematical expression. Thus my theory of tropisms is vigorously antagonized by mystic anthropomorphic hypotheses of the zoologists, and, on the whole, even men like Prof. Wilson of Columbia University do not think very highly of a physicochemical explanation of life phenomena.

You see, at heart human beings are still mostly mystics; if it were not so we should not have any war. If Americans could judge the Europeans as mere machines they surely would not send their youth abroad to be mowed down. But even physicists—men like Hale or Michelson—when they leave the field of physics and deal with human affairs they fall back into the antiquated method of analyzing human affairs from an anthropomorphic viewpoint, and the result is all kinds of fanaticism. The historians are of course still worse, but the difference is not so very great.

Well, we are in the war and our "big businessmen" express frankly the hope that it will last at least from three to five years so that they can safely conquer the markets of Russia and South America. For this reason they are very anxious that Russia also should stay in the war, otherwise there is a danger of Germany getting back its hold on Russian commerce. This is also the reason why even the reactionary papers in America have hailed the Russian Revolution with delight because they felt that the old Czar had a preference for German businessmen. I wish your hopes for an early end of the war were justified, because the war makes me literally sick; but on account of the business situation and the vagueness of what the statesmen really want in the war, America will prolong the war for a large number of years. After that there probably will be a war with Japan, so that I am afraid our generation will never see peace again.

I have seen Taylor[1] who is connected with the food situation in Washington. I am afraid that the neutrals will have to suffer more than they have suffered in the past on account of food shortage. However, people talk in a very cheerful way that the destruction of life and property is really comparatively small still so that there is no need of terminating the war so soon. I am afraid the Germans are responsible for a good deal for having started the war; and the German war-madness again can be traced back not only to the greed of their big businessmen but also to allowing the insane press agents of the Emperor, like Treitschke and Houston Chamberlain, etc., to make them all crazy. I think after the war we shall have to start a conscious enterprise to do away with the old text-books of history, where the minds of the youth are already filled with bloodthirsty ideas and notions of after-glory, which of course is only the miserable glory of kings, emperors, and politicians.

I shall make another attempt to send you some reprints and I wonder whether you will be good enough to let me know by a postcard whether you received them. I think most of the things I sent abroad are lost.

I remain with kindest regards to yourself and Mrs. Arrhenius,

Yours very sincerely,
[*Jacques Loeb*]

Loeb Papers, Library of Congress

1. Alonzo E. Taylor.

Robert A. Millikan to George Ellery Hale, September 21, 1917

Dear Hale:

Ford[1] is just back and brings the word that you will be here in about a month. We shall all rejoice when you are back, for, the work of the Council is growing rapidly and the need of increased personnel is clear.

You will be interested to know that we are now on the chart of the Signal Corps as the Science and Research Division[2] and are trying to get essentially this relation to all of the bureaus of the War and Navy Departments. We have asked for five drafted men to act as the links between the Council and the various branches of the Signal Corps. Noyes and I went to Crozier this afternoon and got him to appoint Hildebrand[3] a Captain to act as one of the links between the Ordnance Bureau and the Council, and I told the General that we should expect five or six such links before long. However, you will get all this in the report of the Executive Committee meeting of the 19th. The engineers constitute the disturbing situation now because there are two other engineering committees which have got to be put into relations with the Engineering Committee of the Council, and this means some diplomatic work. I hope that Dunn[4] and Pupin got a satisfactory arrangement yesterday between the Engineering Foundation and the Research Council. Dunn said Wednesday that it was not necessary for me or Noyes to go to New York to the meeting because he and Carty thought that the situation was pretty well in hand. The Executive Committee appropriated $1000.00, which you asked for, for the San Pedro Station; also the $100.00 for the Botanical Raw Products Committee, though the relations of the Raw Products Committee, the Forestry Committee, and the Botany Committee must be clearly worked out, and I think we shall have to get Coulter, East, Bailey[5] and Pearl[6] together for a discussion of these relations and for defining the function of each of the Committees.

As to the Broca tube of large diameter, your results are entirely at variance with those which have been obtained by the other groups.

The New London Committee tested this with great care and found absolutely no difference between large and small diameters, so long as the tube leading from the thin air chamber was constant in size and small in comparison with the diameter of the disc.

The Submarine Board meets next Tuesday, and at the end of that week, that is about the first of October, I am going to Chicago for a week if I can possibly get away, though work is getting more and more pressing here and it is more and more difficult to escape. On the whole I think the Council is making friends and increasing in influence from week to week.

I have received your telegram in which you think it is impossible to appoint Manning[7] as a member of the Executive Committee. Nevertheless, I have been asking him to attend Executive Committee meetings without saying anything about his status. Of course it is entirely legitimate to have him here in the same way in which we have Mendenhall and Durand.[8] Manning is a very useful member, and I am confident that we have done the right thing in getting him into the inner organization.

I have had the misfortune to lose the key to the desk which you occupied, and although Mr. Chamberlin thinks that there were three duplicates for this desk when we got it, the other two are not to be found. If either you or Miss Gianetti have one, it would be a convenience if you would send it on by the first mail.

Cordially yours,
R. A. Millikan

Hale Papers, California Institute of Technology

1. Tod Ford, Jr., of Pasadena, independently wealthy, had served with the Lafayette Escadrille. He would act as the military assistant in the Paris office of the NRC's Research Information Service.

2. Omitted here is an enclosure, the order of the chief signal officer setting up the division. In retrospect, it seems very strange that Millikan, while wearing his army uniform, successfully acted as the man in charge of research for all the services under NRC. The specific relationship he envisioned did not come to pass.

3. JOEL HENRY HILDEBRAND (1881–) was a physical chemist at the University of California, Berkeley. William Crozier (1852–1942) was chief of Army Ordnance.

4. GANO DUNN (1870–1953) was an influential electrical engineer.

5. IRVING WIDMER BAILEY (1884–1967) was a botanist at Harvard.

6. RAYMOND PEARL (1879–1940) was a biostatistician then at the Maine Agricultural Experiment Station, later at Hopkins.

7. VAN H. MANNING (1861–1932) was the director of the Bureau of Mines whose organization was active in gas warfare research.

8. WILLIAM FREDERICK DURAND (1859–1958) was a mechanical engineer at Stanford and science attaché in Paris in 1918.

Robert A. Millikan to George Ellery Hale, September 27, 1917

Dear Hale:

I want to thank you for the full report which you have sent to McDowell and a duplicate to me, regarding the submarine work at San Pedro. It will be desirable, I think, if Anderson follows your procedure of sending in duplicate reports to this office and to McDowell. This procedure is being followed by the New London and other groups, which have been started by the Research Council, and the advantage of it has already been apparent in the starting of new lines of work in certain places.

Messrs. Shampreux,[1] Babcock and King were officially added to the San Pedro Group at the meeting of the Board, which occurred day before yesterday. I am not sending you this information by wire, because I assume that you are actually using them as members already, and that it is not vital to get the appointments to you instantly.

With respect to the relations of scientific men after the war; although I sympathize fully with Campbell's letter, and share, I think, your own feelings and convictions in the matter, it seems to me that the less that is said now about after-war relations the better. What our relations will be will depend wholly upon the way in which the war comes out, and just as I regard the Paris conference, with the discussion of economic relations after the war, as a very bad tactical blunder, so I should regard any agitation at the present time of the relations of scientists after the war as useless because of the impossibility of predicting the situation, and as unfortunate because it would feed the flames of animosity, which are already sufficiently supplied with fuel, and because it might lead us into a statement of positions which would react badly on us later.

We are looking out for Dr. Campbell's expenses, as you suggested.

Hastily yours,
R. A. Millikan

Hale Papers, California Institute of Technology

1. A. J. Champreux was of the Pacific Telephone and Telegraph Co. in San Francisco. A Berkeley graduate, in 1917 he was in Hale's San Pedro Submarine Station, where he was an expert on induced currents and telephone communication in general.

Jacques Loeb to Maurice Caullery,[1] February 4, 1918

My dear Professor Caullery:

Many thanks for your kind note of January 16th and for your most interesting book on the Universities of the United States. I hope your efforts will bear results in showing your people that science, and laboratory science first of all, can never be too well endowed.

As far as the organization of university clubs after the American patterns are concerned, I should like to call your attention to one point which probably was not impressed upon you. You may have been aware of the fact that it was part of the German system of World Conquest to impress the German mind with the legend of teutonic "racial superiority." All those nations and groups which were considered as dangerous to the exploitation of the Far East, especially the Armenians and Jews, were denounced as vicious, inferior races. Thus Dühring[2] alludes to the Armenians in terms of the "Armenian Vermin." We have seen the result of this preaching by the whole tribe of German anthropologists, literateurs, of the type of Houston Chamberlain, and of the reptile press. The massacres of the Armenians were conducted with the active cooperation of the German officers. Moreover, the massacres of the Jews under the Czar in Russia were received by the Pan-Germans with delight. It happened that a number of American exchange professors and presidents, as for instance President Butler of Columbia University, President Wheeler of the University of California, and others, became the agents of the Emperor and the Junkers in spreading this race propaganda in America; of course, in a quiet way. The Emperor took pains to give to each of the exchange professors a copy of Houston Chamberlain's absurd book, of which good use was made by these professors in America to spread the German gospel of hatred against other nations and races. So eager was President Butler in his propaganda for the German Emperor that he got into an actual quarrel with the late Prof. Münsterberg as to who could have parasitic precedence with the Emperor. Now this will help you to understand the fact that the American University Clubs have silently accepted the strictest form of Prussian anti-semitism, with the result that no man of Jewish descent can be made a member. I hope that in starting university clubs in France on an American model you will not allow this form of Prussianism to creep into your organisations. It will surprise you to know that the administration of Columbia University is about as reactionary as the Russian universities were under the Czar of Russia. Of course, the whole thing is done here not as openly as things are done in Germany or as they were done in Russia under the Czar, but they are being done just the same.

Butler, of course, transferred his sympathies to the Allies when the war broke out but his past has been so checkered that we knew that his sympathies are to be had always by those whom he considers most advantageous to himself for the time being. After the war the decoration of the "Red Eagle 2nd Class" will do service again as it did before the war, when he bedecked himself with that piece of brass.

As far as the general situation is concerned, I think every effort has to be directed toward the spreading of liberal ideas among the people of Germany. You must work with the German prisoners in France in this way, separate the common soldiers from their officers; spread among the soldiers literature like the addresses of President Wilson, and especially the literature published by the German exiles in Switzerland. There is one journal in Zurich, I think "Das freie Wort" is its title, which ought to be subscribed for and spread in large numbers among the German prisoners in France. The change in Germany must come from within but it must be helped by a ferment from surrounding countries. I have reasons to believe that great efforts in that direction exist in Switzerland, and I understand also that efforts are being made with the prisoners in England; but the best thing would be the spreading of a journal like "Das freie Wort" from Zurich, and also translations of Wilson's addresses.

I am going to send some reprints for Dr. Guyenot in your care, since I do not know his address.

I hope you have received the reprints I sent to you recently.

With kindest regards and best wishes,

Yours very sincerely,
[*Jacques Loeb*]

Loeb Papers, Library of Congress

1. The French biologist CAULLERY (1868–1958) whose book *Les Universités et la vie scientifique aux États-Unis* (Paris 1917), which appeared in English translation in 1922, was a favorable review of the U.S. position in science. Caullery used his American examples to call for reforms in France.

2. EUGEN KARL DÜHRUNG (1833–1921) was a German economist and philosopher notable for his attacks on Judaism.

Willis R. Whitney to Robert A. Millikan, February 25, 1918 (dictated February 22)

Dear Millikan:

This is a hard letter to write right, and a lot of it will be wrong, but rather than carry the idea longer, until I can put it clearer, I want to start on G. W.'s birthday and give you at least a worm's eye view of the subject of which you spoke last week. Of course I think you are all wrong, and I guess I said so before you told me what you had in mind. That is merely characteristic of me! But let me now tell you what I think.

Scientific research is the greatest panacea the world has ever seen. Every nation which has used the active study of material truth by investigation has grown in civilization more rapidly than others, no matter what else it may have done. It is perfectly plain to me that America is forced to do more than it has done in the way of learning from Nature. If it does not do it, it will be beaten, even by Japan, in a very few decades. I believe the future holds out infinitely more in the way of new and useful disclosures than the past has shown. But what is very clear to me is that these will come by what we call laboratory work, rather than by metaphysics or exclusively mental gymnastics. Discoveries in engineering, in heredity, in general prophylaxis, and physical and mental properties of matter, new principles, etc., etc. (just that line of things which distinguishes us from wild Indians), really come down to being extended physics and chemistry. Take astronomy and pure mathematics, for example, where we have a few National specimens. They had to go thru this bottle neck. Take the man who studies soil-fertility or diseases of plants, and the one who studies eugenics and diseases of children; the man who studies aeroplanes and the one who studies wireless. They all come thru the narrow neck of chemistry and physics. Now turn your eye on the size of that neck. In a few scattered, unrelated laboratories a few hours of a few days of a few of the weeks of the year, a few illy prepared, illy paid so-called assistants are choking off the natural American traits of physical inquisitiveness. How a man ever got to be a physicist in this country interests me as a puzzle, when I reflect on the opposition he must have met. Fortunately, chemists had the way made a little easier for them because iron, copper, and the other ores simply had to be analyzed, and the process of training the necessary chemists, trained by accident some misshapen analysts, who kept the fires of the science lighted.

But certainly we've got to do better, and its largely up to you. Of course I don't think you are entirely equal to it, or I wouldn't write this, but I should like to have you believe that I don't think I am half as equal myself. Both of us ought to be shot at sunrise, tho, if we don't do all we can to fix things. Think of the rottenness of the conditions which reward the student's guide in physics and chemistry in this country. He gets about $500. a year, and a hundred or more students (many of whom might make good investigators of physical truths) pass thru his hands each year. No real contact with investigating men comes to these students. Its like putting lead into life-preservers. Four of these young assistants get about as much combined pay as one good glass-blower. It would take the pay of four of them to hire a fair mechanic, and that of a half dozen or more to pay a sheet roller in a steel mill.

The job pays just what it did forty years ago. How much farther can it go? And yet that is the neck of the bottle.

What I want to have you see is that whatever tends to increase the pay and the number of teachers of general physics and chemistry in this country should be heartily supported. A good laboratory with one enthusiastic investigator is a wonderful start. Even the five hundred dollars of the assistant increases in this atmosphere. That means better assistants, and that means better product and still better assistants. That in turn means more enthusiastic investigators and better laboratories, etc., ad infinitum. Need I add that that was the one and only grand and good thing that the German government thoroly understood and utilized? Are we Americans going to stay blind to it? I wish we could learn, from their experiments, the good things they disclosed. We who know geography even now recall many European cities largely thru the research laboratories or scientists they contained. Take Leipzig, Berlin, Heidelberg, Bonn, for example. Who would know of Göttingen, Freiberg, or Jena but thru the researches done there?

We don't want a central office full of desks and typewriters in place of experimenting laboratories. We want the people to see that they are most deeply concerned as a whole in the rapid growth of our higher educational institutions. In some form, the people should pay more for greater scientific activity, but the thing most to be guarded against is making of good research men mere executive officers. Who knows what Dr. Woodward, Dr. Day, Dr. Walcott, Dr. Stratton,[1] Dr. Manning, and such leaders might have done in guiding the stream of good students which the country could produce? With all the good they now do where they are, I should a thousand times rather see them directing the training processes by which thousands of equally good men were produced.

Take your own case. Barring the war demands upon you, how infinitely greater would be your value at Chicago, starting a few good men every single year along such fundamental physical lines as have met your predilections, than if you were in an executive position like any of the above men! They may be said to produce material, but you would produce men each year who would in turn produce much more. One is a sort of simple arithmetical progression, while the other is geometric.

What applies to you applies to others, but they are all too few and they are too little appreciated. The job of the Research Council is to inform representative men and Congress. Don't let us try entirely new experiments while we have had before us for fifty years the eminently successful work of Germany in this respect. No matter how we may hate many of her attributes, we must be careful to utilize this one good point, to which the world long paid homage: her broad system of federal aid to scientific education. Why should we fail to produce in many of the cities of our country just such Meccas of learning and research as Leipzig or Jena? The way is so simple that I should think even we might see it.

Yours very truly,
W. R. Whitney

Archives, National Academy of Sciences

1. SAMUEL WESLEY STRATTON (1861–1931) was a physicist, who was director of the National Bureau of Standards. From 1923 to his death, Stratton was president of the Massachusetts Institute of Technology.

Thomas Edward Thorpe[1] to Frank Wigglesworth Clarke,[2] June 22, 1918

My dear Clarke,

I have just received your letter of June 4th. and note what you say respecting reprinting the last table of Atomic Weights without change. I will submit this suggestion to the Council of the Chemical Society and also inform them that you hope to be in a position to make a complete report next year, presumably based upon your fourth recalculation of Atomic Weights, to be published by the Smithsonian Institution next autumn.

In sending in my last report to the Chemical Society I pointed out that it was highly improbable that the existing arrangements for the compilation of the Annual International Reports could be continued. I have received very emphatic statements from Urbain that neither he nor any other French chemist would consent to be associated with Ostwald in this matter, and I think it not unlikely indeed that they would refuse to co-operate with any German representative. A feeling of this kind is perhaps not quite so acute here, but I think it is gradually hardening, and if the whole Chemical Society were polled I have little doubt that there would be a considerable majority against having anything to do with Germany. I do not know how the matter stands with you in America, but I think it not improbable from what I read and hear that the opinion of scientific men with you would be no less strong. It seems a great pity that the cosmopolitan character of science should be affected by such considerations, but really the Germans have behaved so badly in this war and they have been so assisted by their men of science in making it what it is, that some indication of what other men of science think seems to be called for, and one slight indication of the general sentiment would seem to be afforded by all the various powers refusing to associate themselves with Germany for some long period to come. We see a tendency towards this action in all sorts of ways, in political matters, in commerce, trade, shipping, etc., and it is hardly likely that science will be able to resist the same influences. The Council of the Society thereupon appointed a small committee consisting of Tilden,[3] Scott[4] and myself, to consider and advise them as to what action should be taken. There is a strong feeling that the work of the Atomic Weight Committee should not be abandoned for there is no doubt that our action has been beneficial and has been appreciated as tending to uniformity and as keeping the chemical world as a whole informed of progress in Atomic Weight work.

As yet the committee has not met, although we have had an informal exchange of opinions. Under present circumstances there seems no need for immediate action, but I have more than once thought of communicating with you with a view of learning your opinion. I presume you still think it would be desirable to make the organisation international. What have you to say as to what nations should be invited to co-operate, Germany and Austria being out of the question and Russia being no longer available? My own opinion is that America should continue to take, as hitherto, primary direction, and I hope that you will reconsider your wish to be relieved of your present responsibilities. You are so fully identified with the subject that all would wish to see you associated with it as long as your health and strength permit you to take an active part. If, however, you find yourself unable to accede to this suggestion would you indicate how America should be represented? As regards myself, I feel like you that "it is time for someone else to tackle the job," and I have so intimated to my colleagues. There are several men amongst us here whom the chemical opinion of the place would indicate as persons who are likely to be of service on such an international committee, and who would take an active interest in its work.

I should be very glad to hear from you again on this matter of the reorganisation of the Committee and I should like to be in a position to communicate your views to my colleagues, as I am sure your opinion and advice will have great weight with them.

With kind regards to Mrs. Clarke

Yours faithfully,
T. E. Thorpe

Clarke Papers, Smithsonian Archives

1. THORPE (1845–1925) was a chemist, Ph.D. Heidelberg (1869), then emeritus at the Royal College of Science, London (now Imperial College).

2. CLARKE (1847–1931) had been chief chemist of the U.S. Geological Survey from 1893. He is best known for his determination of the elements composing the outer crust of the earth. This letter relates to his chairmanship of the International Committee on Atomic Weights, a post occupied from 1900 to 1922.

3. WILLIAM AUGUSTUS TILDEN (1842–1926) was emeritus at what is now Imperial College.

4. ALEXANDER SCOTT (1853–[?]) was at the Royal Institution possibly up to 1911. Later he was at the British Museum.

Willis R. Whitney to John Johnston,[1] July 29, 1918

Dear Dr. Johnston:

I really think that, unless you are very busy with actual war problems, you might help a lot in clearing up the possibilities for the best method of scientific advance by getting into a closer contact with some of the men or groups, who outside of the Council may still be earnestly trying to do constructive work in science. One reason why I can ask this is that, if I am prejudiced in favor of any one method, it is in the one which most directly concerns the greatest number of people, and should be understood and undertaken in the broadest cooperation of different classes. One of the greatest troubles with us scientists is our exclusiveness and aloofness. Try as we may to see and overcome it, we still have some of the puritanic feeling similar to that which made the magicians and the old alchemists retain their aloofness.

On the other hand, the individual desire to see America advance in knowledge and proper activities is now everywhere. Even a dub can enjoy it. I have heard a little of the recent plans of the Council. I have heard about as much of the plans of the Bureau of Commerce, the Bureau of Mines, the Naval Consulting Board, the men of the Land Grant Colleges, research men in universities, Industrial laboratory organizations, National Institute of Inventors, etc., etc. They one and all want to help in the encouragement of research in some way. One emphasizes the production of the mentalities to do the work, another the encouragement of the existing men by unusual opportunities for work, and others the rewarding of them for service, and still others the selection of certain fields and ways for special effort (mines and Navy, for example).

A Napoleon on research might well marshall these forces and accomplish something. You fellows are hired for such work. You look like him; Millikan dresses like him; Hale writes like him. The present stand of the subject suggests to me a bird's nest, each occupant instead of learning to fly is picking the sprouting feathers from his companions. I've heard enough picking on that Inventors Institute to make me wonder if there isn't someone there who might do something useful. The chances are against it, but I wish you could advise me about it. After I saw you I heard other horrible stories, almost as bad as those referring to Grant and Roosevelt as booze artists. For example, when I mentioned Simon Lake as one of the inventors group I was told that not only Lake, but also Holland[2] (the other submarine man), was just a zero two places east of the decimal point. As I grow older, I get very tired of the persistence of belittlement in good people. The men I think the most of are the first to call an Edison a fakir. Then when I see an Edison work, I say to myself, "I wish I had his powers and his record," and many of my friends have neither. My faith in the powers of folks to really recognize values grows less and makes me doubt myself. I suppose that is why I'm driving you to help me.

But honestly don't it seem possible to you to effect some scheme to get representative men together, at least for partial, if not a mutual education? Couldn't you show the National Inventors Institute where they are weak and sure to fail, and couldn't you take part in the work which the colleges have got to do mighty soon?

Yours very truly,
W. R. Whitney

Archives, National Academy of Sciences

1. JOHNSTON (1881–1950) was a Scottish chemist who came to M.I.T in 1907–1908. From 1909 to 1916 he was at the Geophysics Laboratory of CIW, then from 1917–1918 at the Bureau of Mines. He was secretary of the National Research Council 1918–1919, at Yale 1919–1927, and from 1927 to 1946 director of research for U.S. Steel.

2. LAKE (1866–1945); JOHN PHILLIP HOLLAND (1840–1914), the Irish-born submarine inventor.

Forest Ray Moulton to Thomas C. Chamberlin, August 8, 1918

Dear Prof. Chamberlin:

I beg to acknowledge receipt of your circular letter of August 1st to the Members of the University Senate. The variety of your interests and the energy with which you persue them is a never ending source of wonder to me. I suppose this is what keeps you so young.

I cannot close this letter without giving you a little idea of my work here, in which I know you are much interested. I am in the Engineering Division of the Ordnance Department, and under my immediate control are the programs for all the range firings of the numerous guns and much more numerous projectiles. These range firings are carried out at the Sandy Hook and Aberdeen Proving Grounds. In this problem we spend in the course of a month several millions of dollars. It will be necessary for us to get out more than one hundred range tables.

There is opportunity in these problems for the use of considerable mathematics. I have found that those who had worked on them heretofore were very much lacking in the modern training which is important for successfully undertaking such work. Within three months entirely new methods had been devised which enormously reduced labor and gave very great increase in accuracy. The mathematical end of these problems is undergoing as great an evolution as all other features of them. When I came here there was only one other man engaged on this work and he was a physicist. We now have seventeen of whom ten are Doctors of Philosophy in Mathematics. This work has won the respect and approbation of the regular army men who have heretofore been interested in it, and everything is going satisfactorily to them as well as to us.

You will be interested in my general impressions of work in the army. The problems are of a magnitude and complexity much greater than one could possibly foresee. It seems to me that in view of the extremely rapid expansion, work is being carried forward most satisfactorily, and that another year will show to the credit of this country much more than the past year. In fact, if the war should continue for more than one year after this, I think it is highly probable that we shall be able to show the whole world what it means to organize great industries and go into quantity production. I am doing everything possible to see that the scientific ends of the part of it with which I am connected are correspondingly efficiently managed.

But these interests which absorb every moment of the day and more than half of my evenings, together with the high temperature which we are having, leave me absolutely no time whatever for scientific work. I only hope that when I attempt to take it up again I shall be able to do so with renewed energy.

With greetings to Rollin and other colleagues, I am

Very sincerely,
F. R. Moulton
Major, Ordnance, R.C.

Chamberlin Papers, University of Chicago

Edward C. Pickering to George Ellery Hale, August 24, 1918

My dear Professor Hale:

Your letter of August 9 is received. The problem of International Research interests me very much. I brought the matter to the attention of the Council of the Astronomical Society as you requested. Resolutions were passed, which you will doubtless receive from the Secretary.

It seems to me that there are two questions. I am strongly in favor of an inter-allied research council which should secure from scientific men every aid possible for our successful prosecution of the war, and the continuation of necessary routine observations.

The advancement of pure science internationally is a very different question. This is especially the case with modern astronomy, which stands on a different basis from almost every other science. It is purely impersonal, seeking after truth, independently of individuals or nations. Every consideration should give way to its fulfilment of this object. No ordinary punishment is adequate for those responsible for barbarities contrary to the laws of nations and humanity, yet we ought not to ignore the work of those who, laboring quietly in their observatories, have done their best to extend our knowledge in these terrible times. I hope, therefore, that in Astronomy no definite action in this matter will be taken during war times, until we know the attitude of those whom we once admired and respected.

Hoping that the good news of this morning is the beginning of the end,

Yours very sincerely,
Edward C. Pickering

Hale Papers, California Institute of Technology

[Hale] Memorandum of Interview with President, September 10, 1918

He is *very emphatically opposed* to any resolutions directed against German men of science. Says their passage is best possible way to play Germany's game, as it would give color to the claim that Germany is surrounded by vindictive enemies.

I told him that there would be much pressure for such resolutions at London Conference. Council of N.A., while each member agreed that personal relations with Germans would be impossible for many years after the war, was opposed to passage of any resolutions on the subject. Plan for Inter-Allied Research Council excludes this question, because it is to deal with war problems, and therefore only the Allies could be included during the war.

He is opposed to new Inter-Allied organizations, because of complications encountered by existing ones. France and other European nations have felt the war much more than we have, and are therefore likely to take drastic action that might bind us to do things contrary to our natural intent. Hence he objects to formal organizations of any kind.

I remarked that our purpose was chiefly to secure information, such as could be obtained from inter-allied conferences of scientific men on war problems.

He said that such conferences could be called by such a body as the Paris Academy or the Royal Society, inviting anyone wanted.

I spoke of necessary secrecy, and therefore of necessity of accrediting delegates, pointing out precautions taken here, and fear of military men that leakage would occur.

He replied that the military men were very finicky, and altogether too much concerned about such matters.

I insisted, however, that serious results would be sure to follow without some means of accrediting members present.

He spoke of the importance of having conferences open to anyone, whether members of societies or not, and strongly approved my (incidental) remark that the membership of the N.R.C. would not be permanent, but rotating.

He finally agreed, however, that there must be some body in a position to accredit those who are to attend conferences. I asked whether the N.R.C. would not be the right body to do so for this country and he replied that he thought so. He also agreed that there might be some headquarters in Paris or elsewhere for holding conferences. The one thing he strongly opposed was a formal organization for war purposes—

Hale Papers, California Institute of Technology

George Ellery Hale to Woodrow Wilson, October 15, 1918

My dear Mr. President:

I beg to report that the Inter-Allied Conference on International Scientific Organizations, which had been called together by the Royal Society, opened in Burlington House, London, on October 9th and adjourned October 11th. Delegates from the following countries were present: Belgium, Brazil, France, Great Britain, Italy, Japan, Portugal, Serbia, and the United States. Sir Joseph Thomson, President of the Royal Society, was in the chair.

I am submitting to you herewith a copy of the preamble and resolutions adopted by the Conference (enclosures A and B).[1] With reference to these I beg to offer the following comments. On September 30th the Paris Academy of Sciences had adopted drastic resolutions directed against the men of science of the Central Powers, a copy of which is enclosed (enclosure C). One of these resolutions requests the Allied Governments to require in the treaty of peace that the Central

Powers withdraw from international scientific associations. The Belgian attitude was still more stringent. In place of these resolutions the Royal Society proposed a preamble which, in its original form, contained the clause: "If to-day the representatives of the scientific Academies of the Allied Nations are forced to declare that they will not be able to resume personal relations, even in scientific matters, with their enemies on the resumption of diplomatic relationship, they do so with a full sense of responsibility." The task of the American delegates was to endeavour to secure a modified form of this preamble in harmony with the views expressed by the President of the United States.

After much discussion, the Conference finally adopted the preamble as amended by us (enclosure A). In this form the American delegates believe the preamble to be in complete agreement with your declarations regarding future relations with the Central Powers.

It is not the view of our delegates, or of any member of the Conference, that it will be possible to renew *personal* relations with German men of science for many years to come. You will notice, however, that neither the preamble nor the resolutions adopted contain any statement to this effect. The preamble indicates that after the Central Powers have undergone such changes as your recent messages demand, their men of science will be eligible for membership in international associations. To attempt, however, to renew *personal* relations before the bitter feelings engendered by the war have been removed would undoubtedly tend to postpone the final return to friendly intercourse which we hope may ultimately be possible.

The plan embodied in the "Suggestion for the International Organization of Science and Research" (enclosure D) offered by the National Academy of Sciences was adopted by unanimous vote of the Conference. When this "Suggestion" was submitted to you last spring it received your acquiescence in your letter of May 13th, 1918. It calls for the organization in each of the Allied Nations (and subsequently in other countries) of a National Research Council similar to our own. The ultimate federation of these Research Councils will constitute an International Research Council for the promotion of scientific and industrial research.

The difficulty discussed during my interview with you on September 10th will not arise. You remarked on the danger of joining an Inter-Allied Council formed for the discussion of war problems. The International Research Council will not come into existence until the various National Research Councils have been formed and federated; their organization will undoubtedly occupy many months, and probably a full year. The idea of holding conferences under this international body for the discussion of war problems has also been abandoned. Thus the International Research Council will devote itself to the promotion of research in harmony with the views expressed in our "Suggestion for the International Organization of Science and Research." It goes without saying that the United States should take part in such an organization, as you were quick to realize when the plan was first submitted to you in the spring.

Perhaps I may be permitted to express my satisfaction that our own National Research Council, formed at your request for the federation of the research agencies of the United States, and permanently established by your Executive Order of May 11th, is soon to have counterparts in many other countries. An international union of such bodies should be able to contribute materially to the progress of science and the arts.

In voting on the resolutions passed by the Conference (enclosure B) the delegates of our National Academy of Sciences abstained in the case of the first paragraph of Resolution I. Thus we are free to continue any desirable connections with existing international scientific associations, and at the same time at liberty to join new ones.

As for the proposed new associations, only two are definitely in prospect. That which is to deal with astronomy will doubtless result from the union of such organizations as the Commission on the International Chart of the Heavens, the International Union for Cooperation in Solar Research, and other bodies relating to special branches of astronomy. In geophysics it is planned to form a body to bring into cooperation those who are concerned with special phases of the subject, such as geodesy, terrestrial magnetism, meteorology, seismology, and other branches hitherto dealt with separately. You will thus observe that our guiding principles in the future will be broader cooperation and closer coordination of effort.

The question of expelling honorary members of the various national Academies of Sciences who are of enemy nationality was brought up by the Royal Society. Our delegation opposed their expulsion, and this view prevailed with the Conference. Thus such action is not likely to be taken by any of the Academies.

After completing my work here, I expect to go with Dr. Flexner and Dr. Noyes to Paris and Rome to comply with requests for assistance in organizing the National Research Councils of Belgium, France, and Italy, and to attend a meeting of the Committee of Enquiry (Resolution IV) in Paris on November 26th. I hope to reach Washington soon after the middle of December.

Believe me, Mr. President,

Yours faithfully,
[*George Ellery Hale*]
Chairman of the Delegation of the
National Academy of Sciences
Hale Papers, California Institute of Technology

1. The voluminous enclosures are omitted here. What Hale does not say is that the Conference (and the American delegation) was not acting in accordance with the spirit of Wilson's wishes as given in the interview of September 10.

Excerpt, Minutes,[1] Board of Trustees, Carnegie Institution of Washington, December 13, 1918

[PRESIDENT WOODWARD:] In passing, it may not be out of place to offer some remarks concerning the rise and progress of two novel adjuncts to the Government service, with which your humble servant has been connected. These are the National Research Council and the Naval Consulting Board. They afford examples of the ways in which new organizations arise more or less unexpectedly and grow finally into permanent attachments in governmental machinery. The National Research Council had its origin in the National Academy of Sciences, which has been by law for more than half a century one of the constituted advisory adjuncts of the United States Government but which has been very rarely consulted during this time. With the entrance of the United States into the world war, it was suggested by representatives of the National Academy of Sciences to the President of the United States that an advisory council in respect to the business of research might be advantageous to the Government. This suggestion was cordially approved by the President and the National Research Council resulted, with functions similar to those of the National Defense Council, with which the Research Council became in fact affiliated. Similarly, in the latter part of 1915, the Secretary of the Navy invited a number of technical societies to designate two members each to form what was initially called an Advisory Board on Inventions for the Navy. In addition to the two representatives from each of these eleven technical societies, the Secretary appointed Mr. Thomas Alva Edison Honorary Chairman of this aggregation and subsequently Mr. Edison's personal secretary, Mr. Hutchison, was added to the number. At the first meeting of this Board, its name was changed to Naval Consulting Board. It has met fortnightly with nearly uniform regularity during the intervening years up to the present time, and it will be of interest to the Trustees to know that during the past year and a half, or up to October last, when suitable quarters were furnished by the United States Navy in its new building, this Naval Consulting Board has met in this room.

The histories of these two organizations should be of special interest to the Institution, since they have represented two distinctly different modes of procedure toward common purposes. On the one hand, the National Research Council has proceeded on the supposition that discoveries and advances may be most reasonably expected to arise with those who have already shown capacity to make them. The theory of the Council has been that the best advice to the Government in cases of emergency is most likely to come from experts of repute in their various fields of research. The initial theory of the Naval Consulting Board, on the other hand, was that discoveries and advances are about as likely to come from untrained as from trained minds and that, since the number of amateurs is very large, the best way to secure advances is to set experts at work examining the suggestions and inventions of inexperts. In addition, initially, the Naval Consulting Board was also encouraged to believe that discoveries and advances are developed chiefly by abnormal minds and that it is therefore worth while to set men of proved efficiency and capacity at work scanning the horizon for the scintillations which might otherwise emanate unperceived from exceptional men, who are supposed to be in hiding, or at best more or less concealed behind books and bottles in dingy laboratories.

These two opposing theories are of great interest and importance, not only to the Institution but to all research establishments, since success or failure with them depends upon which of the two rival theories is adopted. The theory which dominated the Naval Consulting Board during the first two years of its existence is by far the more popular of the two and, if appearances are not deceptive, is the one which would receive an overwhelming majority in its favor if put to a popular vote. It is a theory with which the Institution has had to deal daily during the past seventeen years and one against which it must continue to contend for a long time to come. According to this theory, science in America, at any rate, begins with that respectable journal called The Scientific American and culminates in the Patent Office. Science is thus limited to the narrower, egoistic aspects of invention alone. That this fatal fallacy is widely prevalent is perhaps most strikingly witnessed by the fact that it was nearly unanimously approved by the Congress of the United States in recent negotiations with Garabed Giragossian,[2] who had hastily reached the conclusion that the remarkable entity called the ether is an exhaustless source of "free energy."

Happily for the reputations of the members of the Naval Consulting Board, this initial and popular theory was subjected to the tests of plain experience, which proved what is well known in the history of science and what has been demonstrated on a grand scale in the experience of the Institution, namely: first, that revolutionary discoveries, advances and inventions do not arise suddenly or in necromantic fashion; and, secondly, that the poetic process of winnowing vast quantities of intellectual chaff with the hope of securing good grains of truth is the sheerest of futilities. The so-called "wizards" of the Naval Consulting Board produced no epoch making inventions to win the war. They examined about 110,000 miscellaneous suggestions and inventions and found less than 10 of these worthy of application and development.

On the other hand, it should be said that the Naval Consulting Board and the National Research Council have rendered valuable service to the Government in an advisory way concerning a great number of important matters. One of the most important of these emanating from the Naval Consulting Board is a recommendation that the United States Navy have a great experimental laboratory in which new and improved devices may be invented, developed and perfected to the point of effective application. This recommendation has been approved by the Secretary of the Navy and by the Congress of the United States and an appropriation to start the establishment has already been made.

But while the two organizations just referred to have rendered valuable service to our Government in the present emergency, it appears essential to remark that all such advisory or auxiliary bodies must appear quite amateurish for the simple reason that they lack the two requisites fundamental to efficiency, namely, authority to act and coordinate responsibility. Experience with them makes it plain that governments have not yet evolved adequate methods for making use of knowledge now available in the more advanced fields of science. This criticism of existing governmental machinery is rather glaringly brought to light in the experience of the Institution in the production of optical glass and in the production of optical instruments for the Ordnance Office of the War Department. Our Government as such does not yet recognize the existence of an altruistic establishment, and it would probably have been difficult, if not impossible, to achieve the success attained in the production of optical glass if the enterprise had not been financed by the Institution alone.

It may be of interest to the gentlemen here to know about what it has cost the Institution to produce optical glass. I estimate about $175,000. If we had had to do it, confronting the obstacles of governmental machinery, we probably would not have much more than gotten started by this time.

Governmental inability to cooperate with an altruistic establishment was also revealed in the recent optical work undertaken in the shops of the Mount Wilson Observatory. The technical, legal forms of the Ordnance Office are designed to cover only the extreme case of profiteering manufacturers. There was no room in those forms, except in a Pickwickian fashion, for an establishment which was willing to furnish a good share of the men and the apparatus required free of charge. The legal machinery of the Ordnance Office appeared to be designed mainly to prevent rascality and to discourage, rather than to promote, altruism . . .

Archives, Carnegie Institution of Washington

1. A concise summary of Woodward's reactions to the wartime experiences but also reflecting viewpoints which would become widely ingrained among scientists and engineers.

2. Giragossian of Massachusetts claimed the discovery of a method of obtaining inexhaustible "free energy." Both houses of Congress passed a joint resolution granting him patent rights plus other benefits, provided that a committee of scientists certified the veracity of his claims concerning his discovery. Although it was ridiculed by some members of Congress and objected to as circumventing the established procedures of patent law, the resolution passed amid talk of the possible boon to humanity and the general hostility greeting great innovators. The scientific committee curtly disposed of the matter, and the original joint resolution was repealed. Although Giragossian of course denied it, his was basically another perpetual motion proposal. This was the first attempt by Congress to grant a patent right since the case of Charles Grafton Page in the last century. See Robert C. Post, *Physics, Patents, and Politics* (New York, 1976).

Jacques Loeb to Svante Arrhenius, January 23, 1919

My dear Professor Arrhenius:

I do not know how long it is since I have written to you but it must have been before the armistice was signed. It is strange that all that has happened since has not caused more enthusiasm. A year ago I should have considered the establishment of a republic in Germany impossible; today it seems a fact—at least temporarily—and yet somehow I am not elated. The reason is probably that we begin to become conscious of the misery the war has wrought, and also that we are feeling a little troubled about the future. Naturally we are in hopes that the epidemics which have visited us may have been only war epidemics and may be a transitory affair. At present I have had four patients in the house, Mrs. Loeb being one of them, but she is over the worst now and is about again. But we have lost a number of friends and naturally that has caused us a good deal of grief. The worst is, I have a feeling that our whole civilization is in danger. The prices are rising terribly and there is no hope of their reduction. The farmers refuse to be satisfied with the small returns and the long hours of labor of former years, and labor does not wish to see any reduction in wages. The consequence is that professors, and professional men in general, who have no large means are threatened into a mode of living which will render university work practically impossible. The other day there was a harbor strike and the cooks and the deck hands demanded three thousand dollars a year. My son Leonard, who we hope will return from France sometime in March, will probably have to continue his research without remuneration or if he gets a university position he may be lucky if he gets half what the deck hands and the cooks on a tug boat are demanding.

With all that, we are threatened with the worst danger of Bolshevism. While ultimately they cannot accomplish the end, they may, however, succeed in upsetting civilization pretty thoroughly. It is rather a sad outlook for the new year. On the whole it comes down to the stupidity of the business men who rule the world. It was the business men in combination with the feudal militarists in Germany who started the war, and our own militarists are crying for a large navy and a large army if they can get it. Unfortunately, Wilson also insists upon a large navy, which, of course, means that we shall enter upon a militaristic era. Imperialism, however, will end in the future as it has ended in Germany with the total collapse of one of the two belligerents and a half collapse of the other. A second failure of the business man's rule lies in the fact that he did not try to conciliate the laborers in time, with the result that he has fostered the revolutionary spirit among the workingmen who are now demanding things which are incompatible with the continuation of our civilization. This only as an explanation why we are not not over happy here in spite of the happy ending of the war.

I wonder whether you have any news of what is to become of German science. Under the circumstances I do not quite see how they can go on with their laboratories. It must have been a terrible awakening in Germany when they began to realize the collapse. I read recently that Haber received the Emperor's autographed picture and two iron crosses. I am afraid they will not bring much in an auction. If it is true that Haber introduced the poisonous gases I am afraid his name will be blackened with a curse in history. Have you heard from Ostwald? I wonder what he is now thinking about his idea of the Germans as the chosen people and the destiny of the Hohenzollerns to rule Europe.

I am working along hard, amusing myself with the work on gelatin, and I notice on every step that the so-called Colloid Chemistry has been built on a false basis and on poor experiments instead of on thorough work. It seems that almost nothing is true of the assumptions of the colloid chemists. Since I have made use of the measurements of the hydrogen ion concentration (for the introduction of which into our laboratory we shall have to be everlastingly grateful to Dr. Dernby)[1] matters have become very clear and simple. I wonder whether you are getting the current numbers of the new Journal of General Physiology, edited by Osterhout and myself. If you are not receiving the periodical I shall see that you are put on the mailing list. You will see in that journal what the nature of the work has been. In addition I am amusing myself with experiments on the regeneration of plants. My assistant, Dr. Northrop,[2] has at last been dismissed from the army and now I think he can participate in the work of the laboratory.

Bobbie expects to receive his M.D. degree the first week in February, that is about the time you receive this letter, and he will be intern in the Massachusetts General Hospital in Boston for the next year. He is the youngest in his class and seems to be leading his class. He has done very well as a student but now I think his real struggle will begin. Leonard made some kind of invention just before the armistice was signed, and received some money from the French government to work out the invention. This has kept him in France a little longer, but he complains that nobody takes any more interest in war work and he feels that under the circumstances he might just as well return. He hopes that he may be back in March. I only wish that it were so.

The American universities have a very heavy deficit, and moreover are not interested very much in pure science. This is largely a result of business men administration. The trustees of our universities are all business men, who like to see the university run like a department store by a captain of education—a university president—whose interest and sympathies are not with science but with a large display of statistics and expansion, as for instance University Extension, Vocational Training, Schools of Education, etc., so that finally no money is left for scientific work. An interesting book has been published here by a Norwegian, Thorstein Veblen,[3] on "The Higher Learning in America" in which the fatal effect of our present system of university administration by business men is pointed out. The fact that the prices of living have risen while the salaries have not been increased, drives our best scientists into industries.

I hope you will write me soon because I have not heard from you for a long time. Sometime ago I asked my publisher to send you a copy of my book on "Forced Movements, Tropisms, and Animal Conduct" and I wonder whether it has reached you.

With kindest regards from all of us and best wishes for the new year to yourself and your family, I remain,

Yours very sincerely,
[*Jacques Loeb*]

Loeb Papers, Library of Congress

1. The Swedish chemist KARL GUSTAF DERNBY (1893–1929).
2. JOHN HOWARD NORTHRUP (1891–) was a chemist at the Rockefeller Institute until his retirement in 1962; he was awarded the Nobel Prize in 1964. Northrup was one of the models for Sinclair Lewis's *Martin Arrowsmith*.
3. Thorstein Veblen was born in Wisconsin of Norwegian immigrant parents.

Franklin D. Roosevelt to Irving Langmuir, April 24, 1919

Dear Sir:

The work carried out during the war by the Navy Department in developing anti-submarine devices and equipping vessels for anti-submarine operations had an important effect in restricting enemy submarine activities.

The result was made possible by the splendid assistance and cooperation of the many distinguished scientists, engineers, and business men who were in one way or another associated with the Special Board on Anti-Submarine Devices, which had been appointed by the Department to supervise work of this nature.

The Navy Department wishes to express its appreciation of the valuable assistance rendered by you in this connection.

Very truly yours,
Franklin D. Roosevelt
Acting Secretary of the Navy

Langmuir Papers, Library of Congress

11

Biology since 1915: Defining the True Way and Locating Its Home

Jacques Loeb was not a typical biologist of the first quarter of the century, either in the United States or in Europe. He was far too polemical and too philosophical. About him there always hovered the aura of old disputes in German university seminars. But Loeb's great self-assurance reinforced his message of a biology based on experimentation and quantification, explicitly modeled on physics and chemistry. That message could spread because there was great receptivity, here and abroad, even before the turn of the century.[1]

Influential as Loeb was, he hardly encompassed all significant views. The older mode of morphological research still existed; it had partially engendered such notable investigators as Morgan and Wilson. In the hands of such investigators, a kind of symbiosis with physics and chemistry existed. Rather than an almost literal reduction to the laws of physics and chemistry, many biologists used these fields (and mathematics) as tools for developing concepts, laws, models, etc., uniquely biological in nature even if sometimes based upon analogies with the forms of the physical sciences. Morgan's work in genetics is the best example.

At the same time, Loeb's model, the mechanistic world of classical physics, began changing even as he preached its hegemony. A new physics emerged which was neither based on the older mechanics nor deterministic in the same sense. It was a world of ordered complexity where statistics played a greater role than it did in the Newtonian cosmos. Many of the older generation were not comfortable in the new order. Even one of its architects, Einstein, did not like the idea, as he put it, of God playing dice with the world.

For Loeb, the belief in the ultimate reduction of biology of the laws of physics and chemistry was associated with a nexus of other beliefs constituting an ideology of science. Although widely shared, this ideology had a fair number of permutations. Scientists who did share Loeb's deterministic faith might not agree with his views on the importance of theory and its relationship to practice. After World War I Loeb himself departed from the faith in internationalism, as is evidenced by the 1923 exchange with Howell. Although he shared the widespread view of his peers about avoiding politics and public controversy, he did become a member of the National Association for the Advancement of Colored People. Loeb was an elderly man in these years after World War I. Never one to tolerate fools or opponents, an acerbity, verging perhaps on crankiness, appears in his later years.

The Rockefeller Institute was hardly a typical organization, even though it was very influential in the United States. Loeb could scorn the lesser practitioners and the pressures for application as he sat in a sanctuary protected by a handsome endowment. From his viewpoint the organizations fostered by Hale represented an intrusion upon the autonomy of science defined as the absolute freedom of the investigator. Yet the differences with Hale were not as great as the passion of Loeb's prose might indicate.

If the progress of the physical sciences were rendering Loeb obsolete, a similar process was occurring to Morgan at the height of his influence and achievement. The ubiquitous presence of statistics did not bother him. Explicit in the work of Darwin was the element of chance and a resulting concern with populations rather than individual organisms. At the same time that cytologists and embryologists became geneticists looking at cells and parts of cells for mechanisms of heredity, other biologists studying the implications of Darwin's stress on the interrelations of organism and environment started applying mathematical tools to populations. Still others were poking about the chemical constituents of the cell.

Morgan and his school succeeded in demonstrating that an abstract entity, the gene, existed in serial order in chromosomes. The chromosomes with their genes were the carriers of heredity. By a skillful blend of cytological observation and mathematical analysis of experiments, geneticists of the classical period established a body of concepts and techniques defining the mechanisms involved. By these methods Morgan and his students—Sturtevant, Bridges, and Muller—were able quite early to specify exact locations for particular genes. From this point they and others proceeded to map the gene loci on chromosomes. Perhaps the ultimate in this line of research was Bridge's work on Drosophila using its giant chromosomes in the salivary glands. The genetics of corn was another achievement of this thrust.

Morgan carefully defined what his work meant; he knew all too well the precedents of rather grandiose speculations. But the measured care of his words, the austere refusal to leap ahead of his data, proved a hazard. Even those who recognized his achievements might underestimate their generality. As in the December 22, 1916, letter to Loeb, Morgan had to contend with friends who did not quite get the point. Loeb wanted an underlying physical or chemical cause and could not wholly comprehend the revolutionary implications of Morgan's work. In the December 26, 1917, letter to Osborn, Morgan is fighting off a well-intentioned manifestation of an older, speculative style of cosmology. What he insisted on was that his mechanism of the transmission of heredity had a demonstrated validity and a level of generality usable no matter what future research disclosed about the chemical composition of the gene. Nor did the coming of population genetics undermine the achievements of the classical school; here too the basic mechanism was assumed.

By the period between the two World Wars genetics had acquired all the outward signs of a mature scientific field. Its literature had unique terminology and symbols. Technical details prevailed, crowding out tendencies for speculative flights of language. The correspondence between Sturtevant, Demerec, and Emerson illustrates the assumed form and content of classical genetics in private discourse just before the work of Avery on DNA. Above all, the literature had less and less direct reference to controversial issues of the day. Not that all geneticists avoided public controversy or philosophic musings; some writings of this nature persisted, but geneticists began to segregate those efforts from their research findings.

It was a familiar style of science in which broader implications existed only implicitly. Public discourse stressed the purity of the advanced research. It was a style comfortable to Morgan's personality but not to that of Loeb. Private discourse was something else. By World War I both men were hostile to the glib certainties of the eugenicists. Morgan could write vigorously in private about the scientific and logical absurdities of proponents of racial pecking orders. The 1919 exchange between Morgan, East, and Loeb as well as the references to the 1920 Eugenics Congress are among the many evidences of his views. But note that Loeb was willing to give East the right to air his opinions, while Morgan insisted on revisions in the privacy of the editorial process. But the belief in a pure science implied, to Morgan, a need to keep science uncontaminated by public passions.

Perhaps that was essential to avoid the uncritical views of nonscientists or scientists, like Davenport, who were committed to the eugenics program. Yet, there is a sense of futility about all efforts to insulate a science from controversy, to extend a purity of knowledge to the scientists themselves. Much as they might strive for detachment, scientists could not help but reflect, to some degree, their upbringings, their education, and their social and economic environments. Assumptions about human heredity had existed from ancient times. Although crude or irrelevant by modern scientific standards, such views persisted in the thinking of individuals at all levels of society, from near illiterates in rural backwaters to the president emeritus of Harvard and the director of the Carnegie Institution's Station for Experimental Evolution.[2]

As in the case of assumptions about human behavior, a sense of biological determinism pervaded human affairs. Birth, death and illness remained too omnipresent for conservative, liberal, and radical thinkers. For differing reasons each part of the ideological spectrum wanted to stress other factors, other determinisms. But biology could not be exorcised, and it remained a constant temptation when easy answers were needed. Biology often appeared a way out of dilemmas beyond individual will and social institutions. To counter the claims of "nature," a rival set of claims for "nurture" were put forward in the interwar period. By demonstrating eugenics' logical consequences, Adolph Hitler placed scientific racism in the shade. The claims of biological determinism are still with us, however, still raising issues challenging easy assumptions of the effectiveness of man and his cultural mechanisms.

Loeb and many of his peers recognized the need to reply courteously to letters from the public, even while being very firm. The letter to Dreiser (1920) is a good example. Walter Cannon's 1923 letter to Firuski is another one particularly interesting as an example of a reaction to the use of scientific concepts by a layman like Henry Adams. The increasing complexity of science placed individuals like Loeb and Cannon in an awkward position. To the public the contents of science seemed more and more removed from everyday experiences. The concepts and data of science were the nearly exclusive possession of the professionally trained. Yet there was a belief in the importance of science and in the value of its being known to a wider public, both in general terms and even in specifics. Both Loeb and Cannon reacted by writing for audiences beyond their professional peers. The National Research Council concerned itself with developing means of reaching the popular press. Yet, popularization inevitably was significantly behind the research front and necessarily presented an attenuated dose of the real thing. Perhaps this could do, were scientific advances solely of concern to the research community. Sooner or later, new findings directly or indirectly touched the entire population.

Loeb's research on the biochemical roots of behavior, for example, was not simply an abstruse enterprise. When carried to logical conclusions (or extremes), the research program could impinge on public policy, value system, and the health care of individuals. A model of biochemical determinism is discernible in some of the documents of Loeb and Cannon. Even more striking is the implications of the Nobel Prize work of Erlanger and Gasser on the electrical conductivity of the nerves. But while the physical and chemical basis of psychology was being extended, quite different views became apparent: a nonphysiological psychology, Freudian and related psychiatries immune to conventional experimentations, a sociology and an anthropology stressing human cultural factors. An echo of this split appears in the 1931 correspondence of Meyer and Cannon.

Medicine's role in the biological sciences grew. Ross G. Harrison might approve, but Thomas Hunt Morgan apparently did not, judging from the former's letter of December 24, 1919. A product of the graduate education at Johns Hopkins University, Morgan was comfortable in a Columbia University department independent of the medical school. He believed in basic research, and he wanted to avoid the effects of applied pressures, particularly in limiting the area of investigation. No American medical school in 1910 was likely to have sympathized with his interest in Drosophila. In developing the California Institute of Technology Hale and Millikan saw medicine as the vehicle to expand their institution's scope beyond the physical sciences. Not surprisingly, they thought of Morgan as the head of the venture. He declined, urging instead a program in pure biology. In 1928 Morgan launched such a venture at Cal. Tech. By that date, a number of outstanding scientists—such as Cannon, Erlanger, and Gasser—were associated with medical schools.

Consequently, in the interwar years, scientists found a variety of institutional possibilities for biology. To Simon Flexner's call for an independent research institution, a Graham Lusk in 1920 raised the banner of the older tradition of clinical research. Flexner's Rockefeller Institute also studied humans afflicted with pathological conditions, but the studies were divorced from the usual contexts of medical practice and hospitalization. In the exchange with Meyer, Walter B. Cannon had given the limit of his support of the older clinical traditions. Experiment differed from treatment, requiring an abstraction from the total reality. A medical science was reductionist, not holistic. Loeb could not have said it better.

When faced with the arguments of Robbins in 1932 for a science policy following the agriculture pattern, Cannon reacted in accordance with the views espoused by Hale, now widely adopted among the leaders of the research community. Loeb, Hale, and Cannon shared a basic professional ideology. But a subtle difference existed, marking Cannon, the man of scientific medicine, as a harbinger of the future. In the 1934 letter below, Cannon wrote in opposition to Simon Flexner's continued support of autonomous research bodies. It is reminiscent of Graham Lusk but only for Cannon's stress on the university, not the clinic. In 1916 a lecture by Hale so moved Cannon that he wrote a warm letter calling for worthy Progressive-era reform efforts by the National Academy. As it happens, Cannon now calls for more basic research to avoid becoming too applied, too much like the Adolph Meyer of his profession. But unlike Hale, Loeb, and Flexner, Cannon calls for something other than general support of the sciences. He wants a project system, one targeted for a general goal. In such statements of the pre–World War II era are the roots of the principal research-support policies and programs of the years after 1945, the Department of Defense, and the National Institutes of Health.

As in the case of genetics, a sense of biological determinism pervaded the scientific study of behavior. Correct policies were important to Cannon both for knowledge's sake and for human affairs. To individuals sharing these views, biology often represented a way out of dilemmas beyond individual will and social institutions. To such believers, the counterarguments of the social sciences ("nurture" rather than "nature") seemed specious, while the responses from the humanities were sheer romanticism. But the skepticism within science could not be dismissed so easily. In the concluding letter to Flexner, Cannon wistfully wants to convert Morgan to his view of the medical school. It was important because to his peers this elderly experimentalist from Kentucky symbolized their hopes for biology's future.

1. By far the best single introduction is Garland Allen's *Life Science in the Twentieth Century* (New York, 1975).

2. Two articles by Kenneth M. Ludmerer are pertinent: "American Geneticists and the Eugenics Movement: 1905–1935," *Journal of the History of Biology* 2 (1969): 337–362;

and "Genetics, Eugenics, and the Immigration Restriction Act of 1924," *Bulletin of the History of Medicine* 46 (1972): 59–81.

Thomas Hunt Morgan to Jacques Loeb, December 22, 1916

Dear Loeb:

I received your book a few days ago and was very glad to get a copy. I did not even notice that I had to express an opinion in regard to it, but now that you have raised the point there are one or two things which I should like to put into your private ear.

We have been reading parts of your book[1] at our Genetics Club, which meets at my house occasionally. At the last meeting we took up your point in regard to maternal inheritance, which is the theme on which a good part of the book rests. We very much regretted that you had overlooked or were not familiar with certain results on maternal inheritance in the silkworm moth published some years ago by Toyama. You will find the substance of it given in our book on the "Mechanism of Mendelian Heredity," p. 136–137, but I feel quite sure you will want to look up the original papers, reference to which you will find in our literature. The point, briefly, is this: that the eggs of the F_1 moth give the same sort of maternal inheritance, irrespective of which way the F_1 was obtained, that is, by a straight cross or by its reciprocal. In a word, if the dominant character comes in through the father it affects the egg of the next generation in precisely the same way as when it came in through the mother. It seems to us that this is the only experimental evidence to which one can appeal.

The other point that met with some criticism was as follows: the distinction which you constantly make between generic and specific characters as compared with individual, Mendelian, superficial or whatever you choose to call them. This seemed to us entirely reactionary, for we find no evidence that there are such distinctions in actual practice, nor can we figure to ourselves very clearly how a generic character is different from an individual one. I wonder if you realize that you are adopting a view which Jenkinson had proposed in 1913 in his "Vertebrate Embryology."[2] If you will turn to p. 92 you will find the following statement:

"Hence the characters, the determinants of which reside in the cytoplasm, are the large characters which put the animal in its proper phylum, class and order, which makes it an Echinoderm and not a Mollusc, a Sea-urchin and not a Star-fish; and these large characters are transmitted through the cytoplasm and therefore through the female alone. The smaller characters—generic, specific, varietal, individual—

are equally transmissible by both germ-cells, and the determinants of these are in the chromosomes of their nuclei."

Now, Jenkinson had practically nothing to go upon, so far as I can see, for proposing such a view and his own work on experimental embryology was not of a kind, I believe, to warrant any such sweeping conclusions receiving the seal of your advocacy. I beg you as an old friend that you will look into this matter rather carefully, for I think there is involved an important matter. You remember that many people believe that Mendelism is a purely trivial pursuit because it deals only with such matters as the colors of flowers and a few other trifling matters such as hairs, diseases, etc. These people fail, I think, to see the principle that is involved in Mendel's great discovery, quite aside from the actual materials that it happens to deal with. In all this, I know, you agree entirely with me. Therefore, quite aside from the actual merits of the case, your statement, I feel quite sure, will be welcomed by just such people who would like to believe that a mechanistic principle like Mendelism can at best only relate to trifling matters, while the *more fundamental problems of nature* are quite untouched by such mechanistic notions.

There is so much that is stimulating and interesting in your book that I hope you will not mind these two or three friendly criticisms which I venture to make; at any rate, I hope you will take them into consideration when the new edition of your book comes out, which I have no doubt will be before very long.

Looking forward to seeing you next week and wishing Mrs. Loeb and yourself a pleasant Christmas and a happy New Year, I am,

Sincerely yours,
T. H. Morgan

Loeb Papers, Library of Congress

1. *The Organism as a Whole* (New York, 1916), which appeared that year.
2. John Wilfred Jenkinson, *Vertebrate Embryology* (Oxford, 1913).

Thomas Hunt Morgan to Henry Fairfield Osborn, December 26, 1917

Dear Osborn:

Owing to the pressure of other work I have put off sending you an explanation of what I meant when in my last letter I wrote that I regretted that you should have introduced a mystification at the very heart of the matter which seemed to me to undo the laudable attempt to treat the subject of evolution from the point of view of the energy conception, that is, as a physico-chemical problem.[1]

1. In the first place, when you say the "causes of the evolution of life are as mysterious as the law of evolution is certain," it seems to me that you are not living up to expectation. I know, of course, that you use "law" in the sense in which the chemist and physicist does not use it, for I gather you mean only a sort of generalization when you speak of law or a strong personal opinion or bias, although at other times I think you use the expression as though it were an ultimate cause itself. For instance, on page xii of the preface you say: "You may be compelled to regard the origin of evolution of life as an ultimate law like the law of gravitation which may be mathematically and physically defined but can not be resolved into any causes." I can not but think that so long as the causes of evolution are admittedly unknown it is not possible to treat scientifically the process of evolution from the energy standpoint, however much one may be convinced that some day we may be able to do so.

2. On page 7 in the first chapter you say: "In other words, in the origin and evolution of living things, does Nature make a departure from its previous orderly procedure and substitute chance for law?" This contrast between law and chance recurs in a number of places in the volume and is likely everywhere to mislead the reader who is not informed, as I know you to be, on the whole philosophical conception that lies back of these two words. From the energy point of view there is no such distinction between law and chance conceivable. If, however, you throw over the energy conception and substitute a mysterious beneficent being that directs all things that are lawful and introduce a devil-of-a-fellow to mess things up generally and take a chance on coming out all right at the end, there might be some grounds perhaps for such a distinction, but from the energy point of view, as the physicist uses that term, this is nonsense.

3. On page 95 you say: "The chemical or molecular and atomic constitution of the chromatin infinitely exceeds in complexity that of any other form of matter or energy known. As intimated above it not improbably contains undetected chemical elements." I understand, of course, that this is a sort of poetic outburst—not an accurate scientific statement, for as you will admit, of course, we know too little about the chemical composition of chromatin or about its method of action to warrant one in saying that it infinitely exceeds in complexity any other form of matter and energy known.

There then follows a sentence relating to the work on radium from which you have drawn an entirely erroneous conclusion, for neither do the authors (the Hertwigs)[2] make any such claim for their *contagium vivum,* nor does the far more accurate work on radium that has been done since that time which I happen to know about at first-hand bear out any such interpretation. This, however, is merely a passing remark leading up to what I regard as a much more important point, namely, that here and elsewhere you intimate very strongly that there is some innate property (chemical, if you like) of living matter that makes it transcend all properties of other matter. Here, then, we come to the "heart" of the subject. Even by intimating the possibility of such a conclusion you leave the safe fields of the energy conception and roam abroad in the Elysian fields of mystery. This is what I meant in my former letter when I said that I regretted to find that you did not live up to the high conception with which you started, which I infer to be that of the modern school of biology, resting its claims entirely upon the method of experimental analysis which has been so successfully followed out by the chemist, physicist, and physiologist. You must have observed that the physiologist, who of all men is most concerned with action, reaction and inter-action has scrupulously avoided entering upon the wide field of evolutionary discussion into which the zoologist has plunged. He avoids the subject, not because of his inability to realize the intrinsic interest of evolution, but because his whole training has been of such a kind as to put him on his guard against going beyond the verifiable evidence at his command. I think he is right, and while I personally, like yourself, take great interest in the speculative side of evolution and take great pleasure in following it through its devious course, I have a feeling that it is not a fair game to play the energy idea as though by this means we could at present even attempt to explain the problem and then at the critical point intimate that after all there was something about which we know nothing in the machinery that is doing the whole business.

During the Christmas holidays I expect to see Goodale[3] in Pittsburgh and will remember to make more definite arrangements with him for giving the skins of his castrated birds to the American Museum. It is quite true, as you say, that this is a classical material on the subject, and it may be well worth while to have it preserved permanently. His latest results, about which I spoke to you in Philadelphia, have now been published in abstract. In substance they are as follows: that at the time when the testes of the duck are nonfunctional, the nuptial plumage develops, whereas at the time when the summer plumage comes on (you recall that the summer plumage is more like that of the female) his testes are functioning fully. In such ways does Nature pursue her devious course—this is a poetic flight of my own.

Very sincerely yours,
T. H. Morgan

Loeb Papers, Library of Congress

1. Osborn's *The Origin and Evolution of Life, or the Theory of Action, Reaction, and Interaction of Energy* appeared that year, an enlargement of his Hale Lecture.

2. OSCAR HERTWIG (1849–1922), the German embryologist and anatomist, and his brother G. W. T. VON HERTWIG (1850–1937), like Jacques Loeb, had erroneously reported mutations due to X-rays.

3. HUBERT DANA GOODALE (1879–[?]), who was at the Massachusetts Agricultural Experimental Station (1913–1922) where he experimented on the control of plumage color.

Jacques Loeb to Thomas Hunt Morgan, October 29, 1918

My dear Morgan:

The last number of SCIENCE has a review of that stupid and brutal book of Madison Grant[1] by a man who appears obviously to be as much of an amateurish bungler as Grant himself.[2] I remember his having written a eulogistic book on the ruling families in Europe in which he even speaks of families of blackmailers, robbers and liars like the Hohenzollerns as of a "superior strain." I wonder whether it might not be well to point out in SCIENCE that Madison Grant is the pretentious amateur that Houston Chamberlain is, and since we know what that kind of literature has done for Germany, it might be well that this country bases its judgment on "racial superiority" on more substantial ground than mere journalistic propaganda. I should do it, but if it came from me it would appear, of course, as if it were dictated by personal interests. I think the type of literature like that by Woods and Grant should at least be properly stigmatized in SCIENCE.

With kindest regards,

Yours very sincerely,
[*Jacques Loeb*]

Loeb Papers, Library of Congress

1. GRANT (1865–1937) was a lawyer, amateur naturalist, and conservationist. This reference is to a revised and enlarged edition of *The Passing of the Great Race* (New York, 1918), a pseudoscientific and pseudohistorical work of great impact.

2. Frederick Adams Woods almost unquestioningly accepted Grant's position; see *Science*, n.s. 48 (1918): 419–420.

Thomas Hunt Morgan to Jacques Loeb, October 30, 1918

Dear Loeb:—

I quite agree with you that Cattell ought not to have published Woods review of Madison Grant's book or any other review of that book that did not condemn it. However, you fellows when you become editors, have to look out for copy and sooner or later everything is grist that comes to your mill. Cattell knows what I think about Madison Grant and also about Woods but he thinks I take a very extreme position and he told me once, I think, that Woods is not half bad.

The first edition of Grant's book was reviewed by Boas[1] adequately in The New Republic and Muller tells me that it was adversely reviewed by Guerard, L. J. Pub. League of Natural Science. I suppose it ought to be reviewed by somebody who knows something of history and anthropology but unfortunately these are just the kinds of men that indulge, *themselves,* in that kind of thing. All the zoologists could do would be to point out the insufficiency and inaccuracies of his statements wherever it was possible to check them up. As my colleague and your warm, personal friend[2] has fathered the book with an introductory statement, our hands of course are completely tied.

Sincerely,
T. H. Morgan

You might get Cattell to reprint it in Scientific Monthly!

Loeb Papers, Library of Congress

1. Franz Boas, the anthropologist at Columbia.
2. Henry Fairfield Osborn.

Edward Murray East to Jacques Loeb, April 6, 1919

Dear Doctor Loeb:

I send you a copy of a letter I have written to Dr. Morgan. I also feel that I ought to send you a word privately. I do not want to write the same way to him, because though people like Morgan who pride themselves on their broadmindedness, think they are not hurt when their advice is not taken,—they are—just the same.

Now I love Morgan, and I think he is unquestionably the greatest geneticist in the world. I am exceedingly grateful for the time and trouble he has taken in criticizing my MSS. Nevertheless I think editorial criticism of this kind should be confined to a question of actual truth and error.

This book was planned carefully and written carefully. It is not a Masterpiece, not what I should like to have it, but it is about as good as I can do. I wish I could write better—who of us does not. The material that I want in the book is there, however, and there according to plan. The best literary style in the book is in the last two chapters, that is the reason Morgan thinks they are written in "heat." His suggestion that I cut out these chapters, as well as chapters 2 and 10, would leave nothing but a dry technical paper that I could just as well publish in Genetics. You read only the first and last chapters. The first chapter is not satisfactory. I wrote it five times and will have a try at it again. The last chapter you could hardly judge without reading the rest. I have made no radical suggestions, but have only outlined roughly just what follows in an application of the experimental conclusions to the human race. I believe in them and I believe I can support them in argument against anyone.

Questions of fact I shall of course go over carefully, as well as questions of logic. They are, it seems to me, the desirable kind of criticism. Questions of literary style are, between you and me, not Morgan's forte. And anyway outside of actual grammatical errors which presumably will be caught in proof-reading, style is a matter which must remain characteristic of a personality.

Now I am not clear whether it will be left to my judgement what corrections to make, and the MSS accepted on that basis, or whether Morgan will insist that the whole thing be made satisfactory to him. It happens that my book being a Genetical book, the last mentioned attitude could be taken; but I do not see just how that is to be done for most of the remainder of the series.

Will you not be good enough to write me and give me your personal views on the whole situation?

Sincerely yours,
E. M. East

Loeb Papers, Library of Congress

Jacques Loeb to Edward Murray East, April 7, 1919

My dear Dr. East:

I am anxious to have your monograph appear in our series, and I have no doubt that this desire is shared by Morgan. As far as his criticisms are concerned, I should say offhand the following.

First, as far as technical objections are concerned, I think there will be no difficulties. I have not seen Morgan's list, but I have no doubt from your letter that that can all be arranged. As far as the actual existence of atoms and molecules is concerned, I am afraid that I am the author of the objection, because since the work of Perrin,[1] Svedberg,[2] etc., nobody questions any more the real existence of the molecule and electron (you may also look up Millikan's book on "The Electron"). I mention this not in the way of a criticism but because I feel that physicists who will read your book will practically unanimously raise the same objection and for that reason I think you had better change the statement. But all these single points constitute I am sure not the slightest difficulty between yourself and Morgan.

Second, Morgan's suggestion concerning the last chapters, especially in regard to your attitude concerning racial biologists, two attitudes are possible, the one saying that the statements of amateurs or worse, like Houston Chamberlain, Madison Grant, etc., are beneath contempt and that science cannot take them into consideration. I suppose that is what Morgan means by his "agnostic attitude." If you assume that attitude, I think you should come out squarely in your book denouncing these men as making statements for which they have no proof and the only consequence of which can be race hatred, the promotion of which is probably being intended by them. The second attitude is the one you have taken. Now there it becomes a question to what extent the editors should feel responsible for your statements. Personally I am inclined to take the attitude that a man of your reputation and accomplishments ought to be able to assume full responsibility for what he says and I do not think that Morgan or I should feel responsible or insist upon any objection if you feel that you want these chapters to appear in print as they are. I will take up this matter with Morgan and see what can be done. I have no doubt that everything can be arranged to your satisfaction.

Third, as far as your presentation of the evolutionary viewpoint is concerned, insofar as I can judge it is merely a matter of expression where a few changes probably will make matter acceptable to Morgan.

On the whole I would suggest that you may feel that Morgan really means well and that his objections are those which probably might be made by others. He read my book very carefully and he made a number of suggestions, most of which I accepted since they were all right; others, where I felt he was wrong, I ignored. I should advise that you go over the manuscript accepting those of Morgan's suggestions which you think are right, and the acceptance of which means an improvement of your book. As far as the rest is concerned, you let me know and I will take up the matter with Morgan. In all cases you may be certain that Morgan shares my high opinion of yourself and your work.

With kindest regards,

Yours most sincerely,
[*Jacques Loeb*]

Loeb Papers, Library of Congress

1. JEAN PERRIN (1870–1942) was at the Sorbonne. He received a Nobel Prize in 1926. Among other things, he was concerned with structure of matter.

2. The Swedish chemist THEODOR SVEDBERG (1884–1971) was at Uppsala, known for work in colloidal chemistry and the development of the ultracentrifuge.

Edward Murray East to Jacques Loeb, April 8, 1919

My dear Dr. Loeb:

I think that things are cleared up much better by your letter than by Dr. Morgan's.

Of course I have no objection to any sort of revision where it is possible for me to make matters better. In certain places where it is a question of style it is easier to wish to do better than to actually do it. I think that some revision of the last two chapters is perhaps necessary. I am usually apt to state frankly what I think of people like Chamberlain and Grant, but I have considered it better policy to ignore them. I do feel that there are certain conclusions which follow logically from our solution of the questions of inbreeding and outbreeding which are of interest to sociology. Doubtless there are changes which ought to be made even with the attitude that I have taken, but I can't feel that the attitude itself is wrong.

My criticism of the technical objections is that in many cases they were put down so hastily that the sentences seemed hardly to have been read. This, I feel, is the case regarding the chemical atom. I did not state, and do not believe, that there is no physical reality to the atom. What I said was, "The chemical atom is a conception invented for the purpose of simplifying and making useful observed chemical phenomena," and again, "As used mathematically, both the genetical factor and the chemical atom are concepts." Neither Morgan nor yourself seems to see the distinction which I make between the factor and the atom as mathematical concepts and physical realities. The atom had a history of nearly one hundred years as a concept. People wondered what it might be like and invented theories as to what it might be like, but not until within twenty years was there any acceptable physical theory. If you would stop for a moment and see the force of the phrase, "as used mathematically," you could see that no physicist or chemist could possibly criticize my statements. I particularly like the analogy, because it is only physicists and chemists who really appreciate a conception invented for the purpose of simplifying interpretations of physical data. They know that while such a conception must have as much physical basis as there are physical facts at hand and the conception must change with the addition of more facts, nevertheless the entity may be wholly hypothetical. I hope that this will make my distinctions clear to you.

In connection with the whole manuscript it is not that I do not wish to make corrections or that I feel that the material or its form is wholly satisfactory, but that I wish to know whether I am to be the final judge, or not. In spite of our great regard for Morgan's ability, it would be rather monotonous to have the whole series of monographs stamped with a Morgan die. I think that your letter has made this clear.

Sincerely yours,
E. M. East

On reading this over it sounds cold and lifeless, and not at all what I really wish to say. Your letter was so kind and courteous that it gave me the needed impetus to go ahead and endeavor to get things into shape. I am going to talk matters over with Osterhout[1] on Friday and think we can straighten everything satisfactorily.

E. M. E.

Loeb Papers, Library of Congress

1. W. J. V. OSTERHOUT (1871–1964) was a plant physiologist. He was at Harvard 1913–1925, at Rockefeller Institute 1925–1939. He coedited the series of works with Morgan and Loeb.

Thomas Hunt Morgan to Edward Murray East, May 6, 1919

Dear East:

Loeb sent me over the second and last chapters of your book and I have looked them over and returned them to him yesterday.

It seems to me that you have taken out most of the objectionable matter from the second chapter but I agree with Loeb in his view that on the whole the chapter as it stands does not add anything to your book although personally, I don't think it can do any harm.

In regard to the last chapter which I have read with a good deal of care, it seems to me the changes are all in the right direction and I only wish that you had lived up to the revision of the first few pages. I will confess that I should still like to blue pencil some of it towards the end particularly, especially the peroration. As you yourself point out the situation is very complicated, and how far the variability introduced through the crossing of related races arises through genetic recombination and how far it is due to the intermingling of different sets of traditions which may or may not have anything to do with the genetic problem, is I think not sufficiently clear from your context although your context seems always to imply that these things are due very largely or entirely to the genetic background. In this opinion I am inclined to agree with you but I think it is far from being established and ought to be stated with the utmost reserve. You see I have my own pet theories about some of these matters and probably that is why I am so cantankerous.

I am sure your book will be very successful and I think the subject you have is so interesting and the opportunity to get over a very important matter is so great that I hope you will forgive me for arguing about so many different questions, many of which I realize are only differences of opinion, but where there are differences of opinion between you and me, then I think the matter must be uncertain. My own book is now in Loeb's hands and I am waiting for my medicine.

Sincerely,
[*Thomas Hunt Morgan*]

Loeb Papers, Library of Congress

Thomas Hunt Morgan to Jacques Loeb, May 14, 1919

Dear Loeb:—

Thinking over the changes that I have made in the chapter on the maternal inheritance,[1] I ought to tell you that I put in the statement about the theory of the embryo-in-the-rough for your personal edification and expect to cut it out of the proof if it gets that far. Perhaps it might be well to add a line at the end of the quotation from you to the effect that you are using the word species and specific in the sense of specificity of certain constituents of the organization. This will clear you of the reproach that you are dealing with an antiquated topic but will still leave you responsible for the suggestion that the cytoplasm transmits fundamental peculiarities. It is this point that I am anxious to go for, because of its wide spread belief amongst biologists in general for which I can find absolutely no real basis except an emotional one. It is for this reason mainly that I have not hesitated to hold up as examples two of my best friends and a very famous German investigator.[2] I hope the former will forgive me and I shall take my chances with the latter if we meet, as seems unlikely, in another world.

Sincerely,
T. H. Morgan

Loeb Papers, Library of Congress

1. This refers to Morgan's book, *The Physical Basis of Heredity* (Philadelphia, 1919).
2. Conklin and Loeb are the best friends. The very famous German is THEODOR BOVERI (1862–1915).

Edward Murray East to Jacques Loeb, May 23, 1919

Dear Dr. Loeb:

I had really forgotten that your recent letter needed an answer.

I do not mind mitigating my statement regarding the Irish, especially if that hurts Morgan's feelings, although I think it is the truth and never have any hesitancy about speaking the truth. I had some Irish ancestors myself as well as Morgan.

Both Havelock Ellis' careful statistical study of British genius and Galton's English Men of Science[1] bear out the statement. It was not made wildly nor because of prejudice; but, as you say, perhaps in order to keep peace it had better be revised.

Sincerely yours,
E. M. East

Loeb Papers, Library of Congress

1. Francis B. Galton's *English Men of Science* (London, 1869).

Ross G. Harrison to Thomas Hunt Morgan, December 24, 1919

Dear Morgan:

Thanks for your letter of December 20 which interested me very much.

Nichols[1] is an Assistant Professor in the Department of Botany—a young man of considerable energy and ability, who has written a good deal on ecological subjects. The trouble here is just as it is in many other places. Botanists don't care to participate in a general course themselves and they are envious of the zoologists when the latter are successful in teaching general biology. We have had such a course at Yale for a great many years and since we have been in the new laboratory, we have some years had as many as 500 students take it. For a few years after I came here the Botanists gave part of the work. Owing to differences in method and for various other reasons, the course was not nearly so successful as it has been in the hands of our department alone. Last year they became quite restless and we had a good deal of discussion about the question. I often talked it over with Nichols and saw most of the correspondence with outside Botanists and Zoologists. I was able to make some suggestions and advised him to tone down some of his statements. In fact, I was amazed at the bitterness that some of the Botanists showed.

I think that probably the best solution or at least the only one that might satisfy every one, would be to have three separate courses: one in Elementary Zoology, one in Elementary Botany, and one in Biology, allowing the students to choose. The thing would then work itself out, though probably more in accordance with the ability of the respective instructors than with the intrinsic value of the subject.

Your sketch is very good and if elaborated, will no doubt be a valuable contribution. I think, however, that you make a mistake in slurring the medical profession, who can be counted among our best friends and customers. It was they, more than any one else, who demanded a course in biology, rather than Zoology and Botany of the old type. Huxley[2] himself had medical training and knew the problems of medicine, though of course he had much more besides. If you look over the list of men in Germany who have contributed to the development of biological sciences in the last century, you will find that many of the most notable were connected with the medical faculties. They did not devise a course in General Biology, but they did arouse very great interest in it and made some of the most important contributions. I need only mention Johannes Müller, Max Schultze, Gegenbaur, O. Hertwig, Roux, Nussbaum, and Verworn.[3] There is one branch, immunology, which owes practically all of its recent development to men of medical training and yet it is of as much fundamental importance and of as great interest as any other.

I am glad you are going to the Princeton Meeting and look forward to seeing you with pleasure.

Very sincerely,
[*Ross G. Harrison*]

Harrison Papers, Yale University

1. George Elwood Nichols (1883–1939).
2. Thomas Henry Huxley (1825–1895).
3. The physiologist MÜLLER (1801–1858); Schultze, the anatomist; KARL GEGENBAUR (1826–1903), the anatomist; the zoologist OSCAR HERTWIG (1849–1922); the embryologist WILHELM ROUX (1850–1924); MORITZ NUSSBAUM (1850–1915); the physiologist MAX VERWORN (1863–1921). In this field, as in so many others, American scientists were most conscious of the Germans as great contributors to science in the last century.

Jacques Loeb to Richard B. Goldschmidt, February 16, 1920

My dear Professor Goldschmidt:

I was pleased to receive your letter of December 30th. I am very anxious to help you if possible but the animosity in this country is at present a little too strong to discuss your immediate return with anybody. I think in a year the minds will be calmer and then I hope such possibilities will not be out of the question. Have you thought of South America? Those countries are extremely prosperous just now and the Rockefeller Foundation is doing everything to encourage and develop research. I have a feeling that there is a great future for scientists who know America and can adapt themselves to the conditions. If you wish me to I shall see what can be done.

As far as the Journal is concerned, I shall give orders that it be sent to the Kaiser Wilhelm Institut für Biologie. Will you be good enough to arrange with your librarian that the Journal will be made accessible to the members of your Institut.

Your paper was sent in to the printer after you gave it to me, but both Osterhout and Flexner insisted that the appearance of your paper would result in a boycott of the Journal in the United States on account of the intense feeling against the Germans at that time, and especially for the reason that you had been interned as an alien enemy. I had to postpone the printing of your paper but I expect to publish it as soon as these violent emotions from which we are still suffering are a matter of the past.[1]

What you write to me about Germany does not surprise me. I fully expected that the German Junker would try to put the blame for his having ruined Germany—and perhaps the whole world—on the shoulders of the Socialists and the Jews. The Junker has always been a rotten loser, and the cowardice of the German Emperor (who is referred to here as the "coward of Amerongen") expresses the moral status of the Junker. Their anti-Semitic propaganda is only an excuse to cover up their own stupidity and incompetence. An Englishman or an American or a Frenchman would stand with dignity his having been beaten, but the Teutonic Junker has not got that strength of character. If Germany had been invaded and had suffered the way France did, the German government would probably have collapsed in less than a month instead of holding out for four years as the French did.

Can you find out for me how W. F. Ewald[2] is? I have tried to get information about him but I have received no answer from anybody.

If there is anything I can do for you do not hesitate to let me know.

With kindest regards,

Yours very sincerely,
Jacques Loeb

Goldschmidt Papers, University of California, Berkeley

1. Goldschmidt's paper was never published in the *Journal of General Physiology.*
2. Perhaps Wolfgang Felix Ewald who received a Ph.D. in 1910 from Munich for research on tropisms and was with Loeb at Rockefeller Institute before World War I.

Charles W. Eliot to Charles B. Davenport, May 4, 1920

Dear Dr. Davenport:

I read in the first number of the Princeton Lectures,[1] in a lecture by Professor E. G. Conklin, the following statement: "There has been no progress in the intellectual capacity of man in the past two or three thousand years." What do you say to that? At the head of Section III appears the statement, "There is no probability that a higher animal than man will ever appear on the earth." Has the present day man any means of estimating that probability? In the same Section Professor Conklin says "All present signs point to an intimate commingling of all existing human types." Are mulattos and Eurasians considered promising or degenerate products? Does the sentence at the top of the last page, left hand column, state correctly the limitations which eugenicists accept?

Does the lecture as a whole take sufficient account of the discoveries of medical science during the past hundred years?

Professor Conklin's remarks about the ordinary representations of angels are correct, but have not so much pith as Louis Agassiz's statement "a woman with wings and arms would be a monstrosity."

Sincerely yours,
Charles W. Eliot

Davenport Papers, American Philosophical Society

1. "Has Evolution Come to An End?" The full statement of Conklin's position is in his book *The Direction of Human Evolution* (New York, 1921). While agreeing on general points, Conklin came to quite different conclusions than Davenport. The latter, for example, worried about the decline of the old New England stock. In contrast, Conklin thought such views were undemocratic desires for hereditary castes.

Charles B. Davenport to Charles W. Eliot, May 14, 1920

Dear President Eliot;—

Following the receipt of your letter of May 4th I sent to Princeton for a copy of the Conklin lecture in "Princeton Lectures" and have just received it.

"There has been no progress in the intellectual capacity of man in the past two or three thousand years." This does not seem to me to be a scientific statement but a broad generalization, based on a vague opinion. One does not know what the "intellectual capacity of man" is intended to convey. If it is meant there were individual men living two or three thousand years ago who had intellectual capacity as great as that of any modern man the phrase might be accepted. At least, I suppose it would be accepted that the intellectual capacity of the best of the Greek writers and thinkers was at least equal to any writer and thinker of the present time. On the other hand, the sentence is open to other interpretations, which are less obviously true. It seems possible, if not probable, that there has been progress in the proportion of men possessed of high intellectual capacity since that remote time.

For example, there are probably five millions of mulattoes in the United States today. If we classify them with the negro race and compare them with the negroes of Africa and the mulattoes of other countries may we not say that the proportion of "negroes" reaching a certain level of intellectual capacity has increased in the past thousand years? Progress in the proportions of mankind reaching a given level of intellectual capacity has very likely been brought about by the circumstance that that part of mankind with the greatest intellectual capacity has been able to migrate fartherest over the world, to dominate that part of mankind with inferior intellectual capacity, to mate with them and to produce hybrid offspring which have had a higher average intellectual capacity than the lower race and who have, in turn, dominated over the lower race. In this sense it seems probable there has been progress in the proportions of men reaching a given level of intellectual capacity, or in the average intellectual capacity of man in the past two thousand years.

"There is no probability that a higher animal than man will ever appear on the earth." Whereas, in certain other statements made by Conklin, he seems to me to show certain limitations of vision because within the historic time we find that there has been no elevation of the highest intellectual capacity attained by man it does not follow that either man has stopped evolving, or that the lower primates can not evolve something which will be a "higher animal" than man. "Higher animal" is vague enough, perhaps. What is meant is a race containing a larger proportion of the eminently fit for greatest effectiveness in physical, mental and moral output. There is, certainly, a great opportunity for increasing even the highest 1% of man. There is room for improvement of the social instincts and self control, of memory and judgment, of vision and speech, of better feet (that is, more adapted to the needs of the social organization), of higher disease resistance. Just as within the last 25,000 years, perhaps 15,000 years, the Nordics have arisen from the Cro-Magnon and the Grimaldi races and these, in turn, from the Neanderthal man, so in the future, there may arise, perhaps not from the Nordics but from some other of the human races, thru mutation, combinations superior to the best of the modern men. There is room for development of the best to higher levels and it is difficult to deny that such improvement might occur.

"All present signs point to an intimate commingling of all existing human types." It is obvious that the white man is penetrating to all parts of the land surface and, wherever he penetrates, he leaves hybrid progeny. Hybridization is occurring on so vast a scale that it may well be expected to become practically universal, tho relatively slight in the most inaccessible peninsulas and islands. In the latter, it may well be that the human representatives will be able to maintain a separate human type, thru inbreeding and mutation, protected by their isolation. However, it must be recognized that this wide spread hybridization results in disharmonious combinations which are largely lethal. The relatively high death rate among the hybrids tends to retard the commingling processes. This is seen in the high infantile death rate among the colored population of our larger northern cities, a death rate which, in some cases, permits of no natural increase in numbers from generation to generation. We see the dysgenic effect of the discrimination against hybrids in the catello hybrids of cattle and bison, which are maintained only in the face of great difficulties.

"The most that could be expected would be that the standards of the race as a whole would more nearly approach the more perfect specimens of humanity which now exist." In view of the above cited considerations, in view of the possibilities which may become realized it seems difficult to assert that there may not be more "perfect specimens of humanity" than now exist.

It is my present opinion that advances in medical art, at least, are not working toward the increase in the proportion of men reaching a high level of intellectual or physical capacity. The preservation of the "culls" by modern medicine is possibly, if not probably, pulling down the average faster than the increase of eugenical ideals, leading to an increased production of the higher types, can possibly upbuild it.

Sincerely yours,
[*Charles B. Davenport*]

Davenport Papers, American Philosophical Society

Charles B. Davenport to Charles W. Eliot, May 22, 1920

Dear President Eliot;—

Responding to your letter of May 17th, I had not thought to comment publicly on Professor Conklin's lecture. In the matter he discusses mere opinion must play a prevailing part: there are too many unknown factors for scientific prediction concerning the remote future of man.

As to relative viability of mulattoes, as compared with whites and blacks, to which he refers, there is a lack of precise information since Census returns do not distinguish between mulattoes and negroes. It is fair to conclude that northern cities have a much larger proportion of mulattoes than southern rural districts. According to our Census returns the infant mortality rate is higher (24) in northern cities than in southern rural districts, where it is 18. But this difference can not be ascribed solely, if at all, to the direct physical facts of miscegenation.

The environment is probably better for negroes, especially children, in the southern rural districts than in the great northern cities. A further complication arises from the fact that, since the mulatto is on the whole more intelligent than the negro, he secures better sanitary conditions than negroes would in the same locality. However, we have some statistical data that looks toward inferior resistance of mulattoes. This is referred to on page 12 of "Defects Found in Drafted Men"[1] which I am sending you. The detailed evidence is given in the fuller work bearing the same title.

sincerely yours,
[*Charles B. Davenport*]

Davenport Papers, American Philosophical Society

1. *Defects found in Drafted Men* (Washington, D.C., 1920). This was an army study.

Thomas Hunt Morgan to Henry Fairfield Osborn, June 14, 1920

Dear Osborn:

Bateson's attitude[1] toward the kind of people who meet at Eugenic Congresses is that, I think, of many other men working in genetics. The former are excited about all sorts of social propaganda and most of them have little scientific training, and some of them are not scientific at all. The interests of the geneticists are of a very different kind, and they do not care to get mixed up in the sort of business that appeals to the eugenicists.

In this country the distinctions are not so sharply drawn, in part because we are not so sophisticated and are much more naive.

Personally I should rather see the two groups do their work apart, and I think it very unlikely that the men in Europe who have at heart the interests of genetics will care to join the other group. There are other reasons also why it may be better to meet first in a neutral country.

We have settled down here for a quiet year of uninterrupted work. With best wishes for a pleasant summer—

Sincerely,
T. H. Morgan

Davenport Papers, American Philosophical Society

1. See Bateson to Morgan, May 19, 1920, in Genetics Department (University of California, Berkeley) Papers, American Philosophical Society.

Theodore Dreiser[1] to Jacques Loeb, August 26, 1920

Dear Mr. Loeb:

For some time I have been intending to write and thank you for the pamphlets you so kindly sent me—those relative to some of your experiments. Quite recently I read The Science of Human Behavior—by Parmelee[2]—a most involved and poorly presented version of *behavior.* Isn't there some clear readable summary of the Science of Human Behavior—to date—which you can recommend? I am without scientific training and only a good writer can interpret science for the layman. I find no difficulty in reading & following you—but so often writers present their data so poorly that it is slavery to follow them. I will thank you.[3]

Theodore Dreiser

Loeb Papers, Library of Congress

1. DREISER (1871–1945) was the American novelist whose plots sometimes resembled Loeb's descriptions of tropistic behavior.

2. Maurice Farr Parmelee, *The Science of Human Behavior: Biological and Psychological Foundations* (New York, 1913).

3. Dreiser enclosed a clipping about the behavior of goldfish when transferred from a smaller to a larger pool, essentially one of his themes as a novelist.

Jacques Loeb to Ernest Rutherford, August 31, 1920

My dear Sir Ernest:

Many, many thanks for your great kindness in sending me your last papers which I have read and reread. Of all the miracles you have performed, those described in your Bakerian lecture are the most thrilling. The most wonderful part is that it looks as if you were just entering on a new series of scientific conquests. Well, it is good to have lived and read these things even in the midst of all the misery which politicians, generals, and capitalists have settled on the world.

We have spent the summer as usual in Woods Hole. My second son, Bobbie, has been working with me before going to Johns Hopkins Hospital as Resident Physician and we expect Leonard daily for a few weeks' vacation. He has been working on gas ions and may have written you about his work.

We are threatened here with a bureaucratic administration of research by the National Research Council. Hale has gotten the Carnegie money (about 120 millions) practically under his control by having one of the henchmen appointed President of the Corporation,[1] and it is intended that in the new building in Washington (for which he has already the necessary millions) an office be established to which every worker is supposed to promptly announce the subject of any investigation he may decide to start upon, so that he may be advised whether or not his work is a duplication of somebody else's or whether for other reasons he had better change his subject. This information is, of course, "voluntary" on the part of the investigator, but given a good American political machine and the money to back its will I should like to see which young men will dare to stand up against it. The outlook for development of research was never especially good in American universities, but if Washington bureaucracy and politics becomes the de facto and official order one wonders what will happen. The bureaucratic positions of the National Research Council pay well and will be the Mecca of those university men who have political talent and are willing to take orders. And in spite of all this Hale is probably convinced that he is doing a wonderful thing for the promotion of science. I am under the impression that your Bakerian lecture will do more for science than all the National Research Councils in the world put together.

With kindest regards to yourself and Lady Rutherford,

Yours most sincerely,
[*Jacques Loeb*]

Loeb Papers, Library of Congress

1. James Rowland Angell.

Jacques Loeb to Theodore Dreiser, September 11, 1920

My dear Mr. Dreiser:

No exact work has been done on human behavior. The psychiatrists, of course, have to deal with the subject but their work is more or less of an amateurish character. Twenty-five years ago hysteria was à la mode under the influence and leadership of Charcot[1] in Paris, then an era of hypnotism followed, and now we have the wave of Freudianism. While in all of these hypotheses there is some basis of fact, the methods are so amateurish and crude that nothing permanent has or can come of it. Unless we can get exact methods such as I have tried to introduce in the form of the tropism theory in lower animals, we have to be satisfied with admitting our ignorance.

Before the days of Pasteur, medical men were as fertile in inventing theories of infectious diseases as the psychiatrists are now in inventing theories of human behavior. But all these theories of infection were promptly forgotten the moment Pasteur introduced exact methods. So it will be one day in regard to human behavior. This condition of affairs makes one wish that one could come back to life in a thousand years, but alas, such possibilities exist only in mediumistic circles.

With kindest regards and best wishes,

Yours sincerely,
[*Jacques Loeb*]

Loeb Papers, Library of Congress

1. JEAN-MARTIN CHARCOT (1825–1893), French neurologist.

Graham Lusk[1] to Simon Flexner, December 13, 1920

My dear Simon,

You mentioned Liebig today. Liebig was born in 1803 and was professor of chemistry at Giessen at the age of 21. He moved to Munich in 1852 at the age of 49 years where he died in 1873. Wilhelm Ostwald speaks of the latter period as Liebig's literary period. He did little research and stipulated that he should not be troubled with students. He gave evening lectures to fashionable audiences.

His work is the work of early fire and genius culminating at about the age of fifty.

I suppose Liebig supplemented his income from the sale of Liebig's extract of beef. At least both Voit and Rubner[2] received royalties from this source.

Baeyer,[3] Liebig's successor, was reputed to be in receipt of 100,000 Marks annually, even in my day, from patent rights.

I believe that accomplishment in science depends on the quality of the intellectual protoplasm and to the possession of a good conscience, rather than enforced restriction of financial resources.

I am the son of a physician who studied all his life until mid-night or after, and killed himself in the service of others, and I believe that you can still find men who will become distinguished and do service of the highest value without artificially segregating them and eliminating that normal activity which is and has ever been a part and parcel of the physician's life.

The artificial barriers which you are championing are going to chill and kill values of a high order. And you will find this out within a few years. At least that is my judgement.

With kind regards,
Sincerely yours,
Graham Lusk

Simon Flexner Papers, American Philosophical Society

1. LUSK (1866–1932), a physiologist, was a notable investigator of human metabolism.

He was on the faculty of Cornell Medical School. Here he is arguing against Flexner's belief in segregating advanced research in special institutions. His father, William Thomson Lusk, was an eminent obstetrician.

2. The physiologist MAX RUBNER (1854–1932), like Voit and Lusk, was interested in metabolic studies. He applied his findings to industrial hygiene.

3. ADOLF VON BAEYER (1835–1917), 1905 Nobel Laureate in chemistry.

Thomas Hunt Morgan to Richard B. Goldschmidt, February 15, 1921

Dear Goldschmidt;

Your letter came not long ago, and I was very glad to hear from you again, and to know that you are all getting along well. Out here, we do not get much news, and we lead an uneventful existence. Unfortunately, none of your reprints are sent to us, and I have not seen any of your papers since we left New York. I gather you have sent several new ones to me, and I am anxious to read them. We will scurry around, and look up the journals here, and also look them up at the University of California, where we go next Friday to give a seminar on our flies. Most of the Genetic journals are here, but the current numbers are sometimes hard to find. Bridges has a new intersex line that is a wonder, but as it is not yet clear how it works, or what has happened, I better not say much about it.

I sent you a few days ago a copy of the chicken paper. I mean the Carnegie one. I had already sent one from the Biol. Bull. about ten days earlier. I am sorry that the only copy that I had was defaced in bringing [it] out here but it is readable, and, no doubt, it will be much more defaced by the time you have cut it up.

We are having a very fine winter. The spring has come, they tell me, but as we never knew when the autumn came to an end, the change is not striking. A few days ago snow covered the mountains, yet our garden was still full of flowers. After school yesterday Howard went up to the top of Mount Hamilton, where the observatory is, and had a climb of four miles through the snow. A week ago we went to the sea shore, only an hour away, and collected sea urchins, which we brought back here and had plenty of eggs to study for several days. Such are the possibilities of California. I have not been able to get Baur's new edition from New York, where he sent it; consequently I have not written to thank him. Will you explain this to him, kindly, when you see him. Nachtsheim[1] writes that the German edition of the Physical Basis is finished, but awaits certain cuts that I had sent from New York. Also Brachet[2] is having the Mechanism translated (it was left unfinished by Herlandt).[3] This has been thoroughly revised, but how much of the revision will be included I do not know. We have some papers in press and will send them to you as soon as they appear. You will be especially interested in a new stock in which the fourth chromosome is absent. It behaves in a queer way. The absence of the chromosome, predicted from the genetic evidence, has been demonstrated with the greatest clearness by cytological preparations. Also there is crossing over in the IV chromosome in the female. The crosses between the mutants of simulans and melanogaster are giving interesting results. Also the new chromosome map of D obscura, that Lancefield[4] is building up, is extraordinarily interesting in comparison with its chromosome group. But all this will come to you in good time. Perhaps Baur will be interested to hear that the chromosome business is at present "doing well thank you."

With kindest regards to Mrs. Goldschmidt and remembrances to the children.

Sincerely,
T. H. Morgan

Goldschmidt Papers, University of California, Berkeley

1. HANS NACHTSCHEIM (1890–[?]), geneticist at the University of Berlin, had studied with Morgan. The German edition appeared that year as *Die Stoffliche grundlage der vererbung*.
2. The Belgian embryologist A. T. J. BRACHET (1869–1930). Apparently nothing came of this.
3. ACHILLE CHARLES HERLANDT (1850–1920?).
4. DONALD E. LANCEFIELD (1893–[?]), a Morgan student who later taught at Oregon and Queens College.

Thomas Hunt Morgan to Richard B. Goldschmidt, July 5, 1922

Dear Goldschmidt;

Your letter was forwarded to me in England, and I thank you very sincerely for your invitation to come to Berlin. I should have liked very much to have extended my trip and particularly to have gone to Germany. But as I had refused to spend the summer in England, and as I was anxious to get back here to my work and to my family; I made a very short sojourn on your side of the water. In fact, I was just two weeks and three days ashore, and this included a visit to Edinburgh, to Oxford and to Cambridge. However, rushed as I was, I managed to see all the men in genetics and most of the work that is going on. I did not try to convince anyone of the chromosome hypothesis, and while it is true I gave a lecture and three separate talks about these particular carriers, I did not attempt to make converts further than this. In fact I tried to make clear that if they didn't like the chromosomes that was not *my* funeral. However, everyone was most kind and considerate and I had a very pleasant visit. I had been to England several times before and had seen the outside of things, but this time I was taken in and saw the inside. And they do indeed lead a most charming and delightful life. Its contrasts were most marked coming straight to Woods Hole, but fine as English life is, I can not but think that our simple life here is better for our scientific work.

On the way over I met a Japanese teacher, and he told me that you were booked as exchange professor at Tokio next year, so I suppose you have solved that problem and will get your trip to Japan. I am very glad of it and you will get some fine results with the Gypsys.

Here the place is filled completely up and is humming like a bee hive. All my old friends are back and there is a promising crop of youngsters too.

Sturtevant was married in May to Miss Reed, my assistant. This was a by-product of the California trip. They have been in England and Sweden for a couple of months, but returned yesterday. Sturtevant got a very good impression of Nilsson-Ehles[1] work.

We all send greetings to your family. If you go to Japan you will probably pass this way and we shall have a chance to see you. Please give our kindest regards to your wife and remembrances to the children who must be quite big now.

Sincerely
T. H. Morgan

Goldschmidt Papers, University of California, Berkeley

1. The Swedish geneticist, H. NILSSON-EHLE (1873–1949).

Walter Cannon to Maurice Firuski,[1] April 4, 1923

Dear Mr. Firuski:

I did not acknowledge immediately the receipt of Henry Adams' book[2] which you so very kindly sent me because I wished first to have the experience of reading it. After some delay, due to continuing obligations, I have at last had a chance to read it. Parts of it I have read aloud to Mrs. Cannon. The Introduction on The Heritage of Henry Adams is one of the most amusing examples of overweening family pride that I have ever come upon. How any one in his right senses could speak of his relatives as Brooks Adams had in such unmeasured terms as he employs I cannot imagine. Certainly, so far as scientific attainments are concerned, the claims made for the grandfather and the brother are in the highest degree grotesque.[3]

In Henry Adams' discussion of the relation of the second law of thermo-dynamics to the human race, it seems to me that he has merely amplified the problem which arises in the case of any individual who accumulates a vast amount of wisdom, experience, and ability to act effectively, and thereupon dies. The whole procedure seems to be an appallingly wasteful process. All the elaboration of teaching and personal effort seem to have gone for naught—certainly so far as the physiological factors are concerned, the result is merely an increase in the great ocean of heat. There are influences that remain, no doubt, which George Eliot recognizes in her "Choir Invisible," but these would go with the passing of the race. The interpretation of death which the biologists give is that of making way for a new generation, which obviously could not arise if all the old people continued living. Thus results the opportunity for growth, not in the physical sense, but in understanding and in better adaptation, and in improved mores. All this can occur without breaking physical laws. What the purport of it may be is not at all clear. Indeed, the physicists themselves are at a loss to explain why a universe which has existed for infinite time still has such centers of heat, instead of a vast dead even temperature. It seems to me that there still are many elements which we in our limited knowledge do not comprehend, and that the struggle for human betterment may have significance which the physical laws do not take into account. What this may be I do not know. Indeed, it is here, I presume, that one may be justified in an act of faith in order to secure any reason for continued existence.

Again, many thanks for sending me the volume which I shall doubtless dip into again from time to time.

Yours sincerely,
[*Walter Cannon*]

Cannon Papers, Harvard University Medical School

1. Firuski is unknown to us. From the address in Cambridge, Mass., he may have been a book dealer.

2. *The Degradation of the Democratic Dogma* (Boston, 1919).

3. Brooks Adams's introduction does talk extravagantly about John Quincy Adams and Henry Adams as scientists. More precisely, they reflected the intellectual environments of their day, including bits of science and attitudes toward science. Henry Adams and his brothers reacted bitterly to a world in which they no longer had assured positions of power. Science provided a spurious rationalization for Henry Adams's personal trauma.

Excerpt from unpublished autobiography of Francis G. Benedict[1] (International Congress of Physiology, Edinburgh, August 23–27, 1923)

Although the visit to Glasgow and Aberdeen followed the International Congress of Physiology, description of the Congress has been deferred until this point. Great interest had been taken in this Congress since, while it was originally planned to be an *international* congress with *all* nations participating, in the last congress at Paris (the one reported by Professor Miles),[2] the Germans were not invited and on our tour we heard a great deal of discussion pro and con as to the feasibility of holding an *international* congress as early as this, particularly in view of the very irritating situation which had developed in the Ruhr district. In Paris of course we heard very strong opposition to the Congress. Indeed, it was stated that the French would not attend, but subsequently three did attend with no untoward results.

An example of the extremes to which these international complications can develop, even in the innermost circles of scientific laboratories, was the rather bitter opposition expressed in Paris to Sir Edward Sharpey-Schafer,[3] for having invited the Germans, it being stated that he was himself a German, German born, and that he had invited the Germans on his own responsibility without consulting anybody else. I was very glad to point out the fact that, in the first place, he could not be German born or a German citizen, because no one is knighted in England who is not British born, and furthermore that Sir Edward Sharpey-Schafer had lost two sons during the war and to my knowledge was by no means pro-German.

All along the line this matter of the International Congress kept coming up, and we finally heard that, due to the instigation of Professors Frederick and Heger,[4] who, as Belgians, took the invitation of the Germans particularly to heart, the French and Belgian Scientific Societies had passed resolutions, if not forbidding, at least strongly opposing the attendance of any of their members at this Congress. It so happened that finally Professor Richet[5] attended and gave an address at the Congress, also Dr. Laugier[6] of his laboratory, and Dr. Tiffeneau.[7] From Belgium Professor Noyons of Louvain (who, however, was a Dutch citizen) attended.

The masterful way in which this Congress was managed under the most delicate international complexities speaks volumes for the personal character and ability of Sir Edward Sharpey-Schafer, the president. It was a triumph of international diplomacy. The Germans were invited, but Sir Edward asked a vote of the British Physiological Society as to whether he should or should not invite them, and I was told there was a unanimous vote that they should be invited. He did not, I understand, confer with the International Physiological Committee, and on this particular point there may be some theoretical grounds for criticism, but I believe it has not been customary to call this committee together on anything other than to decide the place of the final meeting.

On our tour we kept hearing of different individuals who were to attend, and it was a matter of a great deal of personal satisfaction that I heard subsequently in Edinburgh several, who attended, state that had it not been for my own personal urging and setting forth, so far as I knew, the program, they would not have attended. This was particularly the case with several Germans.

When the Congress began, it was found that about 25 Germans had accepted and had attended. The difficulties which they had to go through in accepting this invitation may be in part realized when it is told that six of them came from Hamburg to Newcastle on a coal steamer, to save expenses. Others told me it would be utterly impossible for them to come because the expense of the trip to and from Edinburgh would cost them more than a quarter of a year's salary. Perhaps one of the most magnificent things that I heard of in Europe was the fact that *every* German who was invited, received a *personal invitation* from some *private* individual in Edinburgh to be his guest during the Congress. Secondly, all the Germans were offered and I think practically all accepted a sum amounting to about 3 English pounds, to help in travel expenses. When one realizes that hardly a house in Edinburgh has not lost some one of its immediate family as a result of the war, the forbearance in inviting these men to their houses as guests is marvellous and well worthy of consideration by the world at large.

It is interesting to state, furthermore, that the Germans who attended were most enthusiastic and appreciative of what had been done for them. As I heard one German say, "We have been entertained almost to the limit of endurance. We have been over fed and over wined and shown every hospitality possible." I did not hear one discontented note throughout the entire Congress from a German except Frank of Munich,[8] who is notorious as a discontented, complaining individual. On one morning I found him tearing about the corridors of one of the buildings, complaining bitterly because his lantern slides would not fit. He said, "My lantern slides are German. They are the wrong size. They are all wrong. Everything German is wrong." I joked with him with regard to the fact that he was getting unduly excited and I was perfectly certain Professor Barger,[9] who had charge of these affairs, would provide him with a suitable slide carrier, which was of course done. Any one who knows Frank would know this was a common state of mind with him.

I also heard not one word of unfriendly criticism with regard to the presence of the Germans, except, I am very sorry to say, in the case of Dr. Cathcart[10] from whom for the first time in my European tour I heard the words "Boche" and "hun." I found subsequently, however, that this was quite common in Glasgow and Cathcart was simply using the natural phraseology of Glasgow scientific circles.

The appearance of Professor Richet was also an international triumph. Although Professor Richet had early signified his intention of attending, as indeed he was almost in honor bound to do, having been the last president in Paris, I was told with every degree of conviction by Professor Heger in Brussels on the last day of June that Richet *certainly would not* attend the Congress. It may or may not be significant that Dr. Laugier, who was the Chef des Travaux Physiologiques in the Sorbonne, was from Richet's laboratory, but Tiffeneau was the professor in the College of Medicine in Paris. Professor Tiffeneau was present with Madam Tiffeneau.

In looking through the list of members, one finds, however, other Frenchmen registered. I had no means of knowing how many of these attended, but undoubtedly more than the three I have cited. The important thing is, if no Frenchman but Richet had come, it was sufficient to break the evil spell, so to speak, and the general consensus of opinion was that the French had made a great mistake in their action. It is quite comparable to the mistake made in the foolish patriotic resolution of a body of physicians in the Rhineland, of whom Professor Abderhalden spoke, who had voted not to treat medically either French or Belgians. The important point is that no one noted the absence of the French and it ceased very shortly to become a subject of conversation.

The meetings were all characterized by a marvellous harmony, considering the rather smoldering discontent and uncertainty that had been going on all through the year. A very tactful selection of presiding officers was made for various sections, so there was a sprinkling of all nationalities with, if anything, a little larger proportion of Germans than others, there being frequently two or three Germans in the course of a morning session. The German representatives were an excellent group, representing physiology in Germany. Their own papers were well presented. There was no undue humility or undue pride. Everything was as natural and as delightful as could be imagined.

Benedict Papers, Harvard University Medical School

1. FRANCIS GANO BENEDICT (1870–1957) headed the CIW Nutrition Laboratory from 1907 to 1937. A chemist with a Ph.D. from Heidelberg (1895), he became an expert on animal calorimetry and respiratory gas analysis in the course of his work. Benedict devised the basal metabolism test.

2. The physiological psychologist WALTER RICHARD MILES (1885–1978) was then with Benedict at the Nutrition Laboratory, later at Stanford and Yale.

3. SIR EDWARD ALBERT SHARPEY-SCHAFER (1850–1935) was born in London, the son of a naturalized British citizen. A histologist and physiologist, he was at Edinburgh University.

4. E. H. R. A. FREDERICQ (1887–[?]); PAUL HEGER (1846–1925).

5. Charles Robert Richet.

6. HENRI LAUGIER (1889–[?]).

7. M. E. P. A. TIFFENEAU (1873–1945).

8. OTTO FRANK (1865–1944).

9. GEORGE BARGER (1878–1939) had come to Edinburgh to a chair of chemistry in relation to medicine.

10. EDWARD P. CATHCART (1877–1951).

Robert A. Millikan to George Ellery Hale, August 28, 1923

My dear George:

I am writing you very briefly, after consulting Evalina, to acquaint you with dreams which are now floating about the California Institute of Technology. Rose,[1] President of the General Educational Board, and Pritchett have both been here within the month and coincide in the general view that if anything is done in Southern California in the field of biochemistry, biophysics, and medical education, it must be done in immediate contact with the present work of the Institute. Rose says that Welch's plant is suffering already from lack of contact with physics and chemistry. He says the Rockefeller Board will not be interested in any medical plan in Southern California which is farther away than across the street at most from the Institute.

We have gone over the whole matter very thoroughly at two dinners attended by Fleming, Robinson, O'Melveny, Chandler, Bridge,[2] Pritchett, Noyes and myself, and the conviction is I think quite general that we can scarcely avoid this responsibility, and that it is incumbent upon us to find some plan for which a medical school, with emphasis upon research, may be started at once in the region.

Rose says it is useless to attempt anything in this connection with less than $10,000,000, and leads me to suspect that if a Mr. Eastman[3] could be found to put in $5,000,000 here, the Rockefeller Board, assisted possibly by Carnegie, might contribute another $5,000,000.

Since Mr. Huntington[4] has already made plans for putting up a hospital either in Los Angeles or on the Huntington property as a memorial to Collis, and as Dr. Bryant[5] is the man on whom he relies for advice in this matter, O'Melveny has cabled Bryant, who is now in Europe, not to come to any decision with Mr. Huntington until after he has discussed matters with the group here; it being thought possible that

Dr. Bryant and Mr. Huntington both might get the vision of something more important than a hospital for Southern California.

Now I wish to suggest just a possibility for your consideration. The whole end of the campus west of the Physics and Chemistry buildings, barring only the Auditorium, might conceivably be given up to biochemistry, biophysics, and medicine. Pritchett, Goodhue,[6] and Fleming have all talked this over and agree that there is ample room. There is also ample room adjacent to Throop Hall for the divisions of engineering proper, and the rest of the campus, clear to Hill Avenue, would be filled with dormitories, faculty club, etc. You have here a picture of three hundred biological men and twelve hundred physicists and engineers living together under conditions not found anywhere in America at present. Pritchett says it would meet the lack all medical schools have felt in the past, isolation from other sciences.[7] Your imagination can work on this plan just as well as mine. If anything is to be done toward its realization there is no time to be lost.

Mr. Huntington is expected back here sometime in September, when a conference will be held with him and the whole group which have been giving this matter consideration. You will be a great factor in the decision, if you are in shape to put your head upon such a matter.

While you are in New York, you may possibly be able to talk it over cautiously with Rose or Abraham Flexner, or both. You may possibly also see Mr. Huntington before he comes back, and too you might chance to see Dr. Bryant, whose interest and conviction is essential. So you will at least wish to know something of what is doing. Your imagination can supply the rest. If it does not bother your head we would like to know what your judgments are as soon as possible. If it does, we do not want it, for you are vastly more important than any school or institute.

I am writing you briefly about this because I thought it would be best to do so. Let me hear from you as soon as you can after your arrival. I realize the danger in all this quite as much as you, but looking at the good of Southern California alone, I can see nothing so important as the intensive pursuit of all the sciences, including the biological, at the Institute. This will leave other schools in Los Angeles free to pursue intensively the humanities, social science, and law. This would be the finest thing that could possibly happen to these other institutions.

Yours as ever
RAM

Hale Papers, California Institute of Technology

1. WICKLIFFE ROSE (1862–1936) was an educator at one of the Rockefeller philanthropies, the International Education Board. He had directed the earlier campaign against hookworm in the South.

2. Five Cal. Tech. trustees: ARTHUR HENRY FLEMING (1856–1940) a wealthy lumberman; the lawyer and financier HENRY M. ROBINSON (1868–1937); the attorney HENRY WILLIAM O'MELVENY (1859–1941); HARRY CHANDLER (1864–1944), a newspaper publisher and landowner; and the physician NORMAN BRIDGE (1844–1925) who was active in petroleum.

3. This is a reference to George Eastman's gift to the medical school of the University of Rochester.

4. HENRY EDWARDS HUNTINGTON (1850–1927) was the heir of Collis P. Huntington, one of the builders of the Southern Pacific Railroad. The younger Huntington endowed the Huntington Library and Art Gallery at San Marino, Calif.

5. Perhaps ERNEST ALBERT BRYANT (1869–1933).

6. The architect BERTRAM G. GOODHUE (1869–1924).

7. Rochester's medical school was also being developed at the main campus. University of Pennsylvania's medical school was adjacent to the rest of the university.

William Henry Howell[1] to Jacques Loeb, October 8, 1923

My dear Doctor Loeb:

In the final session of the Congress at Edinburgh this summer when the question came up as to the next place of meeting Carlson[2] speaking for the American Physiological Society invited them to come to America. The leading spirits knew that the invitation was coming and had discussed the possibility beforehand. They came to the conclusion that unless some financial aid was forthcoming none of the physiologists of Germany and Austria could stand the expense. Consequently the device was hit upon of naming an international committee and leaving the place of meeting in their hands. For my sins I was made the American representative on this committee. At a meeting of the committee it was evident that the next meeting must be either in America or Scandinavia. America was preferred if a fund could be raised to help pay expenses especially for the Germans and Austrians and I was requested to ascertain if this plan is feasible. Among Americans at the congress feeling seemed to be divided. Some e.g. Mathews[3] and Carlson were rather against any subvention for this purpose, while others thought that it would be desirable. While I don't like the general principle I am in the latter group for two reasons.

First, if the next congress can be held in America and it is made possible for them to come in any numbers I am sure that the barriers of international enmity will be broken down completely and international relations of scientific men will be brought back to their normal condition. Owing to Schafer's courage in the matter a very good beginning was made at Edinburgh. The Germans were there in force, aided by a subsidy, and they were entertained nicely. It is true that the French and Belgians for the most part refused to attend, but Richet and two or three others from France came and took an active part in the meetings, while the attendance from Great Britain and this country was large. It was a beginning of a really international congress. In the second place I think that such a meeting in our country would exercise a beneficial influence on the promotion of the physiological sciences and on their relations to medicine and to science in general. I am writing to you about the matter because in the discharge of the duty imposed upon me I have laid the matter before President Vincent in the hope of getting the Foundation interested in the project. I have no idea how he will react to the proposal, but if he considers it at all I assume that your opinion will be asked. I am ignorant of course as to what your feeling will be, but if the matter is referred to you I hope that you will give it serious consideration. It is evident of course that the Foundation can not help to finance international congresses of all kinds but there seem to be special reasons for this case in helping to restore international amity and in the fact that we have a relatively small group to deal with and a group that has given evidence of a willingness to set an example. If you think favorably of the project I wish that you would discuss it with Flexner.

Sincerely yours,
W. H. Howell

Loeb Papers, Library of Congress

1. HOWELL (1860–1945), a physiologist, received his Ph.D. from Johns Hopkins University in 1881. From 1893 to 1931 he was on the Hopkins faculty.
2. ANTON JULIUS CARLSON (1875–1956), physiologist at the University of Chicago.
3. ALBERT P. MATHEWS (1871–1957) was a biochemist at the University of Cincinnati.

Joseph Erlanger to Herbert Spencer Gasser,[1] October 16, 1923

Dear Dr. Gasser:

Your letter of October 3rd comes the day after sending away the reconstructed manuscript. You will have noticed already the difficulty I have had in distinguishing between "conduction with a decrement" and conduction at different rates in the constituent fibers of the nerve. All of our observations point to a gradient of conduction. This involves not alone rate, but also broadening and lowering of the action current, that is "conduction with a decrement." I, nevertheless, feel that the alternative of differences in conduction rate in constituent fibers is worth while hanging on to. This hypothesis clearly accounts for the changes in form and might even be made to account for differences in conduction rate, if these are apparent rather than real and due to the difficulty in recognizing the start of the spread of the conglomerate action current.

I had read all of the papers you refer to in your letter and, since you feel as you do about this subject, I would suggest that certain changes be made immediately on page 46 of the manuscript such, for example, as:

> Insert line 11. "Contrary to the experiences of Adrian and Forbes (J. P. Vol. 56, 1922–301) it has been observed by us to occur in the isolated saphenous nerve also (see figure 13)."
> 7th line from bottom after "accelerating" insert "or at least linear" (Rehorn Zt. f. Allg. Physiol. 17, 49, 1915).
> 5th line from bottom for the sake of clearness add after "in" "the fibers composing."

We have at last gotten the miniature B batteries into workable shape and hope within the next few days to get some more exact measurements of the conduction gradient both in the normal and in the reversed direction, as I wrote you yesterday. The difficulty in getting accurate measurements on amplitude is appalling, for we have to deal here not alone with the overlapping of the escape and action current, but also with some queer alterations in the relation of the base line as drawn on the tube and the bottom of the action current, as indicated by either its start or its termination, or both. In order to be able to get at the "probable decrease in height from the degree in separation" (quoted from H. S. G.) it would seem to me that it would be necessary to assume that all the constituent action currents are of the same height where they start.

We have no news here that can compete from the standpoint of delightfulness with the experiences that you are having. Things are, however, running along quite smoothly.

With kindest regards in which Mrs. Erlanger joins me,

Sincerely,
Joseph Erlanger

Erlanger Papers, Washington University School of Medicine

1. GASSER (1888–1963), a graduate of Johns Hopkins Medical School, shared the 1944 Nobel Prize with Joseph Erlanger for the studies of the electrophysiology of the nervous system. Gasser succeeded Simon Flexner as director of the Rockefeller Institute (1939–1953).

Jacques Loeb to William Henry Howell, October 24, 1923

My dear Professor Howell:

I have not answered your letter before this because my secretary was away on her vacation. I have no definite opinion one way or the other, except that I do not quite see what good it could do to bring a number of German scientists over here at a great expense. If we bring over one or two first-class men who have something new to give, I think I should gladly advise such a step, but to bring a number of more or less indifferent workers over would mean a more political than scientific enterprise and I must confess that I prefer not to mix up the two things.

I have not noticed that the exchange of professors between various countries before the war helped very much in promoting the cause of peace. I think some of the worst sinners during the war were university men who had traveled in the countries against which later they were violently declaiming.

Yours very sincerely,
[*Jacques Loeb*]

Loeb Papers, Library of Congress

Joseph Erlanger to Herbert Spencer Gasser, October 30, 1923

My dear Dr. Gasser:

Again our letters have crossed. It is perfectly obvious that more information is needed regarding the simplest of nerve problems, namely conduction rate, and Bishop[1] and I are working toward that end. We have made no progress as yet because since getting the B batteries straightened out we have been hampered by an unusually prominent escape. We can say that it is not going to be difficult to get good ascending action currents. They seem to be almost as high, as a matter of fact, as the descending. We believe that in a green frog's sciatic we have seen the beta wave in the ascending action current.

We have in our collection enough information to settle tentatively, at least, some of the points you raise. Thus, you may recall that we have three sets of observations on branchless (vagus) nerve (2) and on a nearly branchless (phrenic) nerve (1). Though these data are not as good as they should be, they are much the same as are those from the sciatic.

The fact that the impulse travels faster in a catelectronic area I was familiar with. Indeed, we have one observation on the green frog showing this. I had, however, neglected to fully reread Gotch in Schaefer's text book.[2] There, on page 480 II, is an excellent reason for expecting to find differences in conduction rates in the fibers of the mixed nerve supplying flexors and extensors. This should be referred to in our paper.

I am not yet ready to give up as a working hypothesis slight normal differences in conduction rate as the cause of the change in form of the action current with propagation.

I expect Forbes[3] here tomorrow. He wants to look over our outfit and will spend several days here.

Hadley[4] makes a very favorable impression on everyone. He realizes fully his shortcomings as an educator and for the present, at least, says he merely is the executive officer of the faculties.

We still are without an internist and pathologist, and the Physiological Department without its third assistant. Can't you find someone over there for us? Fenn,[5] I presume, is too far advanced for such a place.

With kindest regards,

Sincerely,
Joseph Erlanger

Erlanger Papers, Washington University School of Medicine

1. The neurophysiologist GEORGE HOLMAN BISHOP (1889–1973) remained at Washington University until retirement.

2. The *Text-book of Physiology* (2 vols. [London, 1898–1900]) had a section by FRANCIS GOTCH (1853–1913), the English physiologist who detected and recorded electric currents in the mammalian brain.

3. ALEXANDER FORBES (1885–1965) of the Harvard Medical School.

4. HERBERT SPENCER HADLEY (1872–1927), an attorney, was president of Washington University from 1923 until his death.

5. WALLACE O. FENN (1893–1971). See his "Born Fifty Years too Soon," in *The Excitement and Fascination of Science* (Palo Alto, 1965), pp. 109–118.

Walter Cannon to Ernest Henry Starling,[1] November 12, 1923

Dear Starling:

As you may have noted, I have recently been employing the heart, isolated from the central nervous system, but left in the body, as an indicator of various changes taking place in the blood stream. It is an admirable signal of adrenal secretion. We have used it for quantitative experiments which have recently been confirmed by Kodama,[2] using Stewart's[3] method. It yields the same evidence for reflex and asphyxial control of adrenal secretion that is obtained when direct splanchnic stimulation is employed.

Recently I have been struck with the possibilities of using the denervated heart in the body as a sample tissue—an automatic mechanism whose action might be indicative of what is going on in other parts of the organism. I am writing to ask you whether you have made progress in putting together the admirable researches which have been performed on the heart-lung preparation in your Laboratory. If the survey of the work is not near completion so that I could soon refer to it, would you be so kind as to give me your judgment on the suggestion which I have just hinted at?

I am especially interested in the evidence which the denervated heart might yield as to variations in metabolism. The observations of Evans[4] show that the rate of beat and the utilization of oxygen increase together step by step with increased temperature. Variations in temperature probably present as general conditions as could be offered. Presumably stimuli which bring the heart into extra activity, e.g., adrenalin and impulses via the accelerators, likewise increase its metabolism to a degree corresponding with the increased rate. Is it true? In other words, if the rate increases for any reason, is it probable that the metabolism increases correspondingly? I recognize that there are other elements concerned than the rate of beat. Obviously the rate may remain very little changed and the degree of contraction might be very largely affected. Do we know enough about conditions in the body to be able to evaluate these two elements? If the blood pressure varies in only a small extent, and there are no changes which would affect the delivery of blood to the heart, might the rate of beat of the denervated organ be regarded as indicating its rate of metabolism?

I realize that I may be asking about conditions which are altogether too complex to be resolved by any present knowledge. There is no one, however, to whom I can turn at the present moment with more assurance of receiving a reliable answer. Any light which you can throw upon my problem I should be most grateful for.

With very kind regards, which you will please share with Mrs. Starling, I am

Yours sincerely,
[*Walter Cannon*]

Cannon Papers, Harvard University Medical School

1. STARLING (1866–1927), physiologist at University College, London.
2. Perhaps the Japanese chemist Sakuji Kodama.
3. GEORGE NEIL STEWART (1860–1931) at Western Reserve University.
4. SIR CHARLES LOVATT EVANS (1884–1968), professor of physiology at St. Bartholomews Medical School, London.

Rollins Adams Emerson to Milislav Demerec,[1] January 13, 1925

Dear Dr. Demerec:—

Please send me Mrs. Trajkovich's address. I want to have a few copies of her memoir sent her. Professor Hutchison writes from Belgrade that he has heard from the Trajkoviches and that he will see them when he gets into Jugo Slavia.

When I get started again on the corn linkage paper, I shall want all the linkage records you have which have not been published. I really prefer that new data be published by the men who get it, so that I will need only to summarize it and cite the references in the general linkage paper. If we save a lot of new data for this paper, it will be very long.

I do not feel that my name should be used with yours on any white-seedling paper. While it is true that I have grown and have had pollinated for you some of the stocks, the plan of the work has been yours almost alone.

I hope that the white-seedling, apparently linked with sugary, is on the end opposite tunicate, so that I can use it as a check in my low-sugary studies. I am tabulating the data on this study now and find it a big job.

Sincerely,
R. A. Emerson

Demerec Papers, American Philosophical Society

1. DEMEREC (1895–1966) was born in Yugoslavia and received his Ph.D. from Cornell in 1923. From 1923 on he was at Cold Spring Harbor where he served as director from 1943 to 1960.

Alfred H. Sturtevant to Milislav Demerec, July 10, 1926

Dear Demerec:

My Oenotheras or a few of them anyway, are about to blossom, and I find myself ignorant of technique. What kind of bags do you recommend (to cover the whole spike, which is the approved technique)? What do you call them, how big are they, and where do I get them? I plan to be in N.Y. some day next week, so can scout around, if I know what I'm looking for.

When do you go to Ithaca, and can't you arrange to go via Morristown and drop in here? I hardly think I'll go myself, as I think when I make a trip it will be to Woods Hole. I noticed a couple of variegated Delphinium in the garden here—one with a dark blue spike and a somewhat paler one. Better come see them.

Flies are going O.K. I still can't decide whether my new singed allelomorph in similans reverts like reddish, or merely overlaps normal. The sn^2 females and also the sn/sn^2 females are sterile; but now I have $\frac{rb\ sn^2}{dy}$, which should give the answer before long. I already know that the thing either reverts or has given a surprisingly large number of crossovers among the daughters (from not-sn fathers) used to carry on the line.

The temperature outside today is running around 30°–31°; but in my cellar laboratory it has just touched 25°. In fact I have the coolest place

I've found around here—last month it was so cool that melanogaster took 18 days for a generation.

I went down to Princeton a couple of days ago and saw Shull and his Oenothera outfit. Shull was very agreeable and pleasant—asked me to come again when he's pollinating, and showed every indication of wanting to encourage me to fool with Oenothera. But neither of us mentioned Renner's[1] name or any of his work or ideas. What do you make of that?

Have picked up some very nice ants, and have larvae and pupae coming along from a wild ♀ of a species of Drosophila I had never seen alive before. Most of today I've spent watching and experimenting with a very rare parasitic ant whose habits and life history are almost entirely unknown.

Sincerely,
A. H. Sturtevant

Demerec Papers, American Philosophical Society

1. The German geneticist O. RENNER (1853–1960). See A. H. Sturtevant, *A History of Genetics* (New York, 1965), pp. 62–66, for a discussion of Renner.

Alfred H. Sturtevant to Milislav Demerec, July 24, 1926

Dear Demerec:

Many thanks for the bags and the advice. I haven't used either yet, as I still have only one type of wild plant in flower. But franciscana will be along in a very few days. I've spotted four distinct forms of wild plants, and will soon be able to cross some of those. One Lamarckiana plant is also showing small buds now, as are a couple of nanella rabricalya. So I should be able to get in a large crop of seed.

The Delphinium is annual all right, and we've marked the dark branches. Will collect open-fertilized seed from both kinds of branches. Some at least of the dark flowers are also off in shape—deformed spur etc.—and I am wondering if they are diseased.

Have you looked over Benedict's data on Nephrolepis? As I remember it, the stuff suggests "multimutation," though I have always supposed it meant that islands of unmutated tissue were still present in mutated branches. As a botanist, you can of course judge of that better than I can.

As to moths, the best laboratory species I know of is Whiting's Ephestia—the beggar that must sometimes get into your corn seed. Goes fast, and Whiting has worked out the technique and published it in the Journ. of Heredity. I did my best last summer to get Whiting interested in trying Harrison's stunt on it—but don't know if I succeeded. Fogg, at Columbia, is also supposed to be doing this, but I hardly think he's to be taken seriously. The real trouble with Geometridae, or even very near them; and it is that family that Harrison's data will pertain to. In fact, except for the nun (same genus as the gipsy), I know of no cases of "industrial melanism" outside of the Geometridae. I'm afraid all that family will take a year for a generation, and will require green leaves to feed on. I think your best plan is to try and get hold of Pink Jesus (= Bug Forbes) when you go to Ithaca. He can give you the best advice and the largest dose of it of anybody I can think of.

It looks now as though singed-2 in simulans is not reverting and never has done so. The best guess is that low temperature (such as we had here in June) greatly increases its tendency to overlap wild-type. Data before June and since all agree that the gene is entirely stable.

The human F_1[1] is now due at any time, but hasn't yet shown signs of coming soon. Mrs. Sturtevant is in very good shape and sends her regards to you.

Sincerely,
STTT

Demerec Papers, American Philosophical Society

1. William Sturtevant, now an anthropologist at the Smithsonian Institution.

Thomas Hunt Morgan to Robert M. Yerkes,[1] February 21, 1928

Dear Yerkes:

Many thanks for your kind letter. I am glad you think that I have not gambled too heavily going to Pasadena. At my time of life it is taking somewhat of a chance making a fundamental change.

I was there two weeks ago for ten days, and not only enjoyed the climate but confirmed my previous idea that they are attempting to do really scientific work with a good program. In the case of biology, much will depend on whether we can collect the men we want. We shall go slowly and not attempt to start with a rush.

I am glad to hear that the primate program is so promising. I am not sure that the material is really better than the fruit flies—but perhaps, with the combination of the primates studying the fruit flies, we have certain advantages.

With best regards to Mrs. Yerkes from Mrs. Morgan and myself,

Sincerely,

T. H. Morgan

Yerkes Papers, Sterling Library, Yale University

1. YERKES (1876–1956) was a comparative psychologist best known for his studies of the great apes. From 1924 to 1944 he was a member of the Yale faculty.

Alfred H. Sturtevant to Milislav Demerec, March 15, 1930

Dear Demerec:

This is not a letter, but only an announcement of a new mutable gene—garnet eye-color in Dros. obscura. One strain gives about 0.5% reversions, another gives about 0.1%, and a newer one (not yet spread out much) seems to give about 2.0 to 5.0%. No mosaic flies seen; reversion happens in homozygous females, apparently not at all in males. No other genes lie close to the locus—the nearest to the left is about 15 units away, to the right 50 or more units. No constant allelomorph known. The eye-color varies in intensity in a pure stock (*not* due to mosaics), and selection seems to be effective in changing the average intensity, though neither extreme has yet been made to breed true. I still don't know if this variability has anything to do with mutability or not.

Anderson and I each have a plant of Nemophila that is mutating from white to blue (corolla), after the Delphinium fashion. I'm afraid, however, that the species is not very favorable—too few seeds per flower.

Best regards from all the Sturtevants to all the Demerec's.

Sincerely

A. H. Sturtevant

Demerec Papers, American Philosophical Society

Adolf Meyer[1] to Walter Cannon, May 7, 1931

Dear Dr. Cannon:

I venture to put before you in the form of a question—not a dogmatic argument—what I feel might establish safe and common ground. Will you pardon my following up our discussion, and will you take it in an inquiring spirit?

While I appreciate what an opportunity you have to work in a planned experimental manner in fields which you can choose for their opportunities, I also see another opportunity. As a worker and a teacher, you can for yourself and your students reserve a place for those who have to work with what comes to them, not as experimental preparation but as persons with a past, present and future and a setting that has problems equally real as your self-chosen opportunities but urgently requiring study and solution not merely according to our own option, but also in their own way open to creative work. I do not believe that any of us would care to belittle the problems you create and solve and use for your synthetic pictures, merely because they are not wholly adequate for what our problems would make us reach for. Why is it that you create the semblance of inadequacy of our effort to meet our own problems just because they could not be reached by your methods or because in your settings you have no special facts that force you to cope with the only data and formulations with which we can attain our results? If I speak of your type of facts as non-mentally integrated or mentally integrated, I realize that we use the same material and functions in either, but in the mentally integrated functioning I know that the full understanding demands a reconstruction of the complete situation in which the activity could occur spontaneously, while you are satisfied to reckon with the experimental preparation for just what it is able to do at the time of the experiment. I should have felt much happier if you had put your difficulty in the form of a request for explanation rather than in the form of a derogative disapproval. You are perfectly in your rights if you ask why we use such concepts as total personality and in what sense. But to expect that a word could save you the trouble to inquire into the facts with which one has to familiarize oneself to make words worth while, is not apt to get one far. Draper got us very far from what might be made simple and common ground. But if physiologists cannot grant the existence of a field in which their methods and data fall short, many "mentalists" might be driven to consider it futile to talk of operatively produced reactions when one has to deal with the disorders of the whole live organism in action; indeed they might feel justified in letting organism go and to talk just of a mind, and when that becomes a bit too plain and inadequate to invent the "unconscious."

In the paper I had prepared, I had in mind to meet exactly this need of depicting the nature of the extension of the situation we have to recognize and adjust to, when we deal with "mentally-integrated" functioning including what you study in more or less isolation without having to bother with the total result, beyond describing it and calling it a name such as rage; i.e., having to study why in the presence of all the structures we get so many modalities of the all or none reaction with which you deal. That is where your methods are insufficient and where our concern begins. I am sorry to have to meet the same rebuff as the ones who indulge in generalizations for over specialized and strange data in our field; I am sure it would be better to allow a ground for the organization of the generally acceptable and controllable facts even if I have to use concepts which do not belong to your sphere of experimentation.

It would be a great thing to attain common ground where common ground can be made readily available, and within the common ground we might define the relative spheres or zones of special concern. If we more conservative and considerate workers are thrown into the same category as those whom we all might equally find fault with—that of generalizing from material demanding an undue amount of hypothesis and new mythology—progress is greatly delayed. Mankind is not going to wait for the settlement of its physchobiological problems by animal experiment with the scalpel or any similar method. Why not grant us the confidence that perhaps we too use our best judgment and experience within our field and as required by our field, without ever being pushed out of the domain of general scientific respectability? It just tends to frustrate a good share of the work of those of us who respect the various spheres of objective science and who are trying to bring human behavior and mentation on as objective a basis as we do. Wherein do you see the difficulty and what are the objections you wish me to do justice to? What is the common ground and where do you suspect me of slipping off?

I have had my troubles with the psychologists à la Titchener[2] and à la Watson.[3] Why should there be a semblance of troubles with the physiologist and the biologist? Why should we have to encourage those who give up all attempt to strike common ground? Is the question beyond the pale of mutual understanding? I do not think so.

The present situation of the treatment of the personality-functions threatens to pass altogether into the hands of the psychoanalysts. There is a fair chance to keep the field out of cults and out of propaganda. But it cannot be done without a getting together of those who are more interested in inquiry than in argument. Do I bore you? Or am I able to make my request for consideration clear? Is it not legitimate to speak in terms of groups or sets of "integrates" requiring their specific rules of the game? And yet a good and intelligible lot of common ground and common principles? And the same approach of determining the facts, formulating them and bringing them to observational and experimental tests? Why should we stop at the extension of the facts in play when the organism acts, reacts, feels and thinks and when the question is not only how but why it does so in one or another way?

There have to be divisions of labor. But there should not be artificial limitations and obstacles. Do I make myself clear?

Sincerely yours,
Adolf Meyer

Cannon Papers, Harvard University Medical School

1. The Swiss-born MEYER (1866–1950) was a psychiatrist and neurologist at Johns Hopkins.

2. EDWARD BEADFORD TITCHENER (1867–1927) was a Briton who studied with Wundt, receiving a Ph.D. from Leipzig in 1892. From that year to 1927 he was at Cornell University and, like Wundt, a believer in an experimental psychology of introspection—influential yet hardly typical of what developed in the United States.

3. JOHN B. WATSON (1879–1958), a Chicago Ph.D. where he was influenced by Loeb's example. He was the founder of behaviorism.

Walter Cannon to Adolf Meyer, May 13, 1931

Dear Doctor Meyer:

Let me thank you for your letter of May 7 which came two days ago and which I have been thinking about since its arrival.

First of all, let me assure you that my spirit is truly an inquiring spirit. I am certain that I must have given you the wrong impression in opening the discussion at Atlantic City. My object at that time was to rouse further consideration of the subject and I purposely put my remarks in such form as to call forth comments from persons whose work I spoke about. Naturally I supposed that we were to have a discussion; and you can imagine, perhaps, my chagrin when the meeting was called to a close without any remarks from anybody except yourself.

As I have read your letter, it has seemed to me that the main difference between your point of view and mine lies in a difference as to what we are considering. My chief interest is in an attempt to explain phenomena by research and experimentation. As I have understood your letter, you have pointed out the necessity which you are under to concern youself with the total situation of an individual in giving him *practical treatment*. Am I right in making this discrimination? I can readily see that a physician who, fifty years ago, had to treat tuberculosis would naturally enough study the social, the economic, the mental and other aspects of his case and would then use his best judgment in advising the patient what to do. I submit that that would not be a precise way to determine the etiology of tuberculosis. The nature of tuberculosis was investigated by neglect of the total situation and by attention to detailed lesions, the occasion for these lesions, and the experimental proof that organisms found in the lesions are capable of producing similar lesions in normal animals.

It seems to me that in a study of the functioning of the nervous system and even in psychiatry the situation is not unlike that which I have just mentioned. We do not know enough about the causes of mental diseases to permit the psychiatrist, who has to deal practically with a given situation, to do more than to study his patient in various relations and aspects and then exercise his best judgment as to what is appropriate treatment. Is there very much hope of advancing the science of psychiatry by that procedure? May we not expect that the experimental method applied to the complicated conditions of cerebral disorder will have results there similar to those that it has had elsewhere? That means, of course, modifying one variable at a time and watching closely the results—naturally the selection of a significant variable is important. May I illustrate by general paresis. I hardly suppose that a study of the total personality, as such, of cases of general paresis would have advanced our knowledge of the mode of treating that disease. Is it not true that only actual tests with acute infections and artificial fever, which brought *proof* of an effective mode of treating the condition, has been helpful?

Perhaps I am wrong in judging that we have been talking about different functions in medicine—that of the practical physician and that of the investigator. If I have made a mistake, I should be glad to be corrected.

As regards investigation, I think that we can find common ground insofar as psychiatrists are willing to recognize the functioning of the brain. I have had a strong impression that the vocabulary of certain aspects of psychiatry has somehow taken the place of attempts to think of mental phenomena in any sort of association with cerebral processes. When I am met with the "ego," the "super ego," the "censor" and the "id" and hear these terms bandied about as if they disclosed a large amount of clear information, I must say that I despair of finding any community of interest between the physiologist and the psychiatrist. That seems to me highly unfortunate. The progress in knowledge of cerebral function may not have reached a point where it can be very useful to the psychiatrist, but after all it is on the neurological side, I think, that causal relations exist; and, as illustrated by recent progress in treatment of general paresis and extraordinarily suggestive observations of the effects of breathing high CO_2 and O_2 mixtures, it is on that side that we may, perhaps, expect most valuable researches. Certainly physiology has gone forward with leaps and bounds by use of the experimental method, and the physiology of the central nervous system is yielding to the experimental attack as other systems have done. It seems to me that by use of scientific methods of observation, analysis, inference, experiment and cautious conclusion there has been a sounder progress in knowledge of the central nervous system from the point of view of its physiology than from the point of view of its psychology. If you agree to that, don't you think that it would be well for psychiatrists to attempt to meet the physiologists halfway and perhaps do their part in trying to formulate the disturbances with which they are dealing in terms which would be understandable to the physiologist?

You must not think for a moment that I am lacking interest in your point of view or that anything that you might write would be unwelcome to me. I am sorry that I ever have given you, even inadvertently, the impression that I would not be hospitable to your ideas and open-minded in considering them.

Yours very truly,
[*Walter Cannon*]

Cannon Papers, Harvard University Medical School

Walter Cannon to Leonard B. Nice,[1] March 16, 1932

Dear Dr. Nice:

I received your letter of March 14 with a report signed favorably by Drs. Greene,[2] Mann,[3] Carlson[4] and yourself, as Chairman. I supposed, of course, that there would not be any formal disposition of this question until there had been opportunity for all the members to consider the pros and cons of the situation. There is little reason for action of a committee, it seems to me, if the members are going to register their previous convictions without regard to any opposing factors. May I state my point of view, so that you will know why I am not in favor of the giving of Federal aid in the form which has been advocated by Dr. Greene.

As Exchange Professor to France in 1929–30, I had occasion to observe the effects of central control of the educational systems. The most striking and deplorable result was the lack of local interest of the people in their own institutions. The responsibility for educational interests had been shifted to Paris. The result was that all the values which come from local and individual enterprise and from competition of one region with another had been lost. As I urged at the meeting of the Society at Montreal, if there is need for support of research in the several states, it is first of all important to educate the State Legislatures regarding their responsibilities and not to shift those responsibilities to

Congress, and thus leave the local legislators wholly uneducated in the relations of research to civilization.

In order to get further evidence on the points at issue, I wrote last year to the American Council on Education at Washington, D.C. and had the following answer regarding the proposed bill "for more complete endowment of State Universities and Colleges of Agriculture and Mechanic Arts."

"The absurdity of the bill does not appear when considered with regard to the consolidated universities and Land Grant Colleges, like, Illinois, Wisconsin, Missouri, Minnesota. The absurdity appears in the nineteen states where the Land Grant College is separated from the State University. This bill would tend to build up a research center in Michigan at Lansing instead of Ann Arbor; in Virginia at Blacksburg instead of Charlottesville; in North Carolina at Raleigh instead of Chapel Hill; in Texas at State College instead of Austin. The bill would thus increase the rivalry which exists in these nineteen states by giving the Land Grant College a club by which to enhance its prestige with the State legislature.

"Another absurdity of the bill appears in giving identical treatment to all states whether they want it or not. This artificial distribution of research centers according to political jurisdiction is on the face of it absurd. The most striking case is that of Washington and Idaho where the two Land Grant Colleges are only eight miles apart, though in different states. This bill would establish two centers, eight miles apart, in that thinly settled region, for research in fundamental science.

"The experience of the Department of Agriculture with State Experiment stations indicates that vastly better returns for research in the fundamental sciences are secured when those researches are directed from the Department of Agriculture and set up as regional studies unlimited by state boundaries. The same policy is followed by the Bureau of Mines, the Forestry Service, the Bureau of Standards and other research agencies. This is the sound process to get your money's worth out of the research funds.

"The proposed bill would violate state autonomy in education by specifying the particular institution in the state that shall be beneficiary. The State Legislature must 'assent to the purpose of said grant' but has no authority as to which institution in the state shall be the beneficiary.

"The authority assigned by this bill to the Secretary of the Interior to withhold from a state its share of the annual appropriation in case the Secretary of the Interior decides that the money has not been properly used is a limitation which will operate to curtail the essential spirit of research."

In consideration of the points made in the foregoing paragraphs, I am quite unwilling to join my colleagues in a favorable report on the bill advocated by Dr. Greene. Accordingly, I am returning the report to you with my name down as dissenting. Since you have not taken occasion to have the Committee consider the points in opposition, I am sending to Drs. Mann, Carlson and Greene a copy of this letter to you.

I shall not be able to be present at the meeting of the Physiological Society in April. If this matter comes up for discussion, I hope that you, as Chairman of the Committee, will be careful to present thoroughly my point of view before the report is finally voted upon by the Society.

With kind regards,

Yours sincerely,
[*Walter Cannon*]

Cannon Papers, Harvard University Medical School

1. The physiologist NICE (1882–[?]) was at Ohio State University.
2. CHARLES WILSON GREENE (1866–[?]) at the University of Missouri.
3. FRANK CHARLES MANN (1887–1962) of the Mayo graduate program.
4. A. J. Carlson.

William J. Robbins[1] to Walter Cannon, April 18, 1932

Dear Dr. Cannon:

Dr. C. W. Greene has recently shown me your letter of March 16 written to Dr. L. B. Nice in which you express your opposition to the plan of Federal aid to research in the fundamental sciences in which Dr. Greene and I have been interested. I am presuming to write this letter to you because I feel that the arguments which you present in support of the position which you have assumed are mistaken ones.

Before discussing the various points which you bring up in your letter, I should like to explain the way in which this idea developed. As you know, through the Hatch, Adams, and Purnell Bills each state receives $90,000 a year for fundamental research in agriculture and home economics. While much of this money has been wasted because of poor direction or lack of a trained personnel, the results are impressive, particularly upon the development of the biological sciences in this country, as much of the first class work in genetics, plant pathology, entomology and related lines has come as the result of the support from these funds. One of the weaknesses in the use of this money has been because some Directors of Experiment Stations have interpreted its use too narrowly with consequent failure, in many cases, to attack fundamental problems.

During the years 1928–30, I was associated with the European Office at Paris of the Rockefeller Foundation and in thinking of the problem of the support of fundamental research in the United States it appeared to me that if the Federal Government were to make appropriations for research in the fundamental sciences such as are now made to agriculture and home economics, the effect would be very notable in the course of twenty years.

The deal I had in mind was the establishment and maintenance in each state of a group of men and women whose primary business would be research in the fundamental sciences. $100,000 available to each state annually for this purpose would maintain a group of fifteen or twenty investigators in each state or a total of between 700 and 1000 individuals in the United States. The subjects with which these investigators would be occupied would be the basic and fundamental subjects such as physiology, chemistry, physics, astronomy, mathematics, botany, zoology, bacteriology, geology, economics, psychology, sociology, etc. Each state would probably not develop all of these equally, but would specialize on some group as circumstances might dictate. The total cost with the complete plan would be about $5,000,000 annually.

The advantages which I saw for this plan were not alone in the productive research which would without doubt result from it, but also in the effect which a group of individuals whose primary purpose is fundamental research would have upon students and their associates in the institutions in which they would be located and upon the people at large. Each group would serve as a center of productive research and the research spirit. Believing as strongly in research and the research attitude of mind as I do, and as I am sure you do, I think that it is of great importance that there should be such centers of research distributed over the United States rather than to have research located in a few large centers. This in brief is the picture which I have in mind.

I note from your letter that you fear that such a plan would result in central control of the educational system with detrimental effects. Aside from the fact that the plan proposed is not concerned with the educational system as a whole, my own opinion is that the possible evil effects of the small amount of control involved are negligible. The freedom which exists under the administration of the Hatch, Adams, and Purnell funds has, from my own personal experience in three state institutions, indicated that the dangers of such control are negligible.

I note further that you feel that research should be supported by state legislatures and that their members should be educated to their responsibilities and to the importance of the relation of research to civilization. It is not so difficult to convince the Texas legislator of the importance of research on the control of cotton diseases, nor the California legislator of the importance of research on the growth of oranges, nor a Missouri legislator of the importance of research on the growth of corn, but we are a long way from being able to convince such bodies of the importance of research in pure mathematics, or in the constitution of the atom or the make-up of the stars. I should emphasize further that research in the fundamental sciences is not of local application, but is of national and international importance. It is, therefore, the business of the Federal government to support such research. I might add that one of the most important methods of educating legislators in the significance of fundamental research would be by the activities of such a group in institutions in their own states.

The response which you received from the American Council on Education, I will consider item by item.

The first objection is without point inasmuch as the writer evidently is under the impression that the Agricultural and Mechanical Colleges alone are mentioned. As we specifically indicated in the preliminary copy of the bill to which you refer, State Universities *and* Colleges of Agriculture and Mechanical Arts are included. In such states as Iowa and Michigan in which the Agricultural College and the University are separated the appropriations would be divided between these two institutions. The division would be made by the state legislature. The effect might be to reduce the rivalry which exists between these institutions in states in which they are separated. It would certainly not intensify it.

The second item is the criticism of giving identical treatment to all states whether they want it or not. May I call attention to the fact that the bill states that the state does not need to accept the money unless it so wishes. I note also that the American Council on Education refers to the distribution of research centers in thinly settled regions. I know of no correlation between the density of population and accomplishment in research. While such a plan as I have proposed might work as well with 40 centers instead of 48, or a distribution on some other basis than state lines, such a proposal would be one extremely difficult to pass through our Congress. Practical considerations demand that the proposal be made on a state basis.

The third item which is mentioned by the American Council on Education is with reference to the experience of the Department of Agriculture with State Experiment Stations in which they argue that the most efficient way to accomplish research is to have a centralized organization in Washington which will control it. In view of your own extreme objections to the effect of centralization, I feel sure that this argument has not impressed you very strongly. Regional studies to which reference is made have little significance in connection with a large part of fundamental research, where regional lines mean little or nothing.

The fourth item, namely that the proposed bill would violate state autonomy in education, is an objection which I can not understand. The bill specifically indicates the state educational institutions and includes not only those now established, but those which may be established in the future. If this does not give to each state the utmost freedom in the disposal of the funds, I can not see what additional freedom could be given.

The fifth item that the essential spirit of research would be curtailed because the Secretary of the Interior has power to withhold the appropriation if it has not been properly used is one to which I have already referred in connection with the Hatch, Adams, and Purnell bills. Experience with these funds has indicated that there is little to fear from such control.

In conclusion may I say that it is difficult for me to understand the active opposition of scientific men to the idea of Federal aid on a state basis to fundamental research. Research in the fundamentals is a national concern, or at least it should be. To my mind it is far better to have this research carried on throughout the nation at institutions where groups of students, teaching faculty members, and investigators are associated together than to isolate it in government bureaus which too frequently are afflicted with dry rot. The state institutions are logical places for such work to go on. The Federal Government spends $12,000,000 on the corn borer, $7,000,000 a year on industrial education, $10,000,000 a year on agricultural extension, billions in an attempt to break the present depression. Why should it not reasonably encourage with $5,000,000 a year the development of research in the several states? What is so absurd in such a plan? Can the American Council on Education at Washington, D.C., or some one else, suggest a more practical or reasonable method of securing the result Dr. Greene and I have in mind?

Yours very sincerely
William J. Robbins
Dean

Cannon Papers, Harvard University

1. WILLIAM JACOB ROBBINS (1890–1979), a botanist. In 1928–1930, he was with the Rockefeller Foundation. At the University of Missouri he became dean of the Graduate School in 1930. Later he was director of the New York Botanical Garden and executive officer of the American Philosophical Society.

Milislav Demerec to Calvin B. Bridges, November 21, 1934

Dear Bridges:

I was very glad to receive your letter describing your idea about the origin of the bands. I agree with you in principle but cannot accept certain details. Double lines are still a puzzle to me. If they originated thru unequal crossing over that would indicate that such crossing-over is a common occurrence and, since those lines are very frequent, that it may take place between any two loci, moreover, that it occurs at random thruout the chromosome. If that assumption is granted, let's do a little figuring to see what should happen. Unequal crossing-over would produce single loci deficiencies in one of the chromosomes and a duplicated locus in the other. Deficiencies would show up as lethals. That they originated thru unequal crossing-over could only be determined thru very special tests and a laborious process. Now let's assume that the X-chromosome has 200 loci and that the frequency of unequal crossing-over for any locus is 1 per 2000 (a justified assumption considering Bar data), then such setting would produce X-chromosome lethals with a rate of 10%, which is not true. Or if it is assumed that the normal frequency of X-chromosome lethals is 0.2% (x-ray controls data) and that the X-chromosome has 200 loci, then to obtain that frequency of lethals thru unequal crossing-over, the rate of that process would be expected to be 1 in 100,000, which is much too low, considering Bar data. It seems, therefore, very improbable that double lines originate thru unequal crossing-over, though I have to admit that this process would readily produce them. I have at present no explanation for their origin. Your hypothesis, however, is susceptible to experimental tests.

Regarding the numbering system for individual lines, it may be advisable to delay the final decision until this subject is discussed by the group interested in the problem. Since you are coming east in a few weeks, that will not be a long delay. As I could gather from a communication received by Whiting, Painter[1] is planning to attend the Pittsburgh meeting. His criticism may be helpful. If the double bands originate thru unequal crossing-over, it would be expected that in some stocks only one would be present and in others two. Two of them, therefore, should have one number only.

I have already analyzed 12 deficiencies, 4 of them affecting a single band only. *Mo* is a heavy capsule to the left of *f* which confirms my genetical results. My data indicate that *fa, spt,* and *Ax* are allels. I shall have final tests before very long.

The map of the chromosome-2, as you have sent it, will go on the cardex panel. As soon as I receive copies of other maps from you, I shall have them all prepared for the exhibits.

The panels for the exhibits are being built. They will be of 3-ply wood with a light molding around. It occurred to me that it may look well if the background is not painted but finished with a light varnish only. That would make the natural wood structure stand out. Would such a background fit your exhibit material? Please send me your answer by air mail.

I am glad to know that you wish to take your panels to Pasadena. I shall pack them separately from ours.

With best wishes,

Sincerely yours,
M. Demerec

Demerec Papers, American Philosophical Society

1. THEOPHILIUS SHICKEL PAINTER (1889–1969) of the University of Texas at Austin did work in cytochemistry and cytogenetics.

Walter Cannon to Simon Flexner, November 21, 1934

Dear Dr. Flexner:

I do not share the fear expressed by Drs. Howell and Dakin[1] that a great Foundation may seriously skew scientific progress by expressing and supporting an interest in some and not in other research projects. The generous contributions of the Rockefeller Foundation to public health, for example, have had, I believe, no demonstrably retarding effect on activities which were not favored—the positive influence was not the occasion for a negative influence elsewhere. In science the genuine investigator is so deeply interested in his own ideas and plans that he is not readily diverted by possible financial support in strange directions; and this would be more and more true as his labors lie distant from the favored fields. Furthermore, a distinction should be drawn between laying emphasis on certain broad ranges of investigation (called "controlling the direction of research") and controlling research. I have as little relish as anybody for the latter. Strictly *ad hoc* investigative activities, such as are seen in governmental and industrial laboratories, are often highly useful but they are likely to be so narrowly conceived as not to have extensive significance. The Rockefeller Foundation is not concerned with such activities nor with planning or controlling research. Through the Division of Natural Sciences the Foundation desires to obtain further knowledge of the factors, mainly biological, which affect human welfare and human conduct, and it is ready to aid investigators whose work is likely to be contributory to that end. The aim seems to me wholly admirable. And the investigators are left free to follow their own "hunches" wherever they may suggest possibilities of progress.

During the discussion at the conference on Friday, November 9, it became clear that the trustees and the administrative officers of the Foundation were hopeful that ultimately the activities of the Division of Natural Sciences would bring forth knowledge important for understanding human behavior, and that the present "sub-fields of interest" (1–9) indicate the extent to which the administration is ready to range in order to obtain that knowledge. From what Mr. Mason said regarding his belief in the prime importance of psychology, and also from the confession of a "major interest" in "experimental psychology, psychobiology or neurophysiology" (see Mr. Weaver's[2] memorandum, page 2) "which bear directly on the general behavior problem and which correlate with the Medical Science Division's program in psychiatry," I infer that the achievement of the ultimate purpose of learning about forces which determine human conduct is much desired.

Although at one point in the discussion regret was expressed that the term "human behavior" had been employed, I must confess that the "ultimate aim" of the officers and trustees of the Foundation deeply stirred my imagination. When we consider that churches, courts and multitudes of social agencies have been trying for an indefinite time to influence the conduct of men and women, without startling success, the possibility that scientific discoveries might prove effective here as they have in other realms (e.g., preventive medicine) seemed to me alluring. The ready correlation of an interest in the biological aspects of conduct with programs in social sciences and in the neglected field of psychiatry was a further attraction.

As appeared in the discussion, two modes of approaching a problem are open. We can seek the hidden thing by beating about the bush generally; or we can look for it first where, according to intelligence based on previous knowledge, we think the chances of finding it are most favorable, and extend the search when that effort shows the need for wider ranging. The argument for the former method is found in the striking and unexpected contributions to human welfare which have come from discoveries in physics (e.g., x-rays), in chemistry (e.g., anesthetics) or in remote realms of biology (e.g., bacteria). That argument, carried out logically, would indicate the importance of supporting *extensively* the basic natural sciences, physics and chemistry, as well as the whole range of biology—on the rare chance that something important might turn up that would affect human behavior. Apparently the Foundation does not intend to spread its support to that degree, for it has stated (Ann. Rep., 1933, p. 197), "A highly selective procedure is necessary if the available funds are not to lose significance through scattering,'' and it has pointed out the ''desirability of increasing emphasis'' on ''the biological sciences, psychology, and those special developments in mathematics, physics and chemistry which are fundamental to biology and psychology'' (Ann. Rep., 1933, p. 198). The principle of limitation is evidently admitted.

Does not the question now arise as to where the limit should be placed? Instances of values appearing from unanticipated sources are highly dramatic and impressive, but set side by side with results obtained by investigators who have ideas and who project them and test them by critical experiments, are they comparable in amount and importance? I venture to suggest that the elaborate fabric of our knowledge of physiology, or bacteriology, for example, has developed much more from carefully considered experiments than from happy chance. And is not happy chance quite as likely to occur in an investigation directed towards a definite goal as in a casual and loosely related study? The line between concentration and distribution of efforts, therefore, I should be inclined to draw at first nearer the center of concentration, and later, if necessary, look for wider support.

If a goal is set up, there is no real reason I can see in not striving for it directly. Human behavior is an almost immediate manifestation of the functioning of the nervous system. I should suggest, therefore, laying a heavier emphasis on the physiology of the nervous system than is obviously prominent in the present program of the Division of the Natural Sciences. There might be included studies on the behavior of animals with experimental lesions of that system, on analysis of the influence of various natural agents (e.g., hormones) on the functioning of the nervous system, on relations in the brain between basal ganglia and cerebral cortex (e.g., in emotional outbursts), on the localization of important functions still obscure (e.g., water and fat metabolism), and on the effects of over-action of parts of the nervous system as shown by special stimulation. There are laboratories in the United States where such studies are being conducted by competent investigators, and where progress could doubtless be accelerated by financial support. I believe also that there are some psychologists who are working carefully with objective methods and whose efforts would justify special consideration. Related to these interests, obviously, would be researches on internal secretions, nutrition (including vitamins), the biology of sex, the physiology of nerve conduction with other aspects of general physiology, and undoubtedly the use of physical and chemical advances as they are disclosed.

As previously noted, such a program would have an intimate bearing on the highly important concern of the Division of Medical Sciences of the Foundation in the urgent and intricate problems of psychiatry.

Good organization would require that the Divisions of Medical and Natural Sciences be coordinated in such manner as to be economically and effectively administered.

Of the four ways of carrying out the program, mentioned in Mr. Weaver's memorandum, the third, "Support of national committees," and the fourth, "Projects," should, I think, be primarily determined by the aim of the Division. If the approach to the problems of human behavior is to continue to be wide-ranging I see no objection to the present distribution of grants. If the approach were more direct I should agree to certain assignments (e.g., internal secretion, nutrition, nerve physiology and the behavioristic aspects of sex phenomena) but I should wish to add others on the nervous system and psychology, as indicated in the previous discussion.

Fellowships and grants-in-aid seem to me incidental to a decision regarding (1) the aim and (2) the projects. Mr. Weaver's statement that questions as to the merits of applications for these two sorts of financial support offer an opportunity for examining new men and new activities struck me as suggestive of good strategy. Continuation of support in these two directions would, in any case, tend towards avoidance of fixed obligations and open the way for recognizing fresh views and enterprises.

I am inclined to accept Dr. Howell's view that the Foundation would be in a stronger position before the public if a group of scientific investigators, in addition to the officers of the Foundation, should engage in the assignment of funds. And if the narrower approach should be regarded as the better way to reach the ultimate purpose of the Division of Natural Sciences, such a group might be especially helpful in the "selective procedure" which has been recognized as necessary. The members of a board of scientific advisers would know the nature of the projected researches, their probable relation to the aim in view, the character of the men who would conduct them, and the reasonableness of the grants made in any case.

There are two possibilities of misunderstanding which I wish to avoid. First, the emphasis which I have laid on the more direct attack on what Mr. Weaver calls the "behavior problem" should not be regarded as indicating confidence in the efficacy of a spear-head thrust. As a biologist I realize the importance of a large strategic plan, and I heartily approve of the extensive interest in experimental biology manifest in Mr. Weaver's memorandum. The suggestions in the foregoing pages are not presented as criticism, but as an emphasis. Second, when an outstanding worker in biological research appears, a man disruptive of established concepts, an opener of new vistas, I should heartily advocate support for him even though his labors might not fit into any scheme.

Money granted to investigators or to university departments seems to me to be much more economically spent than if used to build, equip and maintain research institutes. In university departments, furthermore, young investigators cooperating with the master are trained for research careers and thus the succession of the devotees of science is assured.

Yours sincerely,
Walter B. Cannon

Simon Flexner Papers, American Philosophical Society

1. HENRY DRYSDALE DAKIN (1880–1952), a British-born chemist, was editor of the *Journal of Biological Chemistry* (1916–1931). At this date he was with Merck, the pharmaceutical firm.

2. WARREN WEAVER (1894–1979) was a mathematician, best known as a foundation official. At the Rockefeller Foundation (1932–1959) he was instrumental in the support given just before and after World War II to what is now called "molecular biology" (a term he helped create). See his autobiography, *Scene of Change: A Lifetime in American Science* (New York, 1970).

Walter Cannon to Simon Flexner, November 19, 1938

Dear Dr. Flexner:

My thanks for your kindness and thoughtfulness in sending me a copy of Morgan's "Discourse"[1] would have gone to you sooner if I had not been away in St. Louis. Last night I had time to read the little volume and found myself deeply interested in it. His point of view is remarkably modern. The only feature that is missing is the emphasis on research as an inspiration to teachers, professors and students and as a means of medical progress. I liked especially his tribute to physiology as giving "the clearest light in the cure of diseases," and naturally enough I was pleased by his testimony that "the study of it is most entertaining."

We are in the midst of some rather exciting work just now—evidence of adrenaline in all parts of sympathetic neurones—the significance of which does not yet appear. It would be good fun to have Morgan drop in to see a modern medical school and an active physiological laboratory.

Again with my best thanks and my best wishes,

Yours faithfully,

W. B. Cannon

Simon Flexner Papers, American Philosophical Society

1. A reference, one suspects, to Morgan's Nobel Prize lecture, given June 4, 1934, "The Relation of Genetics to Physiology and Medicine," in *Scientific Monthly* 41 (1935): 5–8. The references to medicine were rather modest.

12 Physical Science between the World Wars: The Abstract and the Reality

The more abstract the topic, the more difficult it is to obtain funding. That is the way it has always been in the sciences and the way it is now. To researchers in the physical sciences, this was especially evident in the years between the two World Wars. Exciting developments in their specialties accentuated gaps between needs and resources. From field to field, even from subspecialty to subspecialty, support was unevenly distributed. As the growth points often were the more abstract topics, obtaining support appeared particularly difficult.

The chemists, as a group, seemed to be best off with their traditional easy accommodation with industry. Yet, when the National Research Council Fellowships were established in 1919, chemistry and physics shared equally on the assumption that they were equally essential. Long before this date, chemistry had acquired many of the characteristics of physics, including even a number of common interests. Some physicists, as heirs of natural philosophy, might look down on chemistry as a kind of superior, empirical kitchen-type operation, but by the twentieth century outsiders had difficulty distinguishing chemists with theoretical leanings from similarly inclined physicists.

Irving Langmuir and G. N. Lewis, for example, although clearly "chemical" in intellectual interests and profession, published, corresponded, and conferred with physicists. As a theorist in a university, Lewis fell into a familiar pattern. His active mind encompassed all kinds of topics but with a particular penchant for those with philosophical or cosmological aspects. Although not so obviously philosophical or cosmological in his concerns, Irving Langmuir continued to seek generalizations in a mathematical form, even as he kept collecting kudos for his many patents. Richard C. Tolman,[1] also trained originally as a chemist, became a notable theorist with a taste for abstruse issues. After serving with the federal program to develop synthetic nitrates in World War I, he joined Hale and Millikan at the California Institute of Technology. Industry and higher education provided stimulating environments for these three men.

Astronomy somehow managed to flourish. Earlier the successes of Newton and the vogue of natural theology had given its practitioners a favored place. Observatories were the first big science installations and remained the largest until the coming of the giant facilities of high energy physics.

Most visible was Hale at Mt. Wilson; in the 1920s he started work on the great telescope at Mt. Palomar. Although there was only a small community of scientists in astronomy compared to the numbers in fields like physics and chemistry, by this time astronomy in the United States was both reasonably supported and intellectually creative.

Physics represented a more complex problem. Ever since Newton, scientists at various times and places have responded with great exuberance and exhilaration to their perceptions of the great pace of scientific advance. As far back as the seventeenth century scientists felt the pace of advance too great for comprehension; the literature was expanding too fast for intellectual control. Because rapid and radical change appeared normal, continuity and routine were glossed over. Almost every generation of scientists assumed that they or their immediate predecessors had participated in something momentous, perhaps even a great intellectual revolution. Although often an exaggeration, there were such great moments of change.

The feel of excitement, of a true revolution, certainly pervaded physical scientists in the early decades of this century. "Classical" physics, going back to the great Newton, was being supplanted with "modern" physics. A complex process with roots in the work of the last century, this ranked as a momentous event in man's understanding of the material world. Of all the figures involved, none captured the imagination more than Einstein, even among the general public who rarely understood the intricacies of the issues nor how Einstein kept aloof from certain later developments in nuclear physics. The spectacular testing of his theory in an eclipse expedition at the end of World War I confirmed Einstein's greatness in the opinion of the public. His later immigration to the United States aroused great public interest and satisfaction. Physics, in the person of the great physicist of the age, had found a new home. Now more than ever this field represented the model for all the sciences.

Looking across the ocean at the intellectual excitement in European institutions, the physics community in the United States successfully campaigned for more support in the first four decades of this century. The arguments offered came down ultimately to two points: the great intrinsic merit of physics, and its obvious potential for utility. The points were strongly held and carefully offered in tandem. Presenting one without the other was frowned upon as inimical to both historical fact and to sound principle. While few might cavil at the intrinsic merits of physics, the notion of utility presented awkward problems. As Einstein's work had not progressed to the point of yielding applications, the shade of Michael Faraday was summoned to stand in testimony. The familiar wonders of electricity and electronics were confidently traced back to his theoretical labors at the Royal Institution. As a historical oversimplification it neatly matched the opposing oversimplification stressing the untutored inventors and their inspired empiricism.

Nationalism played a role in the moves for support of all the physical sciences. Patriotism demanded an end to patronizing by Europeans as well as an American contribution to knowledge proportionate to the country's unquestioned greatness. The arguments for support tacitly paralleled the rhetoric of local boosterism. Catching up to and passing rivals was an easy idea to convey to a sympathetic lay audience.

The true status—the reality beyond the rhetoric—is defined only awkwardly. If discussed on a large scale, such as appraising all of physics (let alone all of science), inevitably one generates a significant element of imprecision, qualitative and quantitative. If done minutely—in terms of a narrow subspecialty—a spurious precision results. Above all, the rhetoric helps to mold the reality as well as to limit perceptions. An outside observer sometimes sees the familiar in a somewhat different framework. Such appraisals of physics in the United States occurs in the September 19, 1925, letter of Manne Siegbahn, a Swedish Nobel Laureate in physics.

Over all, Siegbahn is quite impressed by "the many beautiful results" coming from investigators in the United States. Although no comparison is made with the great scientific powers—only with Sweden—Siegbahn implicitly places physics in the United States in the same class with the research being done in Germany and Great Britain. Two factors catch his eye—one favorably, the other negatively.

Although no specific laboratories are mentioned, Siegbahn is obviously impressed by the great industrial laboratories. In the General Electric and Bell Telephone laboratories were two future Nobel Laureates, Langmuir and Davisson, as well as other notable investigators. The period after World War I saw a notable spurt in industrial research. Yet, Siegbahn's words are perhaps overly eulogistic. Many of the larger industrial concerns lacked laboratories at all or had programs of routine scope. But the leading laboratories in the electrical and chemical industries, like Bell and General Electric, were impressively different from their European counterparts in one significant respect: they combined advanced theoretical research with highly successful developments of practical devices and processes. Unlike Siegbahn's Sweden, there was little stratification of an intellectual and social kind between the scientists of the university and the industrial researchers. The 1926 and 1927 correspondence of Hall, Davisson, and Arnold conveys some idea of the atmosphere of the time.

Unlike the American scientists with their propensity to downgrade their efforts, Siegbahn is also highly appreciative of physics in the universities. Very few American contemporaries would claim that twenty different universities in the United States had "some of the most brilliant men of progressive science." Perhaps Siegbahn was merely being a polite guest. But that positive assertion is made to underscore his negative finding: the universities burden these brilliant men with "lecturing beginners in the elementary things." Siegbahn is echoing the points made by Woodward and his allies in CIW and by such men as Simon Flexner. Implicit in such arguments is an assumption that the leading scientists required deferential treatment separating them as far as possible from the mundane concerns of the world. To some extent, that was actually occurring. Even though the research professorships called for by T. W. Richards were rare, the leading universities and those with ambitions to lead were moving in that direction insofar as resources permitted.

During the interwar years a great upsurge in the physical sciences took place, notably in physics. The National Research Council Fellowships provided support for postgraduate work. Besides the growth in universities and in industry, governmental laboratories expanded their scopes. Aware of the situation Siegbahn complained about, the leaders of the National Academy launched a drive to form a National Research Fund, its primary function to free scientists from routine obligations. Relying on contributions from industry and private philanthropy, the fund expired during the Great Depression. Leading this effort, which was not confined only to physics, were Hale and Millikan. Although opposed by some individuals (as in the blunt 1920 exchange of Hale and Cattell), their policies went largely unquestioned.

One possible source of opposition was largely sidestepped, the engineering community which was a constant source of apprehension. Despite some grumblings, the leaders of the National Research Council managed to effect an uneasy alliance. In Pasadena, Hale and Millikan were outstandingly successful in creating an institution of higher learning on a different pattern than the Massachusetts Institute of Technology. Although funded in large part as an instrument of regional development, Cal. Tech. was science and research oriented, downplaying the routine elements of engineering practice. Implicit in this educational policy was a different social order in which theory and its possessors received deference even from the eminently practical.

In economic terms, physics was a growth industry from the turn of the century, benefiting from both the prestige of its theories and from the assumption of utility. Recent research shows that physicists did very well during the depression years compared to the general population and to other disciplines. After the atom bomb, many of them looked back nostalgically to a simpler age of heroic research untrammeled by governmental restrictions and by knowledge of catastrophic hazards. It is a great myth, derived at several removes from such romantic images as great artists starving in garrets. Romantic images of this sort confused changes in scale with changes in kind.

The mathematicians had a different problem. Theirs was not a mass community, nor was there much appreciation of the utilitarian potential of their talents. As the most abstract of all fields, mathematics was fundamental for all the sciences, at least in the view of its practitioners. Mathematicians like Norbert Weiner and John Von Neumann contributed to the developments in theoretical physics. But this hardly matched the growing utilization of physics and chemistry by industry and government. Nevertheless, mathematics in the United States had developed considerably by the 1920s. It had a certain high-culture prestige, like astronomy, which guaranteed a reasonable degree of support.

But modest support and success only stimulated strongly motivated men like Oswald Veblen to greater efforts. He took a leading role in fostering mathematics in the United States between the two wars. In the first half of the period, Veblen developed the department at Princeton; in the second half, he built the School of Mathematics at the newly established Institute for Advanced Study (see Chap. 13). The 1924 letter to Kellogg is an example of his fund-raising activities. Veblen here repeats in a different context a number of the arguments bandied about in the early years of the Carnegie Institution of Washington. In the brief period of hope for a National Research Fund, Veblen and colleagues like Birkhoff opted for research professorships. Veblen, writing to Kellogg of the National Research Council in 1926, gives a capsule history of mathematics in the United States in the previous 50 years. It is an account composed with an eye to the possibility of a National Research Fund.

The Norbert Wiener letters return us to the ebullient former prodigy, now launched upon a career in mathematics, probably the greatest of that period in this country. As a personality and in his research, Wiener was an original. Yet, in one respect, the course of his career gave him viewpoints at variance with many of his peers, albeit part of an older, ongoing tradition in mathematics. Wiener was bitterly disappointed in not getting an appointment at Harvard, his alma mater (presumably because of anti-Semitism). Instead, he went to the Massachusetts Institute of Technology where he became a most enthusiastic proponent of that organization. His August 3, 1933, letter to Paul de Kruif is an eloquent statement of the role of mathematics in applied work based upon his experiences there. Wiener's views are in contrast with those offered by mathematicians here and abroad stressing the abstract nature of the field. In his own work, Norbert Wiener was as abstract as any of his contemporaries. The outlook he propounded was an older one, shared by such mathematical greats as FELIX KLEIN (1849–1925), sometimes expressed in assertions that physics was too important to leave in the hands of physicists. When applied to engineering, Wiener and those who shared his ideas were proclaiming that scholars could, should, and had to help in the world's work. The world was too complex, even for talented empiricists, and required the rigorous ordering only mathematics could provide.

As Wiener wrote, Ernest Orlando Lawrence was already launched into his notable and influential career in high-energy physics. In 1933 the machines involved were still modest, not yet in the class of the great observatories. By the close of the decade, Lawrence had reached Hale's level. Basic research was the origin and ultimate justification of the expenditures, but Lawrence benefited from the perception of possible medical benefits. The big machines required theoreticians as well as skilled experimenters. Before the war took him away to other tasks, J. Robert Oppenheimer divided his time between the University at Berkeley and Cal. Tech. in Pasadena. In a sense unintended by Wiener, Oppenheimer and other mathematical physicists would exemplify his view of the relation between theory and practice.

1. TOLMAN (1881–1948) was then at the University of California, Berkeley. After World War I, at California Institute of Technology. Tolman was a theoretical physical scientist with a taste for cosmology and philosophic issues.

Irving Langmuir to Gilbert N. Lewis, April 22, 1919

My dear Lewis:

Until a couple of weeks ago I was under the impression that you were still in France, and therefore have not written you in regard to the theory that I have been working on recently.

I am sending you under separate cover the manuscript of a paper which I have sent to the Jr. of the American Chemical Society, and which is to be published in the June number.[1]

You will probably be interested in the history of the development of these ideas. When I read your paper on the "Atom and Molecule" in 1916, I was immediately struck by the very fundamental nature of the ideas you presented, and of their splendid agreement with the general facts of chemistry, so I very soon began to look upon all chemical phenomena from the viewpoint that you presented. It seemed to me that it accounted particularly well with so-called physical characteristics, such as boiling-points, freezing-points, etc. In talking to several other people about the theory, and explaining to them how it would apply to properties of the elements in the two short periods, I gradually began to extend the theory somewhat, especially in the direction of coming to a realization that the tendency to form groups of 8 or 2 was nearly without exception, the cause of the formation of compounds, at least in the case of the first 20 elements. While on war work at Nahant I met Sir Ernest Rutherford, and later, Sir Richard Paget,[2] and I told both of them in detail the importance of your theory, in showing them how it could be applied to the prediction and understanding of the properties of substances.

Early in January of this year Dr. Dushman,[3] of this laboratory, asked me to give a talk at our Colloquium on the subject of "Adsorption." I told him that I thought a very much more interesting subject would be Lewis' theory of the Atom and Molecule, and, accordingly, I read your paper again carefully, and began to study how I could present the matter in a way that it would arouse the most interest. In doing so I was impressed more than ever by the general applicability of the theory, and was surprised that chemists in general seem to have paid so little attention to your ideas.

My interest was then so thoroughly aroused that I spent nearly all of my time in the development of these ideas for about five or six weeks in January and February. I was especially interested in extending the theory to cover all the elements, and to broaden out the theory of valence so as to cover all types of compounds. I think you will be interested to see how I was able to accomplish this. It seems to me there is no field of chemistry where these ideas are not going to bring about radical changes in present conceptions.

Since writing the enclosed paper I have thought over the application of the theory in the field of organic chemistry, and have found, as is really to be expected, that the theory explains, as far as I can see, all of the facts of structural chemistry in a thoroughly satisfactory manner. The stereoisomers of carbon, nitrogen, sulphur and phosphorous, all fit in much better with the new theory than with any theory previously proposed; also the oxonium compounds are fully in accordance with the theory. After having spent weeks in going over the literature I have only found one set of compounds that really puzzles me, and that is, the various hydrides of boron, in which the boron seems to act as though it were quadrivalent. I refer particularly to the compounds B_2H_6, B_6H_{12}, and B_4H_{10}. The only reasonable explanation that I can think of is that two boron atoms form an inner compound with two hydrogen atoms, leaving a pseudo atom with a single octet (like that which I have found for the gases of N_2 and CO_2). This would account for the absence of such compounds as BH_4, B_3H_8, etc. On the other hand, I do not see on this basis how to explain the compound KBH_3O, even if we double this molecule in order to get an even number of electrons. It seems to me that experimental work on the physical and chemical properties of these compounds is needed to give us a basis for the proper theory. Of course, according to the octet theory, it is obvious that no gaseous hydrides of boron should exist.

It seems to me that the theory of the atom and the molecule which we have developed should be capable of throwing a great deal of light on quantitative chemical relationships in such fields as molecular volumes, heat, and free energy, of reactions, dissociation—constants, etc. You will notice that in the case of compounds like N_2 and CO, and again in N_2O and CO_2, quantitative relations are actually predicted by this theory. In the second paper that I am publishing I will show that hydronitric acid and cyanic acid have ions which have similar structures to the molecules of CO_2 and N_2O. As far as I have been able to find the physical properties (solubilities, etc.) of azides and cyanates are the same.

The next time you come East I hope very much that you will stop at the laboratory, for I think we could very profitably spend a day or so talking over the further extension of this work.

Yours very truly,
Irving Langmuir

Lewis Papers, University of California, Berkeley

1. "The Arrangement of Electrons in Atoms and Molecules," *Journal of the American Chemical Society* 41 (1919): 869–934.

2. SIR RICHARD ARTHUR SURTEES PAGET (1869–1955) was at this date technical advisor to the Admiralty.

3. The physical chemist SAUL DUSHMAN (1883–1954) was born in Russia and came to the United States in 1891. A Toronto Ph.D. (1912), he spent his professional career at General Electric's Research Laboratory in Schenectady, N.Y.

Ernest Rutherford to Irving Langmuir, June 10, 1919 (marked "Private")

Dear Langmuir

I have been rather long in answering your letter but you know I am going to Cambridge & I have been kept very busy getting ready for my transfer in both places. In addition, some of my staff were away & I had to buckle into routine harder than usual. The whole military & naval scientific development is held up for the time until apparently Lloyd George returns to settle it. We have hopes that something definite will mature but progress is very slow and in the meantime many of the best men are pulling out of the services to take civilian jobs. Like you, I want to forget the time we have more or less wasted on the warside but I am glad to hear that my work in U.S.A. may have been of some service.

I was hoping to receive your "magnum opus" on atom building but I suppose it will take some time to publish. It sounds very interesting to be able to explain so many things on simple postulates which in themselves look reasonable enough although the theoretical basis is for the moment lacking. I expect you know I am a strong adherent of the law of inverse squares in the atoms for both + & − charges & I feel rather unhappy when new laws of force have to be introduced by the scruff of the neck. The evidence of scattering of α particles seems to me strong support of the inverse square law, while I feel that most of Bohr's work goes by the board if the law for the electron is much more complex for the distances involved in light vibrations. Bohr was over for a day or two recently en route to Holland & told me of the great progress he & others have made in explaining *in detail* the Stark-Zeeman effects. He was of opinion if his work had any substantial basis of truth he had proved the law of inverse squares to one in a million. You will have seen J J T's paper in the *Phil Mag* with its bizarre laws of force to give reality to Bohr's ideas.[1] It doesn't look very hopeful to me. I am of course prepared to believe the electron is something more than a sphere of electricity & in fact I don't see how radiation can be explained unless it has some fairly definite vibratory structure but I have never taken very kindly to the magneton idea & hope it won't have to be invoked to fit in with your ideas. There may be a way of reconciling these apparently diverging points of view. I am a great believer in the simplicity of things and as you probably know I am inclined to hang on to broad & simple ideas like grim death until the evidence is too strong for my tenacity.

So much for my vague ideas—currents Culmus. I have unburdened myself in some papers in the *Phil Mag*—stuff I have had on my chest the last four years. I don't know how things will eventually turn out but the nucleus of an atom attracts instead of repels me—possibly because my character is more $-^{ive}$ than $+$?

I go to Cambridge early in July to try & get my hands in the machinery of the Cavendish. In Cambridge as in every University of this country, students are assembling in shoals & all scientific depts are over-crowded. There have been 200 Naval Officers taking Physics in Cambridge which put the lid on the congestion. There are a good few of your soldiers in Cambridge—I hope enjoying the most wonderful May we have had for a long time.

I have the pleasantest remembrance of my stay & talks in Schenectady. Remember one to Hull, Coolidge, & your scientific chief.

Yours sincerely
E. Rutherford

Langmuir Papers, Library of Congress

1. "On the Origin of Spectra and Planck's Law," *Philosophical Magazine*, 6th ser. 37 (1919): 419–446.

Gilbert N. Lewis to Irving Langmuir, July 9, 1919

My dear Langmuir:

I am very glad indeed to have the complete set of your papers which you were good enough to send to me. You may be interested to know that in one of our recent seminars a considerable part of the time was devoted to a discussion of your paper on Surfaces.[1]

I shall be very much interested in seeing your two new papers on Structure, and I shall not publish anything further in this line until I have seen them. Apparently you have found, as I did, that for the present the easiest progress can be made in the study of elements which I chose for consideration, being not only simpler in structure, but on account of the large number of compounds and the extensive study to which they have been subjected, giving also a much larger body of experimental [data].

It has been extremely gratifying to find that after the extended study to which you have given the matter you have found no one of the numerous and rather revolutionary conclusions of my paper which you have wished to amend. You have been remarkably successfull in applying this theory to a large number of concrete cases, and I do not know anyone who could have done it so well; but to be perfectly candid I think there is a chance that the casual reader may make a mistake which I am sure you would be the last to encourage. He might think that you were proposing a theory which in some essential respects differed from my own, or one which was based upon some vague suggestions of mine which had not been carefully thought out. While I realize what a short distance we have gone towards explaining chemical phenomena, it seems to me that the views which I presented were about as definite and concrete as was possible considering the condensed form of publication. I think if any confusion should arise it would be due perhaps to points of nomenclature. For example, while I speak of a group of eight, you speak of an octet. I think, as a matter of fact, your expression is preferable, and I shall be glad to adopt it, but I should be sorry to see the whole theory known as the octet theory, partly because it raises questions of the sort I have just mentioned, but especially because the octet is no more fundamental to the theory than the electron pair which constitutes the chemical bond. Many years ago, when I first began working on this subject, and before electrons were much known, it was the change of valence by steps of two which seemed to me about the most striking phenomenon which had to be explained by a theory of valence. It was for this reason that I laid particular stress upon the fact that so few compounds are known possessing odd molecules or odd atoms. These are terms which I believe you have not adopted. Did you think of anything better? It is of course important in a new development of this kind that the nomenclature should be as expressive and as simple as possible. Sometimes parents show singular infelicity in naming their own children, but on the whole they seem to enjoy having the privilege.

I trust that you will not misunderstand what I have said, or think for a moment that I am not delighted to see you working in this extremely interesting field. I shall look forward with great pleasure to seeing you in Berkeley, and hope that you will be able to spend more than the one day with us. Do let me know just when you can come and give us as much of your time as possible. We have a number of interesting things to show you, and I should like very much to have our students hear from you.

With best regards, I am,

Yours very sincerely,
[*Gilbert N. Lewis*]

Lewis Papers, University of California, Berkeley

1. It is not clear which of several papers is referred to here.

Irving Langmuir to Arnold Sommerfeld, March 16, 1920

Dear Sir:

After much effort, I have succeeded in borrowing a copy of your recent book on Atomic Structure,[1] and have been reading it for the last few days with very great interest. The results that you have accomplished in accounting for the spectra of elements has been very remarkable.

I notice, however, that you are not familiar with the work of G. N. Lewis on Atomic Structure, which was published in the Journal of American Chemical Society, Volume 38, page 762, 1916.

It would seem to me from the chemical point of view that Lewis' theory marks a very great advance, and is the only theory which is in good accord with chemical facts.

I have been developing this theory of Atomic Structure, and extending it to the heavier atoms and applying it, particularly to chemical valence. That the structure of atoms which we have thus arrived at corresponds to the true structure is indicated by the success that this theory has shown in predicting new relationships in the field of chemistry.

Under separate cover, I am sending you reprints of several articles that I have published in this field during the last year. I am also enclosing herewith the manuscript of an abstract of a more popular lecture that I gave a few weeks ago which may be useful to you as an introduction to the viewpoint that we have developed in this country. This abstract will be published in the April number of the Journal of Industrial and Engineering Chemistry.[2]

In Mr. Lewis' paper on this subject, he assumed that the electrons were stationary in position in the atoms, but if you look thru Mr. Lewis' work, you will find that this assumption was not at all necessary in the development of his theory. As far as the chemical facts are concerned, it is entirely immaterial whether the electrons are stationary or whether they rotate in orbits about certain positions in the atoms which are distributed in space. I have not, perhaps, been sufficiently careful in my own papers to bring out this point, but, if in reading over my work, you will consider that when I have spoken of the "position of an electron," or have shown a diagram, indicating these positions, I mean merely the centers of the orbits of the electrons.

On this viewpoint you will notice that the structure for the hydrogen atom and helium ion, which you and Bohr have developed, is in no way inconsistent with the theory which I have been working on.

In the case of the neon atom, the eight electrons in the outside shell, according to Lewis' and my theories, are arranged in exactly the same way as that found independently by Born and Landé.[3]

It seems to me therefore that you will find that this chemical theory of Atomic Structure will be of use to you in the further development of the theory of spectral lines. The chemical evidence has the advantage that it gives a very definite geometrical picture of the atom. It is, however, qualitative in nature and gives us little or no information in regard to the forces in the atoms. This is the field in which your theory offers particular promise.

I think you will be especially interested in the structure for the nitrogen molecule, as indicated by Fig. 12 on page 903 of the reprint from the June number of the Journal of American Chemical Society. According to this theory, the nitrogen molecule and the carbon monoxide molecule, and the cyanogen ion have similar arrangements of electrons. The striking similarity in the physical properties of nitrogen and carbon monoxide furnish very strong evidence in support of this viewpoint, and the unusual properties of these substances, from a chemical viewpoint, are very satisfactorily explained by the unusual type of structure. You will be interested to know that in a letter from G. N. Lewis, I have learned that he had independently arrived at this identical structure for the nitrogen molecule, but had never published his work. This theory thus suggests that the optical properties of nitrogen gas and carbon monoxide should show extraordinary resemblances. For example, I should expect that the ultra violet absorption bands of both gases should be very closely related. You will notice that the theory also indicates a similarity in structure in the case of carbon dioxide and the nitrous oxide, notwithstanding the fact that the ordinary viewpoint of chemists in the past has indicated very different structures. The agreement in the physical properties, however, is sufficient to establish the correctness of the viewpoint of the new theory.

I believe that the infrared absorption bands of carbon dioxide are nearly exactly the same as those of nitrous oxide. It seems to me that a detailed study of the differences between the absorption spectra and also emission band spectra of these pairs of gases should throw a great deal of light upon molecular structure.

You will notice that one of the outstanding features of our theory is that *pairs of electrons and groups of eight electrons* (called octets) form particularly stable configurations, at least, when these groups are bound by the presence of positive nuclei. It seems to me that this is the most fundamental difficulty at present with Bohr's theory, that it does not explain the remarkable stability of the pair of electrons, nor of the manner in which a pair of electrons function as a "covalence bond" in the union between two atoms.

I notice that in your book you admit frankly that the properties of hydrogen and helium indicate a greater stability for a pair of electrons in the atom than is indicated by Bohr's theory. I feel inclined to believe that these stable groups of electrons will involve some new, but very simple assumption, which is not contained at present in Bohr's theory.

One thing which strikes me forcibly is that according to the chemical evidence, the general arrangement of electrons in atoms follow very simple laws, much more simple than would be expected, if we were really dealing with n-body problems. In the case of an atom having n electrons, your "Ellipsenvereine" and Born and Landé's conception of the coupling between eight electrons, arranged with cubic symmetry, are steps in the right direction, but I believe that still greater simplification will be reached by means of some new, and perhaps more radical assumptions than any that have yet been made.

I am sending you, under separate cover, several copies of the more important reprints, in order that you may give some of these to others in Germany who are interested in this subject, as there seems to have been great difficulty in obtaining these papers in Germany. I would particularly ask you to send a set of the reprints to Prof. Max Born. I do not know his address and therefore cannot send them to him directly.

I hope sincerely that the unsettled political conditions in Germany will not interfere with the progress of your most important work on the Structure of Atoms.

Yours very truly,
[*Irving Langmuir*]

Langmuir Papers, Library of Congress

1. *Atombau und Spektrallenian* had appeared in 1919. ARNOLD SOMMERFELD (1868–1951) was a mathematical physicist at the University of Munich.

2. "The Structure of Atoms and Its Bearing on Chemical Valence," *Journal of Industrial and Engineering Chemistry* 12 (1920): 386–389.

3. MAX BORN (1882–1970), a Nobel Laureate in physics, left Göttingen after Hitler came to power and taught in Edinburgh. The spectroscopist ALFRED LANDÉ (1888–1975) went to teach at Ohio State.

James McKean Cattell to George Ellery Hale, May 29, 1920

Dear Dr. Hale:

I have decided not to print, at all events not at present, the paper that I sent you. The censor by whose decision I decided to abide approves its objects and point of view, but holds that its publication might injure the journals that I edit and that the references to you and some of the rest might be misunderstood. I am thus in the situation to which I particularly object of sacrificing principle to expediency. However, I have myself some misgivings concerning the tone that I adopted and as to whether it would advance the objects that I have in view.

I should in any case wish to be strictly accurate, and apart from publication, should like to be informed of any statement that is otherwise. The one to which you refer must of course be read in its context. I have been fully informed of the circumstances both by Dr. Angell and by the committee of the association. The question in that case is not a matter of fact, but a matter of opinion.

We must recognize the circumstance that the judgment of a scientific man outside the field in which he works is of no more validity than that of any one else. You, for example, believe that our entry into the war was in the interest of civilization; I believe that it was adverse to it. You believe that aristocracy and patronage are favorable to science; I believe that they must be discarded for the cruder but more vigorous ways of democracy. Each of us must follow what he believes to be the better way; very probably our agreement as to ends is greater than our difference as to methods.

Very truly yours,
J. McK. Cattell
Hale Papers, California Institute of Technology

George Ellery Hale to James McKean Cattell, June 3, 1920

Dear Professor Cattell:

Many thanks for your letter of May 29. There were several statements in your MS that did not seem to me to conform with the facts, and others that would certainly lead uninformed readers to a wrong interpretation of them. It is clear, however, from your attitude regarding the case in which Dr. Angell was involved that a discussion of all these points, whether private or public, would lead to no useful result. Least of all would it advance science; in my judgment, if it had any effect, it would be in the opposite direction.

The last paragraph of your letter clearly indicates the futility of such a discussion. To state that I believe that "aristocracy and patronage are favorable to science" while you hold that "they must be discarded for the cruder but more vigorous ways of democracy" is wholly misleading. As a matter of fact, I do not believe in aristocracy and patronage as opposed to democracy, but merely believe in adopting what appears to me to be the most promising means of advancing science and research under existing conditions. If I am not mistaken, you would not object to an appropriation by the Rockefeller Foundation toward the support of work in which you are interested.

As for the entrance of the United States into the war, it is rather difficult to see why your democratic beliefs should lead you to oppose a movement that has done more than any other to unseat "patrons and aristocrats" and to advance the interests of government by the people.

Very sincerely yours,
[*George Ellery Hale*]

Hale Papers, California Institute of Technology

Robert A. Millikan to George Ellery Hale, July 28, 1920

My dear Hale:

There are a great many things which I want to talk over with you, but with 55 students whom I am trying to look after this summer in Quantum Theory, it is hard to find time to do it. There are two matters, however, which ought not to wait longer.

I received your letter and your telegram authorizing me to get Mason,[1] if I could, for the Institute. I am going to have a day with him at Madison within two weeks, and I have postponed saying anything to him by letter until that time. I am not quite certain that it will be wise to approach him even then, though if you think that it is imperative that we get a new mathematical physicist at the Institute beginning in October, there is very little time to delay. The fact is that although I am tremendously attached to Mason, I am a trifle in doubt about his permanent productivity, and I am wondering if an alternative like that which you suggest might not be preferable. It is clear, however, that we could not possibly get anybody from Europe before September. I have been looking over the young European physicists for some time with this very thing in mind, and have been contemplating Debye, who is a Hollander, although working in Goettingen. It might not be possible to detach him, but he is certainly one of the most brilliant of the younger men. I have got to go to Brussels next spring for the Third International Solvay Congress. I have just received a letter from H. A. Lorentz[2] asking me to attend. You may be interested in the program which I enclose. If we are going to get a young man from England or the Continent, it might perhaps be worth while to postpone the choice until next spring when I could confer directly with possible candidates in Europe. We shall certainly not be able to get into our building until a year from this fall, and that is the time at which we may expect full steam to be up at the Institute. If you feel sure enough of Mason to want to have the Institute definitely committed to him, I will present the matter in as persuasive a way as possible two weeks from this week-end, when I shall see him. If you think it is best to do this, let me know before that time. I am delighted beyond measure that Mr. Fleming has been willing to take this step. It seems to me to be a very important one; but we don't want to get the wrong man. I don't think that E. W. Brown would do at all for what we want.

Now, with respect to the new building. I sent on a week or two ago final instructions to Mr. Goodhue, and I hope that we can get the working drawings by the first of September. I am immensely pleased now with both the exterior and the interior of the building. Watson has certainly done his part of the work well so far.

As to the seismograph room, I had thought of that from the beginning, but had concluded that it would probably be better to do that sort of work there in one of the pits arranged for Michelson's experiment, or at least in some detached room where we would be away from possible vibrations due to the machinery and the movement of students on floors of the building. I talked over with Wood[3] the general matter of installing a seismograph at Pasadena, but I did not take up this particular point. If he is still there, his advice on this matter would be of great importance to us. If it is desirable to have the work inside the building, I think we could simply extend without any serious difficulty the room which is to be used for ventilating machinery and adjoin it by a seismograph room without much additional expense. I have from the start been uncertain about this on account of the possible disturbance of the machinery. I am writing to Watson[4] to ask him to make inquiries of Wood regarding it, if the latter is still in Pasadena. If not, there may be others there who have had experience in seismological work, and with whom he could confer. This change could, I think, be made at the last moment, and need not delay the preparation of the final plans. The matter of wiring, however, plumbing, etc., is of so much importance that I suspect Mr. Goodhue will want to have either Watson or myself in his office when the details of this work are being arranged for. Possibly we can get Watson to come on. He could give more time to it than I could, and would do it better. I will make a recommendation regarding this as soon as I hear from Mr. Goodhue.

As you probably know by this time, Edward S. Slosson spent last Sunday here, and we talked over at some length the Science News Service.[5] I hope you had similar opportunity to go over it with him. He is going to be a great addition to the group I think, and I think he ought to be added to the Board of Trustees, or whatever it is ultimately called, whether he is willing to accept the editorship or not. I feel confident that he will not be willing to give up his position with the Independent.

You have asked me to send on any material, typewritten or otherwise, which I have regarding the proposed institute of physics and chemistry at Chicago. I am accordingly enclosing two brief reports, one of which Stieglitz[6] wrote, and the other which I sent to Mr. Judson at his request. May I ask you to return these when you are through with them, as I have no extra copies.

As to the big question of the best place for my permanent work, I did not intend in my last letter to place a final negative upon the Institute's proposals, but I wished to let you know of the changes which are in contemplation here (Professor Michelson thinks most of them are for the purpose of obscuring vision by filling the air with dust), and to make it clear that I didn't wish to be placed in a position of entire commitment to change. The situation here is still somewhat uncertain, and I may feel quite differently by next winter. I am convinced, however, that if I am going to make my chief contribution in scientific scholarship and represent in a worthy way American science, I must devote my time almost fully to a scientific rather than to an administrative job. The work which seems to be opening up now in connection with our experiments on very short waves and its bearing on the structure of the atom (I have an article in the forthcoming number of the Astro-Physical on this subject)[7] is exceedingly interesting, but I cannot make my contribution to it as worthy as it ought to be unless I can have my time largely free to keep minutely in touch with all such work as is to be represented in the forthcoming Solvay Congress.

Cordially yours,
R. A. Millikan

Hale Papers, California Institute of Technology

1. Max Mason.

2. LORENTZ (1853–1928) was at Leiden. He was a Nobel Laureate in physics in 1902.

3. Perhaps HARRY O. WOOD (1879–1958), the first seismologist at Cal. Tech.

4. The physicist ERNEST CHARLES WATSON (1892–) was at Cal. Tech. 1919–1960, dean of faculty 1945–1960.

5. SLOSSON (1865–1929), originally a chemist, headed Science News Service, an organization sponsored by many leaders of the scientific community as the vehicle for public dissemination of reliable news of scientific developments. Tobey's *The American Ideology of National Science, 1919–1930* (Pittsburgh, 1971) discusses the man and his organization.

6. JULIUS STIEGLITZ (1867–1937) was a chemist at the University of Chicago.

7. (With Willard R. Pyle), "The Extension of the Ultra-Violet Spectrum," *Astrophysical Journal* 52 (1920): 47–64.

Irving Langmuir to William David Coolidge, October 11, 1921

Dear Coolidge,

I have just spent four exceptionally interesting days in Cambridge mostly with Rutherford. Perhaps the larger part of the work is related to alpha particles but there are experiments of many other kinds going on. I will describe these in more detail in a general letter to Dr. Whitney. The work of C. D. Ellis,[1] who was made fellow of Trinity College while I was there, is of particular interest to you and Hull. I believe some of his results have just been published in the Proc. Royal Society. He finds that the beta rays from Radium B give two lines when separated into a beta ray spectrum by a strong magnetic field; that is, there are beta rays of only two velocities. If the gamma rays given off at the same time are separated from the beta rays and are then allowed to fall on some other element, delta rays are produced which are also of two velocities, the difference of the two energies being the same as that found for the original beta rays. The delta rays produced from different elements by the same gamma rays differ in energy by exactly the amounts which correspond to the difference in the energies needed to knock out electrons from the K ring as determined from the absorption frequencies. This proves that the beta rays from Radium B are really delta rays produced by the action of the gamma rays on the electrons in the K ring. This makes it possible to determine with precision the frequency of the gamma rays. In this way Ellis finds that the gamma rays from Radium B have a wave length of 0.03 Å corresponding to about 400,000 volts.

Ellis is now working on the gamma rays from Radium C. When these rays fall on different metals, the velocities of the delta rays give lines which prove that the electrons are ejected from the nucleus. The delta rays from lead isotopes are different showing that the work done in ejecting the electrons depends not upon the charge of the nucleus but upon the structure of the nucleus. This line of investigation will lead to an exact measure of the frequencies of the gamma rays and also give a measure of the work done in removing electrons from nuclei. The results so far indicate that the maximum frequencies of the gamma rays of Radium C correspond to voltages of about 3,000,000 volts.

I am writing you particularly because of a request of Rutherford for X-Ray tubes with molybdenum and rhodium targets. He has made every effort to obtain such tubes in England but without success. Bragg finds the radiator type of tube much more satisfactory as he needs no rectifier and profits by the shorter distance from the focal spot because of the smaller diameter bulb. Rutherford wants the tubes for crystal analysis and for other work with monochromatic rays. He says that he is prepared to pay for them. I told [him] that we had given tubes to Bragg except for the cost of the rhodium in the targets.

Rutherford has been unable to get any Pyrex glass for glass blowing work in the Cavendish Laboratory. For example they are using lime glass for making mercury condensation pumps and have much trouble from their breaking. All the laboratories in England complain (the lamp factories also) that the only glasses they can get are very bad and of irregular quality. Rutherford tried some time ago to get some Pyrex glass from Corning but has never received it, and he has never had an opportunity of using it at all.

I told him therefore that I would have some sent to him from our laboratory. I suggest that we send him a box of tubing of various sizes convenient for making glass condensation pumps as well as some cane and some smaller tubing such as he can use for making tungsten seals. Another small box should contain a few sizes of bulbs including a few up to 4 or 5 inches in diameter. I would also suggest sending him a few stems having tungsten seals to show him how these are made and so that he can have a couple to try out. Also send any special glasses used by the glass blowers in making these seals. It is not necessary to send large quantities of any of these things, merely enough for him to get familiar with the use of Pyrex glass and to help him out until he can buy some thru Corning.

October 20, 1921. I have just arrived in Paris, and expect now to go on to Leiden, Berlin, Munchen, Vienna, Buda-Pest, Berlin, Copenhagen, London and to sail from Southampton on the Aquitania on December 3rd. Arrangements have been completed with the German lamp works and Radio companies. So it will now be possible for me to see what progress if any has been made in the last few years. Yours sincerely,

[*Irving Langmuir*]

Langmuir Papers, Library of Congress

1. CHARLES DRUMMOND ELLIS (1895–).

Robert A. Millikan to George Ellery Hale, September 29, 1922

My dear George:

We got back to Pasadena Monday morning, and I have this morning received the good letter which you wrote from Geneva. I also had a card yesterday from Perigord[1] saying that you were not very well. This disturbed me a good deal, but your letter of this morning is somewhat reassuring in that it indicates that the difficulty is not of an altogether *fundamental character* (even though it does *touch* bottom), although a very inconvenient and bothersome one. I hope you have long ago returned to your normal physical activity, and are now enjoying the south of France. I suspect that the difficulties in the Near East may cause you to modify, somewhat, your Egyptian plans. If I analyze this Near East situation rightly, it is another pitiful instance of the result of French jealousy of Great Britain. Isn't it a pathetic spectacle to see the Turk, officered still by German officers, I suppose, setting at naught the whole results of the war, so far as he is concerned, because he has been supplied by France herself with the war materials with which he is able to defy those of the Allies who are still standing by the treaty? It is as discouraging a situation as certain of those which developed in our scientific relations.

I had hoped to send you, before I sailed, a little description of our doings after we parted at Geneva, but decided to wait until I could make some report upon the situations which I found in New York and here. Between about August 8th, when we left you, and September 8th, when we sailed from Liverpool, the Millikans traveled *madly* through Lyons, Avignon, Nimes, Marseilles, Nice, Genoa, Pisa, Rome, Florence, Venice, Bolzano, Innsbruck, Oberammergau, Munich, Jena, Leipzig, Dresden, Berlin, Cologne, Brussels, Haarlem (spending a day there with Lorentz), London, Oxford, and Liverpool.

We had a wonderfully fine voyage on the Montcalm of the Canadian Pacific line, which I recommend to you as getting on the whole the best class of passengers which I have traveled with in crossing the Atlantic, and in carrying them at about half as great expense and quite as much comfort as you have on any of the lines running from New York. The Montcalm is the steadiest boat I was ever on. We did not, however, have any real storms, so that I do not know what she would do in a gale. She did, however, run with perfect smoothness in a moderately rough sea.

I wish you could have been with me in Germany, for I am sure you would have been impressed, as I was, with the exceedingly difficult situation which the fall of the mark has imposed upon the German people. I wish I were economist enough to understand whether that fall has been deliberately brought about by the German government for the sake of reducing the country to bankruptcy, or whether the bankruptcy *preceded* because of the inability of the Allies to come to any sort of agreement as to reparations, and the fall of the mark has been a consequence of this. In any case, unless something can be done to stabilize the mark the world is going to see revolution in Germany and the bringing about of conditions which can scarcely be as favorable for the rest of the world as those existing under the present government. I had no conversations with German scientific men except with Einstein and Westphal,[2] both of whom are of the thoroughly reasonable sort, and both working devotedly to bring about better understandings than now exist.

I spent six days in New York and Washington, and had long conferences with the Research Council group in Washington, including Merriam[3] and Kellogg,[4] who were both just back, and with the engineering group in New York. The two main matters discussed were, first, the question of the establishment of a fund such as you had discussed in the spring and we had discussed in Oxford, and, second, the next move in regard to the Engineering Division.

With respect to the fund: Carty, Kellogg, and Dunn had just had conferences the week preceding with Rummel[5] in which Carty had done all the talking (much to Kellogg's distress), and in which the endeavor had been to show Rummel that money to be devoted to economic and social progress was most effectively spent in furthering science. They all felt, however, that Rummel was merely a go-between, and that nothing was likely to be accomplished so far as the Research Council is concerned unless Raymond Fosdick could be persuaded that the funds in question could be usefully employed by the Council. Another very interesting development, which Kellogg is very enthusiastic about, is that Vincent[6] himself had suggested to Kellogg the turning over of certain funds, the amounts of which were not specified, by the Rockefeller Foundation, to be used for Council projects, Vincent suggesting that the sums turned over for the next two or three years be merely those needed for carrying on the projects themselves (although Kellogg thought that it was not necessary to specify beforehand the exact projects to be carried on). Kellogg understood that Vincent's plan would be to capitalize the whole undertaking after the lapse of a few years.

These two matters bring us face to face with the question which we discussed in Oxford as to whether the Research Council is now in a position to administer and allot funds for scientific researches, and if not what kind of modification it would need to put it in a position to accept funds for such purposes. This is essentially the question which you discussed in the spring, and which I discussed at some length, first with Kellogg and with Merriam, and then with the New York group consisting of Dunn, Carty, and Jewett.[7] Kellogg, as you can imagine, is ready to accept funds right now to be administered by the Council as it is. He says he would not advise more funds for overhead, but wants funds to be administered for specific researches. Merriam, with whom I talked the whole of one evening, agrees with me that the Council as now organized is not in shape to do that, and Dunn, Carty, and Jewett are all of the same opinion. Nevertheless, Merriam, although the most conservative of the group on this point, is not definitely opposed to the plan of finding a mechanism for administering or allotting funds for research. He points out the dangers that we are running by becoming merely one of the one hundred or one thousand research-prosecuting agencies, and wants to see us develop a machinery whereby, if funds develop in very large amounts, we can redistribute the capital as well as the interest so as to keep them down, but he is quite definitely of the opinion that it is *the Academy* and not the Council which ought to have in some fashion the control of the funds in question. He doesn't want even a joint Academy and Council board if it can be avoided, but rather an Academy board with *representation* from the Council, if it is found feasible to make that sort of a distinction.

Dunn, Carty, Jewett, and I met at dinner at the Century Club and discussed for three hours this fundamental question of policy, and although Carty was very cautious, and as usual took up most of the time, we all came to the conclusion that it would be desirable for the Council or the Academy to develop machinery for administering funds for specific researches, or better, for allotting funds for specific researches, but we did not have the time to formulate any opinions as to just what that machinery should be. Dunn, Carty, and Jewett are all strongly of the opinion, however, that *the Academy* should exercise the dominating influence, if possible. I ought to add, too, that in order to get around certain influences which have got to work in Dunn's mind, it is very vital to avoid tying this matter up in any way with the Seashore plan, for he has already decided that the Seashore plan is nonsense, and therefore anything that has Seashore's name in it starts off at a disadvantage so far as Dunn is concerned.

We were unable during the week in which I was in the East to formulate any definite plan, and you see from what I have said that there are difficulties in the way of formulating one because of the fact that Kellogg, who is mainly responsible for both the Vincent and the

Rummel moves is perhaps going to be impatient because of the reticence which Merriam, Carty, Dunn, and Jewett all feel about a plan of getting relatively large funds as soon as possible to be administered by the Council itself as it now is for research projects. I think, however, that some good was done during that week in getting ideas cleared up on the question of accepting funds at all for allotment to specified research problems, and the next move is, I think, for Kellogg and Merriam to formulate something which will embody their joint ideas. Kellogg feels that this question is the main one which the Council has got to decide this year, and every one feels that it is of the utmost importance that it be decided rightly. I expect that some sort of action will be taken between now and April, if not between now and Christmas. Dunn and Carty both feel that in spite of their suggestions it is wholly unlikely that the Rockefeller Foundation or any other such group would be willing to allot to the Research Council or to the Academy funds without studying pretty carefully the machinery for expending funds wisely which we now have and I think they would say that we ought to develop better machinery than we now have before we make any special effort to persuade them to allot funds.[8]

The second big Research Council problem lies in the field, as usual, of the Engineering Division. Rand,[9] who is at present, I think, a bit antagonistic to the Council and who is at present the boss of the engineering ward in New York, recently put through a plan to make Flinn[10] devote his whole time to the Engineering Foundation, and to give up entirely his connection with the Engineering Division of the Council. The plan of engaging a man to devote himself exclusively to the Engineering Foundation was one which Rand and Swasey had discussed together, I think, also Mr. Adams,[11] and which Rand put through with considerable opposition when he had made the choice of Flinn for this man. The great weakness in our relations to the engineers in New York now lies in the fact that there is nobody on the Engineering Foundation who knows much about the workings of the Research Council. We originally counted on Dunn, Carty, and Pupin, but none of these men are now on the Foundation, and Jewett tells me that they are all now too far from the working machinery of the engineering group to exert much of an influence.

Although the letter to the Research Council which accompanied Flinn's resignation from the Engineering Division was very nicely phrased, the actual situation now is, as you at once see, that all the activities which are under the name of the Engineering Division, of which Flinn has been chairman, are likely to be taken over by the Engineering Foundation, and Kellogg is now definitely saying, "What is the use of our trying to do the work of the Engineering Foundation?

Would it not be better to drop the Engineering Division altogether?" I pointed out the danger of our losing contact with the engineers entirely and thereby undoing all that has been done during the past two or three years toward getting the engineering men interested in the Council. That this has been done is evidenced by the fact that Jewett, who is, as you know, on the Engineering Foundation, tells me that the Engineering Division is much better known in New York than is the Engineering Foundation. Indeed, I suspect it was because of that fact that Rand persuaded Swasey that it was time to make some strenuous effort to have the Engineering Foundation function in another way. My solution of the whole situation was to make another effort to get Mr. Adams to take the chairmanship of the Engineering Division, and although Dunn evidently felt it was a hopeless endeavor it was agreed that he, Merriam, and Kellogg should visit Adams as soon as Merriam could be apprised of the plan, and try to persuade him to take the chairmanship of the Engineering Division. If he will do it for a year I should hope that Jewett, after he has been relieved of the very large responsibilities which he now has as President of the Society of American Engineers, would succeed Adams as chairman of the Engineering Division, and act as the link to keep the Council and the Engineering Foundation together.

You will see from the foregoing that we had quite a strenuous time mulling over certain Research Council problems while I was in the East. You will be interested to know, too, that I stood on the corner stone of the new building, and indeed walked over the whole basement. It has been progressing nicely, and is going to be a good deal larger and more imposing affair than I realized.

I was glad to get back to Pasadena and breathe again this invigorating air which you lauded so highly when you set up your first siren song about California. It seems to have stimulated *you* too much, hasn't it? I hope the deadly monotony of Europe will bring your nerves back to their normal quiescence.

The high tension laboratory is as yet only a hole in the ground. Mr. Fleming estimates that it will cost all of about $125,000, and therefore be a charge to the Institute itself of something $20,000 or $25,000 before we get through with it, $105,000 being the amount contributed by the Company. Also, the estimates on the library are $60,000 instead of $50,000, which we had planned, but I understand from Mr. Fleming that Dr. Bridge is willing to assume this extra cost.

We are starting in this fall with an excellent group of men, and the Faculty Club, which is located in the house which stood upon the property to the East of the Institute, bought before you left, is now an attractive feature of life here. Tolman and Darwin are both living there, as well as a number of the younger men.

The thing that worries me most is our apparent inability to get hold of any gratings for our ultra-violet work, either through Anderson or Michelson. I am writing Michelson to-day, but am seriously contemplating having Julius build for us a small grating engine for the ruling of the small gratings which we need. Our problem is an entirely different one from yours at the Observatory. Julius estimates that he could build a grating engine for our purposes in about three or four months time.

You say you wrote me a letter to Brussels, but I failed to receive it. Also you ask me to send my expense account for the League meeting from Brussels back to Brussels. I suppose I could work that out fairly accurately if it is desirable to do so, but I am quite content to let the matter go as it is, since I was not an authorized delegate and I see no special reason why my expenses should be paid. They would be of the order of 1600 francs, which would presumably have been reduced to something like 600 had I been alone.

Mrs. Millikan joins me in affectionate greetings to both yourself and Mrs. Hale, and in the earnest hope that you are now getting back into fighting trim.

Cordially yours,
R. A. M.

Hale Papers, California Institute of Technology

1. PAUL HÉLIE PERIGORD (1882–1959), French-born historian, obtained his Ph.D. at Minnesota in 1924. From 1919 to 1924 he was professor of European history at Cal. Tech. From 1924 to 1932 he was at UCLA.

2. The physicist PAUL WESTPHAL (1882–[?]) was at this time in the Prussian Kultus Ministerium. After 1934 he was at the Technische Hochschule Berlin.

3. JOHN C. MERRIAM (1869–1945), a paleontologist, was Robert S. Woodward's successor at CIW. In 1919 he was chairman of the National Research Council. Before then he was on the faculty at Berkeley from where he organized a state-wide research council.

4. VERNON LYMAN KELLOGG (1867–1937), zoologist. From 1919 to 1931 he was chairman of the National Research Council.

5. BEARDSLEY RUML (1894–1960) was the recently appointed head of the Laura Spellman Rockefeller Memorial which supported social science research for the welfare of women and children.

6. George E. Vincent.

7. FRANK BALDWIN JEWETT (1879–1949) was a physicist who served as vice-president in charge of research and development for A. T. & T. He was president of the National Academy of Sciences 1939–1947.

8. This was one sign of a problem which for some years plagued both the academy and the council. The latter acquired a life of its own not always in accord with the wishes of the scientists in the academy. This produced a crisis over the Science Advisory Board early in the New Deal.

9. CHARLES FREDERICK RAND (1856–1927) was a mine owner. He was on the Board of the Engineering Foundation 1916–1926, and its chairman 1920–1925.

10. The civil engineer ALFRED DOUGLAS FLINN (1869–1937) was chairman of the Division of Engineering of the National Research Council 1920–1923. What is at issue here is the tendency of at least some of the engineers to develop programs and positions independent of the leaders of the scientific community. Hale, an M.I.T. graduate, was conscious of the potential power of the engineers as a group. He was also sincerely convinced that both science and utility would suffer from a split between the creators and the applicators of knowledge.

11. COMFORT AVERY ADAMS (1866–1958) was an electrical engineer at Harvard, chairman of the Division of Engineering of the National Research Council 1919–1921.

Clinton Davisson to Owen Richardson, October 3, 1922

Dear Owen:

Having written you one letter today over Arnold's name, I am spurred to write one over my own. No need to tell you I've been meaning to write for a long time. We were glad to have the card from Lilian and to learn that you were having a pleasant time [in] Wales. It reached me about the time I was returning from Maine. I was up for three weeks and Lottie and the youngsters were up for all summer. At last we are able to send you a few pictures of our cottage and a few of our house here. We are again possessed of a camera. There are also some of the baby—I think you haven't had any before. She is a very strong determined little youngster who would boss the whole family if she could. We are all very fond of her, and Jim and Owen are both very careful with her. The latest news is that she is to have a sister (or brother) in January. We are all well at present—except for colds—Lottie and all of the youngsters have been having them.

I have done the only work that has been undertaken in our laboratory on thoriated tungsten—and that hasn't been much. I made one or two attempts to transfer the thorium from one filament to another—but nothing much seemed to happen. I never tried very seriously, however. I have a dim recollection of seeing some mention of such experiments somewhere, and tried today to find the references—also without success. When I was working with Th.W. I did some work on the change in c.p. by using the Th.W. filament as a grid between an ordinary W. filament and a plate. The Th.W. grid was used cold at various stages of activation and always kept at negative potentials so that it was not bombarded. I think that this method can be made to work. We got results at times that seemed to be about right. At other times the results were erratic. These were both side issues to the main problem which was to find if the G.E. had anything on us in the matter of an efficient source of electron emissions.

I have a paper coming out, in the next Phys Rev. probably, with L. H. Germer[1] on the work function of pure W. We think it is a fairly thorough job, but not much else. At present we are writing up the results of similar measurements on a coated filament, and I am more or less stuck in trying to interpret certain behavior of the filament. The results are these—by measurements on the cooling effect at 1047° K we find a work function corresponding to 1.79 volts. With the emission steady at 1047 we change suddenly to other temperatures. At the new temperature the emission is never steady. If the new temp. <1047 the emission increases, approaching a steady value asymptotically—if it is >1047 the steady value is below the initial. Both initial and steady values give straight lines when plotted in the usual way. The initial values line checks exactly (within 1 or 2%) with the cooling measurements. The steady values line is 23% lower (in 2). What I am trying to make out at present is why the steady values give a line at all—and what the line means. I wish I could talk it over with you.

Germer, at present, is working on the velocity distributions problem. He is doing the work for a thesis at Columbia. He is using the cylindrical arrangement. The filament is heated on rectified current and the iR drop in a resistance in series is used to run the filament off to a high positive potential during the heating. This does away with the mechanical interrupter. From preliminary results it looks as if he is not going to get the factor two that turned up in . . . ing's[2] results.

The scattering work is going along very well. We have investigated Pb and Mg since I wrote last. For Pb we observe as many as four very definite maxima of scattering at voltages around 500, in the range 15 to 135° from the primary beam. With Mg the distribution is simple at all voltages (up to 1500) and changes with voltage in accordance—in a general way—with the formula I worked out. Only in a general way, however. I'm examining what effect the change in mass with velocity will have—not much it appears. All sorts of people seem to be discovering that the inverse square field about the nucleus doesn't extend off to infinity. I wrote Mr. Schenland a letter and sent him a copy of the note we had in Science. Have you noticed that Perrin and Rogers have observed similar departures for alpha particles scattered 1 or 2 degrees by argon.

When up in Maine this summer I worked out a rather interesting problem that may possibly have been worked out by Laue[3] in his lengthy paper on the distribution of electrons in enclosures—or you may have worked it out at some time and thought it not worth publishing—I suspect from something in your book that this is the case. The problem is to arrive at your celebrated equation without assuming in the thermodynamic argument that the density of electrons is so small in the region outside the metal as to leave it practically uniform throughout. By assuming a parallel plate arrangement—one plate emitting and the other reflecting the density distribution can be worked out. The pressure at the reflecting plate is less than it would be if the density were uniform and there is e.s. energy in the field to be included in the thermodynamic argument. Finally I get

$$\frac{1}{Po}\frac{d\,Po}{d\,T} = \frac{W}{RT^2}$$

where Po is the density at the hot surface. Making the distance between the plates infinite, I get the distribution published by you in 1903. What started me on the calculation was an article by Karl Compton in the Jour. of the Franklin Inst. in which he came to some very weird conclusions about electron atmospheres.

Well, I think that I have told you of everything that is worrying us at present. I often wish that our families had some way of getting together once a year or so. It would be a great help to me to talk over some of the these things with you, but this would be only incidental, of course, to the pleasure of seeing you all again. Our best love to Lilian and yourself and to "The Nipper" and to Jack and to "Baby Mary"

As ever
Davy

Richardson Papers, University of Texas at Austin

1. LESTER HALBERT GERMER (1896–1971) was from 1925 to 1961 at Bell Telephone Laboratory, a collaborator of Davisson's. The paper is "The Thermionic Work Function of Tungsten," *Physical Review* 20 (1922): 300–330.
2. Name unclear.
3. MAX VON LAUE (1879–1960), Nobel Laureate in physics 1919.

Richard C. Tolman to Gilbert N. Lewis, January 25, 1923

Dear Gilbert:

At a recent meeting of our physics colloquium Darwin[1] reported on an article by Bragg, James and Bosanquet (Phil. Mag. 44, 433, 1922).[2] Probably you have already seen the article, but since it is kind of a bothersome one from the point of view of the cubical atom, I thought I would call your attention to it. The work was on the scattering of X-rays by sodium and chlorine atoms, and seemed to show that the amount of light scattered in different directions could not agree with a distribution of electrons around the nucleus, such that the number of electrons in successive shells could increase say from 2 to 8 going out from the center.

At the close of the colloquium I tried to point out that the whole method was based on the idea that the electrons were taken as scattering the X-rays in accordance with classical ideas, which hardly seemed justified to me when taken in connection with the introduction of the quantum theory for all the other interactions between light and electrons which we know anything about. Both Darwin and Millikan, however, seemed to feel that my remarks were of no importance at all.

Although the Bohr atom seems to me inescapable in the case of monatomic hydrogen and ionized helium, I do not feel that this necessarily means an abandonment of the cubical structure for other atoms, especially when they are in their lowest quantum state. In many ways, it seems as if all but the valence electrons in an atom were sort of "dead," especially when we consider how few substances are strongly paramagnetic.

I am having an enjoyable time here this term, and am much more cheerful than when I first arrived. I have taken your advice and am doing quite a little experimental work myself on photochemistry, the real old-fashioned test-tube and beaker sort of stuff. I also have three men working on experimental problems and am keeping track of some of the other men too. As for theoretical work, Darwin and I are writing a highly mathematical thing together upon the motion of an electron under the combined action of a central nucleus and an applied magnetic field. It is similar to the usual considerations of the Zeeman effect, but includes the effect of the square of the magnetic field strength, a matter of some experimental significance, we believe.

Edward tells me that I have been sending you insults. Of course you realize that we never insult people unless we envy them their superior attainments and qualities. This is true Freudian dope, and I assure you it is true for the case in question.

Please give my regards to Mary and Miss Gray. With best wishes, I am

Yours very sincerely,
Richard C. Tolman
Prof. of Physical Chemistry and
Mathematical Physics

Lewis Papers, University of California, Berkeley

1. CHARLES GALTON DARWIN (1887–1962), a physicist, was the son of George H. Darwin.

2. "The Distribution of Electrons around the Nucleus in the Sodium and Chlorine Atoms."

Ernest Rutherford to Arthur H. Compton, August 14, 1923

Dear Compton

First of all allow me to congratulate you on your appointment to Chicago. I was very pleased to hear of it and wish you all success in your new sphere of work. I congratulate you also on the fine lot of work you have turned out the last two years. I have, of course, been greatly interested in your later theory and experiments on the scattering of X rays which is undoubtedly very important and may help to clear our notions of the quantum theory. I saw Ross' paper in the Academy with the photograph of the scattered rays. I was interested enough to try at once when I saw your paper[1] whether there was any sign of a shift of a line after 100 or more reflections from a silver plate but of course detected nothing.

We have had a very bright and warm summer. I have been very busy preparing for the Liverpool meeting of the British Association where I am President. I saw Darwin the other day and he said he saw you on his way back.

We are all well and the grandson flourishes. I hope your boy is going strong.

With kind regards to Mrs. Compton

Yours sincerely
E. Rutherford

Arthur Holly Compton Papers, Washington University

1. P. A. Ross of Stanford. The paper is "Change in Wave Length by Scattering," *Proceedings of the National Academy of Sciences* 9 (1923): 246–248, a confirmation of Compton's work.

Oswald Veblen to Vernon Kellogg, February 11, 1924

Dear Dr. Kellogg:

My experience this year has made me rather acutely conscious of the fact that the needs of mathematical research have not yet been brought to the attention of those whose position enables them to have a view of the strategy of Science. This, I think, is chiefly the fault of the mathematicians themselves, who have too easily assumed that an outside world which cannot understand the details of their work is not interested in its success. That such an idea is erroneous has been well illustrated by the generous action of the Rockefeller Foundation in providing funds for Research Fellowships in Mathematics of the same type as for Physics and Chemistry. This was done immediately, and apparently as a matter of course, when the need for such fellowships was pointed out. This experience, as well as much evidence of a less tangible sort, of the friendly interest in mathematics, leads me to hope that it may be worth while to draw attention to the fact that we are now in a situation where another very important step of a similar sort may be taken.

Mathematical research is done almost entirely by university and college teachers. A mathematics department in an American university has to deal with an enormous mass of freshmen, a very large number of sophomores, and with extremely small numbers of juniors, seniors and graduate students. The situation is entirely different from that of a European University, which has to deal only with the last class of students. The subjects taught to freshmen and sophomores are taken up in the Lycee's and Gymnasia. Under our conditions, the men responsible for the conduct of a Mathematics department are obliged to give their primary attention to providing instruction for the freshmen and sophomores. This obligation is due not merely to the number of men who have to be dealt with but also to the intrinsic importance of such instruction.

Nevertheless there has been a great development of mathematical research in this country. Twenty or thirty years ago there were very few men doing such research and they were receiving very little consideration from the Universities. Now they are very much in demand. A man with good mathematical gifts and normal personal qualities has little trouble in obtaining as good a position as is available under our system. But when he obtains it he has a teaching schedule of from nine to fifteen hours a week as compared with three hours a week for his colleague in the College de France, for example. Moreover, he becomes tremendously interested in this teaching; he sees the manifold ways in which it could be improved, and he plays his part in the committees and other administrative devices which are trying to do the obvious tasks of the university in a better way.

He was preferred to other men when appointed, because of his scientific distinction. But just because he has a sense of responsibility and reacts in a normal way to his environment, it is only a small fraction of his energy that goes into research. The university authorities never know the difference (it does not show in the number of his publications, only in the quality) and give him his rightful share of respect as a loyal member of the community.

So we have arrived at the stage where we recognize ability in scientific research as a basis for university appointments but not as a primary occupation for the appointees. This statement is not strictly true in sciences like Physics and Chemistry for the universities which have great laboratories usually recognize the absurdity of maintaining such plants without a respectable output of research. It is brilliantly untrue in Astronomy. But in Mathematics it is true almost without an exception.

The way to make another step forward is obvious. Indeed it has already been partially recognized by the Rockefeller Foundation in establishing a series of Fellowships in various sciences which afford opportunities for research to men of promise at the outset of their careers. What remains to do is to find a way of assuring the continuance of their research to men who have already proved their ability. This is already provided for, to a certain extent, in the laboratories of the experimental sciences, but, as already indicated, there is no provision in Mathematics. To provide it, there are at least two ways which would be justified by the actual amount of mathematical talent in the country. The first of these would be to found and endow a Mathematical Institute.

The physical equipment of such an institute would be very simple: a library, a few offices, and lecture rooms, and a small amount of apparatus such as computing machines. There should also be provision on a small scale for stenographers and computers. But the main funds of such an institute should be used for the salaries of men or women whose business is mathematical research. These people should, however, be provided with the equivalent of the routine work which is always present in laboratory sciences. This work could consist, for example, in editing a mathematical periodical or in preparing a new edition of the Encyclopedia of Mathematics. The latter enterprise would be a very large one but would be tremendously important both for pure mathematics and for its applications. The members of the Institute should also be expected to give lectures to advanced students in their own fields of research.

Such an institute, in my opinion, could operate successfully either in conjunction with a university or as an entirely separate institution. In either case it would treat mathematical research as a profession. There are plenty of men in the country who have shown that they are capable of living up to such a position.

The second plan which I have in mind is essentially that followed by the Royal Society in the Yarrow Research Professorships. It consists in establishing and endowing a number of research professorships which are awarded to individuals who have shown in their own environments that their impulse to research is a vital one. The appointees are not moved to new places. The only difference brought about is that they are freed from all other obligations and thenceforth paid for devoting their energies to research.

In our country it would be advisable actually to limit the amount of teaching or other routine that a research professor is *allowed* to do. He should not be allowed to give more than two or three lectures a week. Perhaps also he should not be allowed to accept more than a limited number of research students. With such restrictions, I think that one of our philanthropic foundations could carry a number of research professors on its salary roll and be confident that no better use could be made of its funds.

The second plan has the advantage that it could be tried out by gradual steps. The mathematical institute has the advantage that it would provide a definite nucleus for mathematical research and foster cooperation in a field that has been treated in the past in perhaps an unnecessarily individualistic way.

Yours sincerely,
Oswald Veblen,
Chairman, Division of Physical Sciences

Veblen Papers, Library of Congress

Percy W. Bridgman[1] to Gilbert N. Lewis, June 21, 1924

My dear Lewis;

I was glad to get the copy of your reply to Campbell;[2] it seems to me to adequately meet the situation and to show the errors in his position. At the same time it has given me a better idea of your own position about U.R.U. than I had before, and I should like to set down here my general reaction to your whole position.

In the first place there are a few minor matters to mention. I was glad to learn of the existence of Sundell's paper, of which I had not previously heard; I will look it up as soon as I again have access to a library. With regard to my use of Einstein's criterion of the "smallness" of a number as a test of the probability of a relationship, it would be better to say instead of "small" a number "which it seems might probably result from purely mathematical manipulations." I think the rest of my test suggests something of the sort as the meaning. At any rate, I have myself emphasized that the idea is only very rough, and the use that I have made of it is as a criterion as to whether further and more exact examination of the problem is likely to be worth while.

Used in this way the idea cannot lead far astray. Most of your statements about dimensional analysis I can subscribe to, but as I wrote in my last letter, I am not sure that Mrs. Ehrenfest's[3] analysis of the situation is entirely satisfactory, and that there are not differences between the model and dimensional methods when dimensional constants are introduced by the integrations. However, I have as yet heard nothing further from Mrs. Ehrenfest, so that I cannot be definite about this.

Now to come back to the main question of U.R.U. I believe that the main difference between us is as to the extent to which we believe nature to be simple. You believe that all the laws of nature of wide applicability are simply connected with each other. By "simple" you mean simply expressible in terms of human mathematical analysis. It would seem to me very strange if our mathematics, which for many phenomena is such an extremely clumsy tool, should turn out to be so peculiarly well adapted for the correlation of *all* the broad universal relations. My feeling is that the simplicity of nature is a much more restricted thing; that there are different domains within which the relations are simple, but that the different domains need not necessarily be simply connected in terms of ordinary mathematical analysis with each other. I recognize that apparent simplicity may arise in different ways; we may have such simplicity as we have in the behavior of perfect gases arising from the statistical effect of enormous numbers, or we may have (perhaps) an entirely different sort of simplicity such as the inverse square law between two electrons. I see no reason why the simple relations in one of these domains should be simply related to those of the other. In fact the whole history of your theory conforms exactly with this point of view. I should expect that there might be domains of related phenomena in which by the proper choice of units the experimental constants could be simply expressed. For each of these domains there would be a set of "rational" units, but the sets of units would be different in the different domains and there would no such thing as "*ultimate*" rational units. What I have in mind is made more definite by the behavior of the gravitational constant. This is admittedly of fundamental significance, and I find it most difficult to see why, if your theory is right, this constant proves so obstinate. If I have not made a mistake in my calculations, the gravitational constant may be made unity by fixing your last disposable unit, which demands that the unit of mass be taken as 4.2×10^{10} gm, a mass which appears to be without physical significance. On the other hand, if the gravitational constant is in someway connected with the whole universe, then we should expect to be able to choose an entirely different set of rational units such as to make it unity. In fact, if Einstein's cosmical speculations are right, we can make the gravitational constant unity by choosing for the unit of mass all the mass in the universe, for the unit of length the semi-circumference of the universe, and for the unit of time the time required by light to describe this semi-circumference.

I find it very significant that your theory has yielded only two results. I should expect it to be very much more fruitful if correct, in fact, a veritable open sesame. The two results that it has achieved seem not unnatural in view of the general considerations above. In detail, I tried to indicate in my book that a simple relation between Stefan's constant and your other quantities might be expected, but I have not yet had a chance to think sufficiently about the chemical constant to have any very definite ideas; evidently the possibility of discussions like those of Tetrode and Sackur indicates a connection between the elements. It may be that you have here merely one of those numerical coincidences that have sometimes proved so disastrous in other fields.

When I have time, I am going to see if I cannot pick out other sets of rational units for other restricted domains of phenomena; I suppose that you would rather that I make the discovery than make it yourself.

A year ago this spring I had a conversation in Washington with E. Q. Adams that shed considerable light on your position. He said that when a man was within a hundred yards of the end of a race he was willing to sprint a little. Your theory of U.R.U. constitutes the sprint. I, on the other hand, do not believe that we are anywhere near in sight of the finish, but that there are vast amounts of significant structure beyond anything we have penetrated and that we are very far from being able to formulate the whole of nature simply.

I suppose that you will not admit much of this as a fair statement of your views; it will be interesting in any event to hear your comments.

Yours very sincerely,
P. W. Bridgman

Lewis Papers, University of California, Berkeley

1. PERCY WILLIAMS BRIDGMAN (1882–1961) was a physicist who received the Nobel Prize in 1946 for investigations of high-pressure effects. For his generation he was unusually interested in philosophical questions arising from science. He proposed a view known as operationalism, sometimes mistakenly equated with pragmatism. Bridgman's entire career was at Harvard.

2. In 1923 Lewis published a call for a program to reduce the number of "arbitrary" physical constants to produce "Ultimate Rational Units." It produced a spasm of protest. NORMAN ROBERT CAMPBELL (1880–1949), a British physicist with the British General Electric, wrote on the philosophy of science, particularly on theory construction. Bridgman's book, *Dimensional Analysis,* appeared in 1922. For Lewis's position, see his "Physical Constants and the Ultimate Rational Units," *Philosophical Magazine* 6th ser. 45 (1923): 266–277; and "Ultimate Rational Units and Dimensional Theory," *Philosophical Magazine,* 6th ser. 49 (1925): 739–750.

3. Paul Ehrenfest's wife, Tatyana A. A. Ehrenfest, originally a mathematician, collaborated with him in several studies.

Clinton Davisson to Owen Richardson, July 7, 1924

Dear Owen:—

Please accept my tardy congratulations on your appointment to one of the Yarrow professorships. I don't know much about them, although I remember glancing over an article concerning them in Nature some time ago. I presume they are very attractive or you would not be making the change. I've seen several items about you in Science and in Nature and understand that King's College is still to be your headquarters. I also heard that you would retain the Wheatstone professorship as well. I noticed the place listed in a "Help Wanted" column, however, so imagine that's wrong. Whatever the arrangement is we all wish you no end of success.

I laid aside your Proceedings article on Electron Emission (Apr 1) for more careful study and am only now getting at it. The argument up to Sec. 3, which is as much as I have worked over thoroughly, is really very beautiful, or satisfying or whatever it is that arguments are when they are all that they should be. I haven't yet come to the part in which you discuss our experiments on tungsten. From the first reading I gathered the idea that you didn't altogether agree with our interpretation. Have I told you that we also have results for coated filament. We have had these for a long time but I've never been satisfied that we had got the correct meaning—or the full meaning out of them. They are not so simple as those for tungsten. I think, however, that we will publish them soon whether we understand them or not. Germer called me on the phone while I have been writing this. He has just returned to N.Y. after being out sick since a year ago April. He is starting to work again tomorrow.

One interesting thing we have been doing here recently is measuring the point to point emission from a tungsten filament. The essential parts of the apparatus are shown in the figure. The parts of the cylinders C_1 are separated by about 0.25 mm. What gets between them is collected on C_2 and measured. The cylinders form a rigid system which can be slid along from one end of the filament to the other. It isn't an ideal arrangement as secondary electrons are always passing from C_1 to C_2 or the other way about depending on the potentials. However, it does well enough to verify one's notion that more electrons are emitted from the middle of the filament than from the cool ends, etc. The effect of

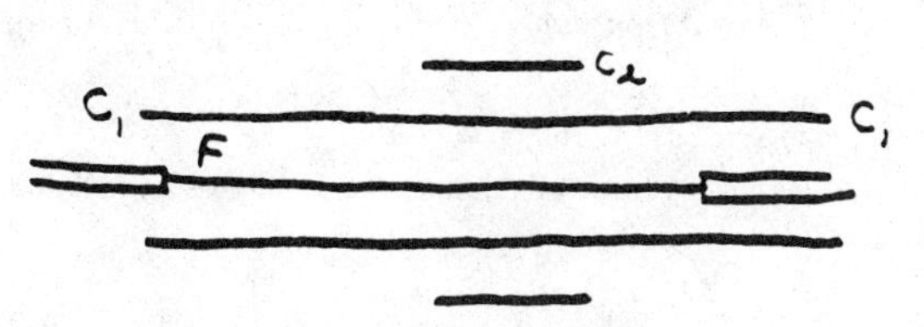

space charge limitations is rather interesting, but the really interesting thing is the rather marked importance of the Thomson effect. The emission except at the center is not independent of the direction of the heating current. When the emission is great the emission from the negative end of the filament is greater on account of the variation in heating current, but for all except these large emissions it is the positive end of the filament that is the hotter. This is apparently due to tungsten's having a positive Thomson coefficient. The value of the coefficient doesn't seem to [be] readily deducible from the results as it seems rather hopelessly tied up with the coef. of thermal conductivity.

But enough of physics. Lottie and the youngsters are in Maine as usual at this season, and are all doing very well according to reports. May and Oswald drove up in our Ford. They took a week to [do] it and arrived without mishap on July 2. It's marvelous how they did as they had only just learned how to drive and neither of them was what you might call expert.

Has Darrow[1] (Dr. K. K.) been in to see you? He is spending a couple of months in Europe and expected when he left to stop in to see you. I should have written long ago to tell you of this. Gibson[2] of the W.E. in England told me in a letter recently that he had talked with you over the phone, I believe it was, and that you had received the coated filament but had not yet found time to use it.

How are Lilian and the youngsters? I'm thinking we'll have to make a trip to England sometime during the next decade to see all of you again.

I'm sending you some pictures. A part of them with May and Oswald was taken rather early in the spring. The others were taken only a few minutes before the family set off for Maine. My mother who is visiting me this summer is in some of these—also the priceless "Helen."

With Much Love to Each and Everyone of You

Davy

Richardson Papers, University of Texas at Austin

1. DARROW (1891–), a physicist at Bell Telephone 1917–1956, was secretary of the American Physical Society 1941–1967.
2. Perhaps Earl S. Gibson who was an inventor of telephone exchange systems and other equipment for Western Electric.

Norbert Wiener to Phillip Franklin,[1] July 22, 1925

Dear Phil:

I have been advised by L.[2] to publish copiously on my work on light theory before going on to my book. My present program of writing for the coming year stands as follows.

(1) Outline sketch of my program (for Göttingen Nachrichten. Already under way for Courant)[3]

(2) Note on same topic for Grenoble meeting

(3) Computational method for non-periodic harmonic analysis. For our journal (?)

(4) Generalization of well-behaved functions to cover needs of light theory. For Lichtenstein & Ztschr (already ordered)

(5) Applications of probability theory. For Annalen

(6) Schrotteffeht. For our journal.

(7) More on Nörlund's stuff. Already ordered by Courant for Annalen.

(8) Problems of polarization, of light-cones, diffraction, refraction, etc., handled by my method. Still in dim and distant future. Won't go to Transactions, anyhow.

I wish I had a couple of good graduate students to do a part of the dirty work. It is a virgin field. I've been looking at the work of Schuster and of v. Laue. It is good stuff as far as the insight goes, but is terribly hampered by the lack of anything like a decent mathematical tool-kit. The whole jargon of big infinitesimals and small infinitesimals needs setting in order, and my methods will do it. Heymans[4] was too much of a damn fool last year to see where my work was leading, so I suppose it's no use trying to collaborate with the physicists. If I only had good enough relations with Harvard to ask for the loan of one good graduate student for a doctor's thesis! I don't want to do the Nörlund[5] stuff, and do paper after paper on one line, but I have nobody to wish the detail on to, and I can't write my book until I have disposed of the minutiae.

Congratulations on the kid! also on the Acta paper![6] When I say here that the kid's name is David, they think you are thinking of the great Göttinger David.[7]

I intend to have a shingle hung out from our office

2-171
WIENER & FRANKLIN
Wholesale & Retail
Mathematicians & Exporters
Sole European Representative:
L. Lichtenstein
Berlin & Leipzig

The senior partner is finding his business trip so promising that he is going to establish a branch factory for the continental trade at Göttingen, provided he can interest American capitalists in the plan. He has already taken orders to the full extent of the production of the factory, and for any further development it will be necessary to increase the staff. There is a good chance for one or two young apprentices, who wish to learn the analysis business from the ground up. We can also employ an apprentice or journeyman physicist or two. Of course, we have to meet the competition of older and better known firms, but the quality of our product has received favorable attention in the foreign and domestic trade. I advise the cultivation of the foreign market, as there we can compete on a more even basis, since the whole export trade is relatively new. This is particularly useful, since I find that some of our competitors have been spreading false rumors as to the honesty, novelty, and quality of the firm's product. As the home styles follow the foreign styles, the way to the home market lies through the foreign market.

As to the state of the market: differential geometry seems rather quiet, and some of the principal operators have deserted it for other securities. Real and complex variables continue firm, without much change. Analysis situs has a bull market. Bull operators have been very active in difference equations, also. Quantum theory continues speculative, with chances of a sharp rise, but the market contains a lot of wildcat stock. Hilbert, Brower,[8] & Co. are doing well with mathematical logic.

Your brother and partner
Norbert

Wiener Papers, Massachusetts Institute of Technology

1. FRANKLIN (1898–1965) was with Wiener at M.I.T. A mathematician, he had been a friend of Wiener's since they had served at Aberdeen during World War I. He received his Ph.D. from Princeton in 1921.

2. The mathematician LEON LICHTENSTEIN (1878–1933) of Leipzig, a cousin of Norbert Wiener.

3. RICHARD COURANT (1888–1972) continued his career at New York University after the Nazis came to power.

4. PAUL ALPHONSE HEYMANS (1895–1960), a Belgian who received a doctorate in physics from M.I.T.'s Physics Department 1926. He returned to Belgium to teach at Ghent but also embarked on a career as an industrialist and banker.

5. The Danish mathematician NILS ERIK NÖRLUND (1885–[?]).

6. "Functions of a Complex Variable with Assigned Derivatives . . . ," *Acta Mathematica* 47 (1926): 371–385.

7. Hilbert.

8. L. E. J. BROUWER (1881–1966) of the Netherlands.

Norbert Wiener to Phillip Franklin, July 25, 1925

Dear Phil:

Yesterday I saw L. again, & got a lot of fatherly advice from him. Went on a hike with a gang of Leipzig professors & Koebe of Jena[1]— who is a queer duck. Took the 11:07 P.M. train for Basel. Pleasant Swiss student in compartment—he is lending me this letter paper. Not much sleep sitting up. At breakfast noticed familiar face in dining car—Einstein. Went up to him and reintroduced myself. We had a pleasant 45 minute chat. He was in a third class compartment. Very simple and unassuming. Told me of his latest stunt—he gets gravity & Maxwell's equations at one swell foop by an unsymmetrical minimization problem. Coming out in Berlin academy. He is tickled to death about it, and has distant hopes that the secret of the electron and of quantum theory may be contained in it. He asked about my line, and seemed really interested in my light and Brownian movement work.

He is not going to travel much in the future, so will probably not come again to the States. He gets too tired, and wants the time for his own work. He is awfully nice & genial. I didn't stay too long so as not to tire him, but will probably look him up later again today.

(Special Scoop from Wiener scientific news bureau, unlimited) He doesn't lose much sleep over the Miller[2] experiments.

Tomorrow I shall be in Grenoble. Einstein asks me to give his compliments to Langevin.[3]

Let's hear all the home news!

Your Schwäxer
Norbert

Wiener Papers, Massachusetts Institute of Technology

1. The mathematician PAUL KOEBE (1882–1945) went to Leipzig in 1926.
2. DAYTON C. MILLER (1866–1941), a physicist, was at Case Institute for most of his career. He made several experimental attempts to determine the apparent relative motion of the earth and the ether. His results, announced that year, purported to refute Einsteinian relativity.
3. The French physicist Paul Langevin (1872–1946) of the Sorbonne.

Norbert Wiener to Bertha Wiener,[1] August 28, 1925

Lieber Berthachen!

I'm having a nice time at the meeting. Miss Payne[2] is having me to dinner with her family in London, after which we shall commence a tour of exploration of its environment from the standpoint of the theater. We had a reception here last night by the Mayor. I sat with a bright young Hindu physicist (from Bangalore!) and a travelling fellow of the Harvard zoology department, of Alsacian origin, who is one of God's own jackasses. He talks loudly and reverently of the Mechanistic Gospel of the Holy Jacques Loeb, and doesn't seem to have the least idea that mechanism (in the Newtonian sense) is as dead as a doornail in physics. His scorn for vitalists is as withering as his respect for statistics is complete. I myself don't think the present day vitalists have much to offer, but I certainly don't think that the mechanistic explanation of living processes is anything more than a one-sided schematization—and even as that, is only a pious hope at present. What disgusts me is that any one with any pretensions to scientific standing should have such an atrophied imagination as to regard the dogmata of science in the year of our Lord 1925 as *endgültige* gospel truths rather than as *zweckmässige* hypotheses. If only the supporters of the Ape weren't fundamentally of the same religious mentality as the partisans of the Diety!

By the way, Ornstein[3] of Utrecht gave a paper on dispersion under quantum theory—50% Q.T., 50% Classical Mechanics, emulsifier badly needed. There is much disquiet on the part of the Kelvinian generation, and I don't blame the poor old dears. *There is no consistent basis for the physics of the present day.* The man is yet to be born, probably, who can put together a decent theory of radiation. When that comes, one A. Einstein will step into the background—as he admits.

Schuster—the periodogram & white light man[4]—is interested in my junk. I have also got in touch with a statistician who has done a very similar stunt independently, but has not published or used it. I lunch with him today. When in London, I am going to look up a guy at the Air Ministry, who is *the* periodogram expert, and see what a practical man's opinion of my generalized periodogram is. I think it knocks spots off the old method, particularly for meteorological work.

Between you & me & the doorpost, barring the relics of the Kelvin generation, I don't think the gang here has any edge on our crowd at all, at all. Cambridge, Mass. is decidedly on the map, as Miss Payne—who is becoming quite a patriotic American—is finding out. In ten years, the scientific center of the world will be somewhere off the Newfoundland banks; in twenty years, it will be—well, say in somebody's back yard at Springfield, Mass.

There is a considerable lack of flexibility in the crowd—Göttingen is a much homier place (English for gemütlich). I haven't seen anyone as nice as Courant—or Harold Bohr,[5] for that matter. And as for Lichtenstein—you know when I was at Leipzig I went into his class, and he was Dad all over again in his strict but sympathetic handling of the students. It is funny how the stiff German professor type is vanishing.

Love
Norbert

Wiener Papers, Massachusetts Institute of Technology

1. Norbert's sister.

2. Perhaps the astronomer CECILIA HELENA PAYNE-GAPOSCHKIN (1900–1979) who came to the United States from Britain in 1923 and was at the Harvard Observatory.

3. L. S. ORNSTEIN (1880–[?]).

4. Edgar Herman Joseph Schuster.
5. Harald Bohr (1887–1951), the Danish mathematician, was the brother of Niels Bohr.

Manne Siegbahn to Ivar Kalberg,[1] September 19, 1925

My dear Mr Ivar Kalberg,

You wish me to write down some of my observations and impressions from my visit to the United States. I am sorrow that just now, at the beginning of the academical year, I have not very much time left for such a purpose but I will try to send You a few lines.

It had been my hope for many years to go abroad and see how the americans are managing their scientific work as I felt from studying the many beautiful results coming from Your country that we european had a lot to learn from Your methods. So I started my journey with a great deal of curiousity. I will mention that I had formerly had the advantage of looking around in all the leading european countries and so I had knowledge not only of the scientific methods in my own little country.

Now the scientific work is not an insolated phenomenon of human doing, but has connections in all directions. To study it and to understand the reasons why there are differences in the european and the american way of managing it, one is pressed to consider many other things than science. Especially there are two: one is the general methods of education, another is the industrial life, which has a great bearing on that branch of science, physics, that I went to study.

I will begin by saying a few words on the last point. In the relations between the industry and science there is a very strong difference between Your country and ours and I am sorrow to say that you are in a much more favorable situation in that respect than we are. I think it is easy to give a probable explanation of these facts. In our country the science had got its complete organization long before there existed any industry. Our Universities had many hundred years of tradition when the industry began to put its seal upon the life. Our science looked upon the rising industry as something very materialistic, with which it should have nothing to do. And the industry replied by considering science as a kind of philosophical sport. No one has gained by that state of things which has lasted long enough. In these days there are signs of changes to better and more rational conditions in this respect. But it will take long time before it is a generally accepted fact that science and industry not only need each other but that no real progress is possible without cooperation.

In the United States the evolution of the scientific Institutes has taken place simultaneously with the development of industry and many of your best scientific laboratories has been born under the auspices of the mighty industry. The industry has been called for at the start of these laboratories and so a cooperation has from the beginning been established. Your big factories have their own Research-laboratories where scientific work of the highest quality is produced. At the meetings of the scientific organizations the men from both kinds of laboratories meet for discussion of their common problems. Partly due to the fact that only very few of our factories are big enough to bear the costs of a research-laboratory, partly due to a disbelieve in the value of spending money for such a purpose, we have in our country practically no such laboratories. So we also miss this connexion between science and industry, which in America has already shown to be of such a big importance.

The understanding between science and industry in this fundamental question was one of my strongest impression from my visit at the United States and by that fact I am convinced that your country is far ahead in the competition of progress.

When speaking of advantages I must also touch what I think is disadvantageous. If a leading man of business spent a considerable part of his time and his forces in writing invoices You should call him queer. But such a thing would never happen as a business-man earns money on his work and pays someone to do that kind of work. A leading man of science is not in general in that position. Most of them are paid not for there research-work but for the lecturing they do at some educational institution. Both things must be done but no-one will be in doubt what of these two things demands the best qualified man. You will find many hundred of good educators on only one man with that special shape of the brain that enable him to widen the limits of human knowledge. It is a bad economy to let such men spoil most of there energy on a work that could be done by lot of others. For poor countries it may be necessary. The United States is not in that condition. On my visit to about 20 different Universities in America I found some of the most brilliant men of progressive science strongly occupied by lecturing beginners in the elementary things. Their work, known all over the world, had been done on the short time left and with the forces not already spent in simple rutin-work. What should have been the results, had these men been able to spend if not all at least most of their time and forces in research-work and in the stimulating lecturing of only the advanced students. It seemed to me that what the american universities need more than anything else is that last-mentioned kind of professors-ships in a relativ number much higher than they have now.

I am
Yours sincerely
Manne Siegbahn

P.S. Will you kindly correct my english before putting it in print.

Compton Papers, Washington University

1. Kalberg remains unknown to us. Nor is there any clue in the Compton Papers why a letter addressed to him ended up in that location. We have not found any evidence of this in print.

George Ellery Hale to Harry Manley Goodwin,[1] January 31, 1926

My dear old boy,

I have had some real adventures, terrestrial and celestial, since I left you that Sunday evening, and since reaching home the celestial ones have kept me so busy that I have been able to do nothing but rest after the daily round. This resting has been greatly helped by reading the adventures of Gallatin fils, which somehow remind me in spots of the "sermons" I sent you from Washington! Thanks heartily for these genial glimpses of the good old days. While my own adventures have been less strictly in harmony with the best traditions of diplomacy, they have been quite exciting enough for this old man.

After a talk with Hoover in Washington I started for Chicago armed with a letter to Rosenwald,[2] which said Hoover hoped he would chip in with his customary generosity, but named no amount. Bill thought this letter a very poor start, and my feet got still colder when Rosenwald's secretary told me I could not see him until the Monday after Christmas at best. So I had to postpone my return to Washington, which had been set for Sunday. On Monday, I was given a chance to go out to see Rosenwald in his remote lair on the West Side, and I started with real trepidation. He received me cordially, however, and gave me time enough—about an hour and a half—for my spiel. Hardly had I opened than he emphatically protested against endowments, which he doesn't believe in *at all*. In fact, he said he was a well-known crank on the subject, and held forth on the evils of setting up perpetual funds for the unknown future. Soon after I had politely dodged a dangerous discussion he casually remarked that he had just promised three million dollars for an industrial museum in Chicago, and we then engaged in an entertaining discussion of museums, during which he presented me a book on the subject, which I treasure as a memento of an up-hill climb. Museums having been adequately reviewed I got back to my topic, and finally reached the point of presenting Hoover's letter. It suggested a gift in annual installments, which Rosenwald heartily commended. In fact, he prepared the way for me by saying that a man would far rather give *ten thousand* a year for several years than to set up a capital sum of the same total amount. I saw my finish and sparred for wind, turning to a cordial agreement with him on the personal merits of Hoover and his value to the nation as stated by Root, whom Rosenwald also admired to the full. Thus reunited, we got back to the subject, and I expatiated on the large calibre of Hoover and the scale on which he does things. This led very naturally to an explanation of why we got Hoover for chairman—obviously the only man who could swing a *national* enterprise which would mean nothing unless carried out on an adequate scale. Slowly, as I built up the imposing picture, and made its scale apparent, I endeavored to make ten thousand a year look like thirty cents, and at the same time I spurred up my courage to spring my demand. Finally, I got it out—one hundred thousand a year for ten years! I had meant to ask for a capital sum of a million or its equivalent, but saw that this would be hopeless. There was a slight pause, and finally Rosenwald said that of course he couldn't decide that day. I assured him he could take his own time, but came around to the trustees meeting to be held in a week, and the great advantage of having an offer to start with. The fact was I didn't dare leave him, so I sparred again for wind, and finally he remarked that he couldn't think of being *alone* in making such a gift. At last, he suggested that he might agree to be one of five to give one hundred thousand a year apiece for ten years. I took him up without losing a second and got away before anything else could happen. Never did I go through a tougher job or feel greater relief when I got into the open air. The temperature was about zero or not far above, but my feet were no longer cold as I made for the the elevated on my way back to town!

I wired Hoover, asking him to try to get one or two to match Rosenwald before the meeting. He did wire Robinson here, asking him to tackle Huntington—this on his own initiative, as I knew only too well that such an appeal would be useless. Robinson did also, and luckily made no move.

Meanwhile we had had a good Christmas, though it was marred by the illness of G. E. H. II and the colds of his two sisters. G's trouble, which involved a slight fever and a sore arm, was mysterious, but on the Sunday after Christmas, when it was four below zero, the doctor suddenly decided he must go to the hospital, where he was operated on that night—osteo-militis (spelling?), with much pus about his wrist bone, but fortunately none within the marrow. He got through it splendidly, and when I came back through Chicago he was again at school and the little girls as lively as ever. As for Jack, the year old Christmas baby, he was the very picture of health and perfection.

In Washington I saw Hoover again and had a long talk with Mellon[3] at the Treasury, though I did not tackle him for a gift. Hoover's plan is to get Mellon to get two hundred thousand a year for ten years from his industries. I arrived in New York New Year's eve and spent the following days getting ready for the trustee's meeting, which went off very well, with excellent speeches by Hoover, Micholson, Welch, Carty, Millikan, Pritchett and Root. The next day, according to plan, Millikan and I tackled Gifford[4] (now President of the A.T. & T. Co.) for two hundred thousand a year for ten years. This was a pleasant interview, with much jollying on both sides—very different from the Rosenwald affair. Of course Gifford could do no more than agree to put our request before his executive committee. Carty wires that it is to come up next Wednesday, but with the condition that we must get Hoover's total of twenty millions (two million a year for ten years). Imagine the job ahead! Daniel Willard[5] of the B. & O. has gone after the railroads for us and Hoover has written J. D. Jr. a request signed by himself, Root, and Hughes.[6] All the above is confidential, as we are still uncertain about the A.T. & T. and not ready to give out the Rosenwald offer. Of course we have plans for an attack on the General Electric Co. (Owen Young was at the meeting, and showed much interest), the Westinghouse Co., insurance companies, &c. I shall have to do some more work in April, when E. and I expect to start east a day or two after the wedding on April 6.

You can imagine what a relief it was to get *home* and back to the quiet of my lab. I found everyone well (though E. had had a light attack of the prevalent grip), and the "bride and groom" in the highest of spirits. The weather was glorious and has been ever since until yesterday, when the greatly needed rain arrived. My 13 ft. spectroheliograph was about ready, and I decided to make a decisive test at once of the oscillating slit device I worked out here at the house, with the same optical parts in wooden mountings, two years ago. Now everything is beautifully mounted, with electric controls of coelostat &c. and a motor to focus the 12 inch Kenwood object. glass (borrowed from Y.O.) that forms the 2 inch solar image on the slits. I can change over to a 6½ inch image in a few minutes, but the small image is just what is needed for this job. The presence of many large and active spot groups is another great advantage at present—when I tested the scheme before there was hardly anything on the sun. Imagine my delight when, with the second slits set on Hα in the first order, I found I could see practically as much of the bright and dark flocculi on the sun's disk as

I can photograph with a spectroheliograph! Since then I have been enjoying the most wonderful views of the sun I have ever seen. The bright prominences can of course be seen around the limb, and frequently, where they overlap on to the disk, they can be made out there as dark structures. The delicate small dark structures all over the disk, and the long dark prominences, can all be beautifully seen. Several times the vortex structure has been plainly visible running into spots, though the larger vortices more than 2 mm in diameter—the width of the window below the oscillating slits—of course cannot be seen as a whole. A great eruptive outbreak a few days ago in a large spot group was the most extraordinary thing I have ever watched, with its brilliant jets and cloud masses, and its rapid changes in form. But the most important thing is the ease of moving the slits across Hα while observing and thus seeing the extraordinary changes in the bright and dark flocculi. Here is where the method surpasses the spectroheliograph, where any given photo. must correspond to a particular setting of the second slit on some part of Hα. Now I can pass in a moment to slit positions which show the ascending or descending gases, bright and dark. The other day Hα was greatly distorted in some places, and by moving the second slits far from the normal position of the line toward the red, I could see the dark forms of the rapidly descending gases, instantly running the slits back to observe the behavior of the other gases moving at different velocities. The pleasure of at least *seeing* the flocculi in action, and of observing all these diverse phenomena, was so great that I couldn't keep away from the instrument, and everything else had to slide. Naturally a number of improvements in both spectrohelioscope and spectroheliograph have occurred to me, and we will soon build a rotating disk spectrohelioscope with over 400 slits, which will give a perfectly steady image, without flicker. A moving picture camera can probably be used with this, to record all the changes in the flocculi caused by shifting the slits across Hα. But the best thing is the *certainty* that some work worth doing will now justify the existence of my new lab.

Brusted's reports are not very encouraging, and I fear his chances are not too good.

Evelina joins me in much love to you all and the hope of seeing you in April.

As ever
G.

Hale-Goodwin Correspondence, Huntington Library

1. GOODWIN (1870–1949), a physicist at M.I.T., was a close personal friend of Hale's.

2. JULIUS ROSENWALD (1862–1932), board chairman of Sears, Roebuck & Co., and a major philanthropist. His best remembered philanthropies were those to improve the South, particularly the lot of the blacks.
3. A. W. MELLON (1855–1937), the financier, was Secretary of the Treasury.
4. WALTER S. GIFFORD (1885–1966).
5. WILLARD (1861–1942).
6. Charles Evans Hughes later became Chief Justice of the U.S. Supreme Court.

Oswald Veblen[1] to Vernon Kellogg, April 7, 1926

Dear Dr. Kellogg:

This is an attempt to comply with the request in your letter of April 2. The period of the last fifty years includes the later work of Benjamin Peirce[2] which had to do chiefly with the classification of linear associative algebras.

It also includes practically the whole work of J. Willard Gibbs who is undoubtedly a mathematician who has been produced by our country. His most important contributions were to thermodynamics and to statistical mathematics. A very large part of thermodynamics, and particularly that part which constitutes the foundation of modern physical chemistry, was his creation. He is perhaps best known for the formulation of the law of mass action and of phase rule, but doubtless the most important thing which he did was to grasp the subject of thermodynamics as a whole and make it effective for purposes of application to chemical problems.

Another important figure is G. W. Hill[3] in mathematical astronomy. His work was chiefly on the theory of perturbation, and his name is associated with the infinite determinants whose theory was afterward developed by Poincaré and others.

Peirce, Gibbs and Hill belong to the period previous to the organization of systematic current work in mathematics and the cooperation in mathematical research which has been brought about by the American Mathematical Society. The last thirty years seem to be characterized rather by large numbers of industrious and substantial contributors rather than by outstanding heroic figures. This is perhaps because we are still in the midst of the epoch. Twenty-five years ago one heard very little about Willard Gibbs. The development of mathematics on extensive scale in this country was brought about by a series of waves of interest in new subjects. The first of these was the introduction of the English school of algebraic geometry by Cayley and Sylvester[4] in the early days of Johns Hopkins University. The impetus given at that time to the study of this subject has continued to the present time. One might mention the names of Coble, Snyder, White and Morley.[5]

A similar but even more intense wave of interest, finite group theory and its applications to algebraic equations was started early in the nineties. A great many important contributions in this field were made by E. H. Moore, Dickson, Cole, Miller, Blichfeldt, Mitchell and others.[6] While group theory no longer seems to occupy the whole horizon as it once did, there is still a considerable current of thought in this direction.

Modern analysis and function theory may be said to have been imported into this country by Osgood and Bocher, and they and their followers have created a function theoretic current which is one of the most important elements in the mathematical stream. With this belongs logically the work on Calculus of variations initiated by Bolza and continued by Bliss[7] and many others.

Another important wave of interest came along about 1900. The subject this time was the logical foundations of mathematics and, chiefly, the foundations of geometry. In this field we must mention E. H. Moore, E. V. Huntington, and R. L. Moore.[8] It will perhaps interest you that it was in this subject that I got my start.

The total effect of these waves of interest was to transplant practically the whole of the European mathematics to the United States. Of course many other subjects of perhaps even greater importance were being imported and studied during the same period. In any case, we can say that from 1910 onward all branches of mathematics have been represented in the United States and American mathematics has been keeping step with that of the rest of the world. During this last period it is perhaps fair to mention Birkhoff as having made important contributions to the theory of differential equations and dynamics, as well as to the theory of difference equations. In the same period Dickson and Wedderburn have made important contributions to the theory of linear associative algebras and related subjects. Dickson has also made other important contributions to algebra and has published his Monumental History of the Theory of Numbers.[9] E. H. Moore has developed out of the theory of integral equations and related subjects a very general theory which he calls General Analysis and which represents perhaps the most abstract type of mathematics at present in existence. Another very abstract branch of mathematics, the theory of point sets and analysis situs has been cultivated by R. L. Moore, Kline[10] and others. Combinatorial analysis situs, a subject which is becoming more and more important because of its applications to geometry and function theory, has been cultivated by J. W. Alexander, Lefschetz, Chittenden, Morse,[11] Birkhoff, and myself.

Finally it should be said that the theory of relativity has had a great influence in stimulating renewed activity in differential geometry and seems to be bringing about some development in the direction of applied mathematics. Mention should be made in this connection of Eisenhart, Kasner[12] and Birkhoff, as well as of a number of younger men who have been interested in what Eisenhart and I call, the Geometry of Paths.

I fear this is rather a disorderly mess but it is the best I have time for this week.

Yours sincerely
Oswald Veblen

Veblen Papers, Library of Congress

1. This is like the letters Hale received before World War I in which U.S. contributions in a field are summarized. It relates to the proposed National Research Fund.
2. The Harvard mathematician PEIRCE (1809–1880).
3. GEORGE WILLIAM HILL (1838–1914), a mathematical astronomer, spent much of his career at the Navy's *Nautical Almanac*.
4. The British mathematicians ARTHUR CAYLEY (1821–1895) and JAMES JOSEPH SYLVESTER (1814–1897). Sylvester taught at Johns Hopkins University from 1876 to 1887 before returning to a chair at Oxford.
5. ARTHUR BYRON COBLE (1878–1943) of the University of Illinois at Urbana-Champaign; VIRGIL SNYDER (1869–1950) of Cornell; HENRY SEELEY WHITE (1861–1943) of Vasser; British-born FRANK MORLEY (1860–1937) at Hopkins.
6. LEONARD EUGENE DICKSON (1874–1954) of the University of Chicago; FRANK NELSON COLE (1861–1926) of Columbia; GEORGE ABRAM MILLER (1863–1951) of Illinois; HANS FREDERICK BLICHFELDT (1873–1945) of Stanford; HOWARD H. MITCHELL (1885–1943) of the University of Pennsylvania.
7. GILBERT AMES BLISS (1876–1951) of the University of Chicago.
8. E. V. HUNTINGTON (1874–1952) of Williams College; ROBERT LEE MOORE (1882–1974) of Texas.
9. Published in 3 vols. in 1919 by the Carnegie Institution of Washington. JOSEPH HENRY MACLAGEN WEDDERBURN (1882–1948), born in Scotland, came to Princeton in 1909 and remained there until his retirement in 1945 with time out for service in the British army in World War I.
10. JOHN RITT KLINE (1891–1955) of the University of Pennsylvania.
11. Born in Russia, SOLOMON LEFSCHETZ (1884–1972) was at Princeton. EDWARD WILSON CHITTENDEN (1885–1977) since 1925 had been at the University of Iowa. MAR STON MORSE (1892–1977) would join Veblen at the Institute for Advanced Study.
12. LUTHER P. EISENHART (1876–1965) was dean of the faculty (1925–1933) and dean of the Graduate School (1933–1945) at Princeton; EDWARD KASNER (1878–1955) was at Columbia University.

Clinton Davisson to Edwin H. Hall,[1] May 18, 1926

Dear Prof. Hall:

I must apologize for my long delay in answering the letter you addressed to Dr. Arnold[2] on February 10th regarding the interpretation of the measurements Mr. Germer and I made some years ago on the thermionic work function of tungsten. Dr. Arnold wrote you, I think, that he had asked me to answer your letter for him.

I was occupied with quite different matters at the time, however, and found it impossible to think about thermionic emission. But recently I have got back to this subject, and on reading again your letter and the attached manuscript find myself in agreement with much that you say. I agree with you that in reckoning the heat absorbed during emission one must think of the condition of the conduction electrons as they enter the element of filament from which they are eventually emitted, rather than their condition just prior to emission. This is a matter that has occupied me recently and it seems to me that in doing this one must take account of the kinetic energy transported into the element by these electrons as well as the kinetic energy which they eventually carry from it—that is, if, as in your manuscript, one writes θ to represent the thermal energy carried from the filament by an emitted electron, he should also include a term θ' to represent the thermal energy, if any, carried into the element by this electron.

Mr. Germer and I are writing a note for the Physical Review in which we correct the value given for what we called, in our original paper, the "thermodynamic ϕ", for the so-called "Schottky effect".[3] We overlooked this in reducing our data, and the fact was pointed out a year or so ago in an article by Dushman. We also have another go at interpreting the results and decide that Tonks and Langmuir are wrong in their recent article in finding from them a value for the heat of charging of tungsten.[4] It seems unlikely to us that anything concerning the heat of charging can be found from experiments such as ours in which the surface charge remains constant.

Yours very truly,
C. Davisson

Hall Papers, Harvard University

1. HALL (1858–1938) was a student of Rowland's at Hopkins. At this date he was emeritus at Harvard. He was the discoverer of the "Hall Effect."
2. H. D. ARNOLD (1883–1933), director of research at Bell Telephone Laboratories, was a physicist. He was a 1911 Ph.D. from the University of Chicago.
3. "A Note on the Thermionic Work Function of Tungsten," *Physical Review* 30 (1927): 634–638.
4. "Surface Heat of Charging," *Physical Review* 29 (1927): 524–531.

H. D. Arnold to Edwin H. Hall, February 7, 1927

My dear Professor Hall:

I read with considerable interest your letter concerning the work of Davisson and Germer and the possible applicability to it of the theories which you have so long promoted, but I regret that I am not sufficiently in touch with the technical details of this work to be able to express any opinion on the matter myself. I have shown your letter to Dr. Davisson, who is in the best possible position to consider it, and he has written a brief note commenting on it, a copy of which I am enclosing.

Of course our commercial interest in this matter of thermionics is dictated by our desire to get electrons copiously and cheaply, but we naturally believe that the best road to this end must include a thorough understanding of the broad facts of electron emission. Dr. Davisson's work has lain principally along these lines and has already resulted in very interesting information. He has assured me of his interest in reading your letter, which I may add to my own thanks for the trouble you have taken in the matter.

Yours very truly,
[*H. D. Arnold*]

Hall Papers, Harvard University

Oswald Veblen to John von Neumann,[1] October 15, 1929

Dear Neumann:

With the return of Weyl[2] to Zurich, the Jones chair of mathematical physics in Princeton is vacant, and we are very desirous of having lectures on that subject during the current academic year, if possible. It has therefore been decided to offer you a lectureship in mathematical physics for the second term of this academic year. The second term begins on February 5th, 1930, and the lectures end approximately June 1st.

The stipend to be offered for this term is $3,000. plus an allowance of $1,000. to cover travelling expenses, so that the total emolument would be $4,000. The duties would be to deliver a course of lectures (2 or 3 lectures a week) on some aspect of the quantum theory. The lectures could be either elementary or advanced, according to your own preference. If you indicate to me that this sort of an engagement would be acceptable to you, an official invitation will be immediately forthcoming.

I hope that this will seem to you a good occasion to visit America and in particular to make the better acquaintance of mathematicians and physicists in Princeton. I would certainly give great pleasure to my colleagues if you could accept. There are several men here who have followed your work with much interest.

One other question. Considerable interest has been expressed in the work of Dr. Wigner[3] who has collaborated with you in some of your recent papers. Would it not be a good idea to invite him to lecture here at the same time as yourself? If this sort of arrangement would be desirable from your point of view I should be glad to suggest it to our authorities. In any case we will be obliged if you will let us know some of the relevant facts about Wigner.

With best greetings and with the expression of my personal hope that you will be able to accept this invitation, I am,

Yours sincerely,
Oswald Veblen

P.S. I am sending a copy of this letter to you addressed to the Mathematische Institut in case your private address has changed.

Veblen Papers, Library of Congress

1. VON NEUMANN (1903–1957), a Hungarian mathematician, came to Princeton University in 1930; in 1933 he joined the Institute for Advanced Study. From 1954 until his death he was a commissioner of the Atomic Energy Commission.

2. HERMAN WEYL (1885–1955) left the Jones Professorship to assume a chair at Göttingen, 1930–1933. Then he returned to the United States to join the Institute for Advanced Study.

3. EUGENE P. WIGNER (1902–) did join the Princeton faculty. He received the Nobel Prize in physics in 1963.

John von Neumann to Oswald Veblen, November 13, 1929

Dear Professor Veblen!

Please excuse my answering your kind letter so late. It has reached me only this morning, through a delay caused by some irregularities in the delivery of my mail. (I was lecturing during the summer in Hamburg, and besides have changed my Berlin address.)

I feel greatly honoured by the invitation you kindly have extended to me, and should be very glad to be able to follow it. But some personal affairs to which I have to attend in the course of the next week have to be arranged before I can give a definite answer. I will write again within a week.

Today I have spoken to Dr. Wigner. He too would very much like to come to America and especially to Princeton. It would greatly please me if Dr. Wigner, whom I personally greatly esteem and with whom I have been closely connected by scientific work, could also come. Dr. Wigner is 27 years old, a Hungarian, and in the spring of 1928 he was habilitated at the Berlin "Technische Hochschule."

Hoping to see you very soon again

Very sincerely yours
J. v. Neumann.

Veblen Papers, Library of Congress

Irving Langmuir to Gilbert N. Lewis, July 2, 1930

My dear Lewis

I have just spent a delightful evening reading over your two papers (Phys. Rev. June 15 and Science June 6) and can't help writing you a note to tell you how much I admire your wonderful analysis of these fundamental subjects,[1] and how much pleasure I have taken in reading them. I only wish I could talk these things over with you as we did so often when you were in Boston.

I like your treatment of "time's arrow" infinitely better than Eddington's.[2] Last fall I had a long talk with Polanyi and Bonhoeffer[3] (while they spent a week-end with us at Lake George) in regard to the symmetry of time, with inclinations on our parts to reach conclusions much like yours but not so well formulated. Polanyi told me particularly of some views which Ehrenfest had put forward in some meeting a few years ago in which Ehrenfest had maintained that there is *no arrow* to time. The apparent running down of the universe is only the result of the "extraordinarily improbable" state which the universe has been in during a long time prior to our entry on the scene. Ehrenfest is known throughout Europe as the "Human catalyst" for he stimulates clear thinking on the part of everyone he comes in contact with. He has written me that he will be in this country during the summer, stopping here in August and that he will be in California in the fall. I know you will enjoy discussion with him.

The theory of the fundamental complete symmetry of time is a wonderful generalization that should have far reaching consequences. You will remember I think, in 1916, at Washington you and Tolman and I had long discussions in connection with some foolish hypotheses I had been led to make by a sleepless night, but they were based on a principle of reversibility which I had repeatedly found to be of service in chemical problems especially in much unpublished work on kinetics of reactions on solid surfaces.

Later I was enthusiastic over Einstein's 1917 paper but was troubled on analyzing it to find that his induced radiation was not completely in accord with it, altho I was gleeful over his proof that all radiation emitted in one act by one atom must pass *undivided* to *one* other atom. But I had not thought the thing out well enough to be able to contribute any thing useful. It is thus a particular satisfaction now to see this whole matter cleared up so perfectly.

I'm spending all this summer writing up Part II of the paper on Discharges in Gases[4] with Compton which promises to develop into quite a book. Experimentally we're getting some interesting results with discharges in neon dependent on the formation of meta stable atoms by absorption of resonance radiation at distances of some 30.cm from the source, in neon at 1 or 2 mm pressure.

I do wish we could see you here again if you come East.

Yours sincerely
Irving Langmuir

Langmuir Papers, Library of Congress

1. "Quantum Kinetics and the Planck Equation," *Physical Review* 35 (1930): 1533–1537; "The Symmetry of Time in Physics," *Science*, n.s. 71 (1930): 569–577.

2. ARTHUR STANLEY EDDINGTON (1882–1944) was a British astronomer and specialist in relativity.

3. The Hungarian physical chemist MICHAEL POLANYI (1896–1976), at Manchester. K.F. V. BONHOEFFER (1899–[?]), a physical chemist at the University of Frankfurt am Main.

4. Apparently never published. Compton is K. T., presumably.

Gilbert N. Lewis to Irving Langmuir, August 5, 1930

My dear Langmuir:

I was delighted to learn that you were interested in my ideas regarding time and that you approved. I know that there has always been some speculation on this subject but it seemed to have died down, when about five years ago, I was forced to some very definite conclusions on the subject while I was preparing my paper on the nature of light. Since then, it has been a question of formulation and it was not easy for a person brought up in the ways of classical thermodynamics to come around to the idea that gain of entropy eventually is nothing more nor less than loss of information.

While recently a number of scientists have thought about the symmetry of time, many seemed to have arrived at a conclusion opposite from mine, being influenced, I presume, by some current cosmological speculations. I am afraid I am a complete sceptic regarding most of the prevailing ideas of the universe. I really do not think that the universe is in an improbable state but rather that the present condition is probably a good sample of what always has been.

I am just sending off to the *Physical Review* a paper on a quite different fundamental principle, which I call The Principle of Identity.[1]

It is perhaps a little brazen for a person who knows so little about quantum mechanics to write on this subject, but in any case, I am not the only offender and I hope at least that my paper will prevent others from saying some of the very foolish things that have been said recently.

With many thanks for your letter and with best wishes, I am

Yours very sincerely,
[*Gilbert N. Lewis*]

Lewis Papers, University of California, Berkeley

1. "The Principle of Identity and the Exclusion of Quantum States," *Physical Review* 36 (1930): 1144–1153.

Ernest O. Lawrence to Frederick G. Cottrell,[1] July 17, 1931

Dear Dr. Cottrell:

I am hastening to let you know that the experiments on the production of high speed protons have been successful beyond our expectations. About a week ago we finally got the magnet built by the Federal Telegraph Company set up and during the past week we have been producing high speed protons. The highest speeds so far produced correspond to 750,000 volts. This upper limit was set by the 150 watt high frequency oscillator at our disposal. Yesterday we went down to the Federal Telegraph Company and procured a 500 watt short wave power oscillator and are setting it up this morning. This alteration should allow the production of protons in excess of a million volts at once. We have been simply amazed at the intensities of the high speed proton currents. The currents are so large that our electrometer system is too sensitive for them and we guess that the currents are of the order of magnitude of 10^{-8} of an ampere. There is no doubt that we can increase these values considerably.

This week of experimental work has demonstrated unquestionably the practicability of the method and I am convinced that with the aid of the big Federal Telegraph magnet we can go right on up to 20,000,000 volts.

The work has advanced to an exceedingly important stage and the greatest difficulty now confronting us is no longer of an experimental nature, but one of finance. It seems that it is going to be necessary to build a separate building for the large magnet, and accessory equipment will be required, such as a fifty-kilowatt oscillator, which will require something like $10,000 more. This will be necessary if we are to proceed immediately towards the goal of 20,000,000 volts. I was talking to Professor Richtmyer[2] at Stanford yesterday, and he was very enthusiastic over our present results and their importance; and he suggested that we attempt to get money from the Carnegie Foundation. He said that they have special funds for special research projects, and that the way to approach them is through an informal broaching of the matter by a third party. I, of course, immediately thought of you. I presume you got the copy of the letter from Dr. Ernst[3] and suppose that the matter is receiving consideration on the part of the Chemical Foundation. Any ideas you may have along this line, needless to say, would be gratefully received.

I am going east again on July 25th and will be away until August 17th. I am glad that the present work has proceeded to this successful stage so that I can leave with a somewhat eased mind, although of course I should like to be right here to push further developments. However I must go east for unavoidable personal reasons. Possibly you may suggest my calling on someone in New York on the trip.

With cordial regards and highest personal esteem, I am

Sincerely yours,
Ernest O. Lawrence
Professor of Physics

Lawrence Papers, University of California, Berkeley

1. COTTRELL (1877–1948), a physical chemist. With the patent rights for a precipitator of particles from gases, Cottrell founded the Research Corp. to support scientific investigations. Its funds, for example, supported E. O. Lawrence's early work on particle accelerators.

2. FLOYD K. RICHTMYER (1881–1939) was a visiting professor at Stanford in 1931. He had been at Cornell since 1906, serving as dean of its Graduate School from 1931 to his death.

3. Perhaps the chemist ROBERT CRAIG ERNST (1900–) of Louisville University.

Richard C. Tolman to Albert Einstein, September 14, 1931

My dear Professor Einstein:

I was very glad to find your letter of June 27th waiting for me after my return from a vacation, and also to receive the reprints on the cosmological problem and the unified field theory.

When I first saw your proposed quasi-periodic solution for the cosmological line element, I was very much troubled by the difficulties connected with the behaviour of the model in the neighborhood of the points of zero proper volume. The remarks in your letter, however, pointing out that the actual inhomogeneity in the distribution of matter might make the idealized treatment fail in that neighborhood, seem to me very important. Lindemann's[1] suggestion, that the formation of planets might have something to do with the small volume, is also interesting. In addition I think that it is pertinent to remark that from a physical point of view contraction to a very small volume could only be followed by renewed expansion. Hence all in all I am feeling much more comfortable about this difficulty, and indeed have just sent an article to the Physical Review discussing among other things the application of relativistic thermodynamics to quasi-periodic models of the universe.[2]

With regard to the question of setting the cosmological constant λ equal to zero, I think that there are a number of arguments in favor of it but one fairly strong one against it. On the one hand, by giving λ the definite value zero, the fundamental equations of the theory are simplified, the conclusions drawn from them are rendered less indeterminate, and it becomes no longer necessary to inquire into the significance and magnitude of what would otherwise be a new constant of nature. On the other hand, since the introduction of the λ-term provides the most general possible expression of the second order which would have the right properties for the energy-momentum tensor, a definite assignment of the value $\lambda = 0$, in the absence of an experimental determination of its magnitude, seems arbitrary and not necessarily correct.

In a separate envelope, I am sending some reprints of the note on quantum mechanics which was published in the Physical Review by yourself, Podolsky and me.[3] Copies have been sent to the regular mailing list of our Physics Department.

I am also enclosing a reprint on the application of thermodynamics to the problem of the entropy of the universe as a whole,[4] which I hope you will find interesting. I think it shows that in relativistic thermodynamics certain thermodynamic processes can take place at a finite rate and yet reversibly without increase in entropy, in contradiction to the conclusion of classical thermodynamics that reversible processes must take place at an infinitesimal rate. And I believe that this might be important for cosmology.

Please give my very best regards to Frau Einstein, Fraulein Dukas and Dr. Mayer, and accept for yourself my warmest feelings of admiration and affection.

Sincerely yours,
Richard C. Tolman

P.S. You will be interested to know that we had another sitting with Upton Sinclair's medium and again nothing happened.

R. C. T.

Einstein Papers, Institute for Advanced Study

1. FREDERICK ALEXANDER LINDEMANN (1886–1957) was a physicist who headed the Clarendon Laboratory at Oxford. A friend of Winston Churchill's, Lindemann (Lord

Cherwell after 1941) acted as the Prime Minister's scientific advisor during World War II.

2. "Theoretical Requirements for the Periodic Behavior of a Universe," *Physical Review* 38 (1931): 1758–1771.

3. Einstein, Tolman, and Boris Podolsky, "Knowledge of the Past and Future in Quantum Mechanics," *Physical Review* 37 (1931): 780–781.

4. "On the Problem of the Entropy of the Universe as a Whole," *Physical Review* 37 (1931): 1639–1660.

Ernest O. Lawrence to John D. Cockcroft[1] and Ernest T. S. Walton,[2] August 20, 1932

My dear Sirs:

I want to thank you very much for the reprints of your epoch making experiments on the disintegration of the elements by high velocity protons, and I hope you will continue to send me accounts of your work in the future. Under separate cover I am sending you reprints of the work of myself and my coworkers on methods for the acceleration of ions which you may find of some interest.

At the present time we are attempting to corroborate your experiments using protons accelerated to high speeds by our method of multiple acceleration. We have some evidence already of disintegration, though as yet we can not be certain. Unfortunately our beam of protons is not nearly as intense as yours although of higher voltage. Whenever we obtain some reliable results of course we will let you know promptly.

All of us are looking forward to the visit of Professor Fowler[3] with us after Christmas and hope that he will be able to tell us a great deal about your work.

With cordial good wishes, I am,

Sincerely yours,
Ernest O. Lawrence
Professor of Physics

Lawrence Papers, University of California, Berkeley

1. JOHN DOUGLAS COCKCROFT (1897–1967) became a Nobel Laureate in 1951 for the first nuclear transformation by artificial means. From 1946 to 1959 he was head of Britain's Atomic Energy Research Establishment at Harwell.

2. WALTON (1903–) shared the Nobel Prize with Cockcroft. From 1934 to 1974 he was at Trinity College, Dublin.

3. RALPH H. FOWLER (1889–1944), theoretical physicist at Cambridge.

John D. Cockcroft to Ernest O. Lawrence, September 17, 1932

Dear Sir,

Many thanks for your letter of August 20th. We have naturally been very interested in your methods of multiple acceleration and have always considered that in such methods lie the main hope of working with ion energies of over a million volts. We shall therefore be most interested to hear whether you can overcome the difficulties of the small current values to which you refer.

We intend here for the time being to limit ourselves to work below a million volts since we shall not have the space for higher voltage work with our present methods. We have very little new to report since our last publication. We have added Magnesium to the metals from which we can get scintillations and have had negative results from gold and chlorine.

We are all rather puzzled by the results on the heavy elements and we now propose to study these rather more intensively. Lithium, boron, fluorine, sodium and aluminium seem obviously to be proton-alpha particle transitions. The remainder require a good deal more work before we can have any certainty.

I look forward to visiting your laboratory within the next year or so. I have a long standing engagement to visit friends at M.I.T. and should hope to include California.

Yours Sincerely,
J. D. Cockcroft

Lawrence Papers, University of California, Berkeley

Ernest O. Lawrence to Frederick G. Cottrell, September 22, 1932

Dear Dr. Cottrell:

I was glad to get your letter with all the eastern high voltage news, but I was sorry to learn that you are not going to be able to come out here this winter. I am glad to hear that Van de Graaff's[1] work is coming along so nicely and particularly that he had made thorough tests proving that his vacuum generator is entirely practical. Harry Barton[2] writes me also that his pressure outfit is working satisfactorily and I hear that Prof. Davis[3] at Columbia is building a Van de Graaff's machine. Out of all this I am sure we can expect some valuable developments of Van de Graaff's original efforts. Of course the most exciting of all is his big project at Round Hill. I can hardly wait to hear what happens when he applies the large spheres to a vacuum tube.

I was glad to hear about Tuve's[4] recent activities and I'm sorry that he has to mark time because of needing housing for the sphere. Tuve is doing excellent work; his design of a high voltage vacuum tube I believe is to form an integral part of the successes achieved with the Van de Graaff generator. Van has developed a source of high voltage and Tuve has developed a tube which will stand the voltage. The two should work together beautifully.

I suppose you know that the President turned over to us the old Civil Engineering Laboratory (the frame building behind Gilman Hall) to house the magnet and provide space for all our activities involving high speed particles. The Regents have given the place the official name of The Radiation Laboratory of the University of California. There is quite a group now working with me on these problems and already we have been getting some very interesting results. Using the small proton accelerator (the outfit in LeConte Hall) we have just completed preliminary experiments on the disintegration of lithium. A report on these experiments, which form a confirmation of Cockcroft and Walton's results, has been sent to the Physical Review in the form of a letter to the editor, presumably it will come out in the October 1st issue. In these preliminary experiments, proton voltages up to 710,000 were used and now we are going ahead to higher voltages. Livingston[5] is busy with the big magnet outfit and is at the present time generating 2,000,000 volt hydrogen molecule ions. He has just succeeded in overcoming some vacuum difficulties which have harassed him for some time and I predict that in the next few weeks he will be producing hydrogen ions with energies up above 3,000,000 volts. When he gets up in the 3,000,000 volt range he intends to stop and bombard various elements with them before going to higher voltages. Coates[6] has our new mercury ion tube working very nicely delivering 2,800,000 volt mercury ions and at the moment he is studying the production of x-rays and secondary electron emission by these high voltage ions. Sloan, Livingood and Exner[7] are busy with the new x-ray tube. In a short while they will have made tests on the new tube and then will proceed on the construction of one for the Institute of Cancer Research. I believe this gives you an outline of our activities in the Radiation Laboratory.

With cordial regards and hoping to hear from you again in the not distant future, I am

Sincerely yours,
Ernest O. Lawrence
Professor of Physics

Lawrence Papers, University of California, Berkeley

1. ROBERT J. VAN DE GRAAF (1901–1967), physicist, was the inventor of an electrostatic high voltage generator.

2. HENRY ASKEW BARTON (1898–) was director of the American Institute of Physics 1931–1957.

3. BERGEN DAVIS (1869–1958).

4. MERLE A. TUVE (1901–) was with CIW's Department of Terrestrial Magnetism; in 1942–1946 he was director of the Applied Physics Laboratory at Johns Hopkins.

5. MILTON STANLEY LIVINGSTON (1905–). See his *Particle Accelerators: A Brief History* (Cambridge, 1969).

6. Wesley Coates worked with Lawrence from his senior year, 1930, until 1933. From January 1934 he was at Columbia but working on his thesis for a Ph.D. at Berkeley. On March 20, 1937, he died in a cyclotron accident at Berkeley.

7. DAVID H. SLOAN (1905–) is now emeritus professor of electrical engineering at Berkeley. JOHN JACOB LIVINGOOD (1903–) has been at Argonne National Laboratory since 1952. FRANK METCALF EXNER (1898–) was then in cancer research. From 1942 until 1963 he was at the Honeywell Research Center.

Ernest Rutherford to Gilbert N. Lewis, May 30, 1933

My dear Lewis

I was delighted to receive your concentrated sample of the new hydrogen isotope in good shape, and we shall certainly take an early opportunity of examining its effects in our low voltage apparatus which Dr. Oliphant[1] and I have been using the past year.

I have been enormously interested in your work of concentration of the new isotope with almost unbelievable success. I congratulate you and your staff on this splendid performance. I can appreciate the extraordinary value of this new element in opening up a new type of chemistry. If I were a younger man I think I would leave everything else to examine the effects produced by the substitution of H^2 for H^1 in all reactions.

Next, I should like to congratulate Lawrence and his colleagues for the prompt use they have made of this new club to attack the nuclear enemy. Cockcroft showed me the letter of Lawrence giving his preliminary results which are very exciting. These developments make me feel quite young again as in the early days of radioactivity when new discoveries came along almost every week, for it is a double scoop not only to prepare this new material but also to have the powerful method of Lawrence to examine its effects on nuclei. I wish them every success in their work and as soon as we can arrange it, I will try out the effects we can observe at our low voltages.

I am glad to know that you enjoyed the visit of my son-in-law, Fowler. He is back again and I am glad to say his children are all flourishing. Cockcroft is leaving today for the U.S.A. and I understand he will visit you in California. Apart from his scientific work proper, he is an electrical engineer on whom we all rely.

With best wishes to you all and good success to your labours; and my most grateful thanks for your splendid gift.

Yours very sincerely,
Rutherford

Lewis Papers, University of California, Berkeley

1. Mark Oliphant (1901–) is a physicist born in Australia. From 1937 to 1950 he was at Birmingham University and returned to Australia in 1950. He was governor of South Australia 1971–1976.

Norbert Wiener to Paul de Kruif,[1] August 3, 1933

My dear Mr. de Kruif:

I have read several of your books of scientific biography with great interest, as well as Lewis' "Arrowsmith," to which you contributed so much of the background, and I have recently been following your story of Kettering in the S.E.P.[2] As a practicing scientist myself, I naturally set a high value on the popularization of science and of the circumstances under which science is carried on. I wish to congratulate you for your pioneer work in a little-exploited field. May I then, without giving offense, submit to you a few questions and criticisms?

First of all, you leave the impression in the mind of the reader that applied science, followed with an eye to the benefit of humanity, is perhaps the most worthy of all pursuits, but that pure science, the fruit of mere curiousity, is likely to be dilettantish and fruitless. This impression may not be intended by you: it may be due to the fact that you are writing of Kettering, who is primarily a practitian. I feel, however, that is your own opinion, and I wish to protest against it.

Here I had better state my own position. I am a professor of mathematics at the Massachusetts Institute of Technology. During the term, I am in daily contact with classes of young engineers, teaching them the calculus of the first two years, the function-theory and differential equations of the third, and conducting an advanced course on Fourier series. I have the closest relations with colleagues in the engineering departments, particularly in electical engineering, in which department I have supervised master's and doctor's theses. I am not an engineer, but I have sufficient acquaintance with electical engineering to have done a little mild inventing in the field of filters and networks. On the other hand, the great bulk of my research lies in the purest of pure analysis. My master, Hardy of Trinity College in Cambridge, regards all applications of mathematics with a contempt which I do not indeed share. Landau of Goettingen has scarcely more use for practice. It is with these men that my scientific bonds are closest.

It is my conviction that there are two ways of approaching scientific research. You may have a problem, and look for an answer, or you may have an answer, and look for the problem that it fits. The first is the path of applied research, the second that of pure research. Both are good methods in the proper hands, but there are men who tend naturally to one or the other. The way of pure research is opposed to all the copy-book maxims concerning the virtues of industry and a fixed purpose, and the evils of guessing, but it is damned useful when it comes off. It is the diametrical opposite of Edison's reputed method of trying every conceivable expedient until he hit the right one. It requires, not diligence, but experience, information, and a good nose for the essence of a problem.

Above all, it requires a sense for the body, the "thickness" of an idea. There is a certain school of scientists, particularly in mathematics, who value all generalizations equally. The Chicago school has a bias this way. On the other hand, some men have a flair for important ideas. (By the way, nobody knows how to define an important idea, but any scholar who is worth his salt knows one when he sees one.) Riemann's[3] collected writings fill two thin volumes, but every paper that he wrote started a new field of mathematics. Cayley's collected works bulk with the Encyclopedia Brittanica—give me Riemann!

Work of this sort is rather slow in paying dividends. Newton's investigations took a hundred years to penetrate into engineering. Riemann's work is taking sixty or seventy, but its effects are already being felt. Even Lagrange's[4] work is barely coming to its own in an engineering sense. Yet the vast structure of modern mathematics is a tool which no engineer worthy of the name can afford to ignore, which no leader of engineering can afford to wield clumsily.

Our Vice-President and Dean of Engineering, Vannevar Bush,[5] is as well-rounded an engineer as we have anywhere, here or abroad. He has an unusual knack with tools, and his basement is a young machine-shop. He has devised inventions and practical gadgets without number, in the fields of electical, hydraulic, mechanical, and mining engineering, and in navigation. Perhaps his chief interest at present is the development of machines to perform the solution of differential, difference, and integral equations—the old dream of Babbage come true, with modern technique. His differential analyser represents a marvellous degree of success in this direction. The inspiration of his program is the philosopher and pure mathematician Leibniz;[6] its tendency and result, the opening to the engineer of field after field of mathematics, which the tediousness of computation has hitherto closed to practice. Dr. Caldwell,[7] Bush's right-hand man in this program, describes himself as a "monkey-wrench mathematician," and I can bear witness how exacting the demands on his mathematical powers really are.

There is no M.I.T. school of mathematicians. My colleagues are of the most diverse mathematical interests and origins. Struik is a geometer from Leiden, Schouten's former assistant.[8] Franklin has a Princeton degree, and is equally at home in analysis situs, in mechanics and in pure analysis. Douglas[9] has solved the Plateau problem, and is one of the pioneers in the geometry of paths. Hopf[10] is a Berlin man, an authority in dynamics, in the theory of differential equations, and in astrophysics. There is however one belief we hold in common: the best applied mathematics can only be developed with the full use of the technique of the best pure mathematics, and the best pure mathematics may draw its inspiration from problems arising in physics and in engineering. We best serve the engineer by following our own bent and curiosity wherever it may lead us, understanding the engineer, and appreciating the scientific interest of his problems, but in no wise regarding ourselves as his servants.

We have very little use for the organized research which is the fetish of the present day. It may be a necessity, especially in a commercial organization, but it is a cramping and crippling necessity. We work in whatever direction our work points, and though we discuss and collaborate, we never apportion research between us. There are those among us who feel the needs of humanity in a very direct and personal way, but in our work, we do not think at the time of some particular need of humanity. If I may speak for the group, this is our philosophy:

"Mathematics is a subject worthy of the entire devotion of our lives. We are serving a useful place in the community by our training of engineers, and by our development of the tools of future science and engineering. Perhaps no particular discovery that we make may be used in practice; nevertheless, much of the great bulk of mathematical knowledge will be, and we are contributing to that bulk, as far as lies in us.

"Moreover, a clearly framed question which we cannot answer is an affront to the dignity of the human race, as a race of thinking beings. Curiousity is a good in itself. We are here but for a day; tomorrow the earth will not know us and we shall be as though we never were. Let us then master infinity and eternity in the one way open to us: through the power of the understanding. Knowledge is good with a good which is above usefulness, and ignorance is an evil, and we have enlisted as good soldiers in the army whose enemy is ignorance and whose watchword is Truth. Of the many varieties of truth, mathematical truth does not stand the lowest.

"Since we have devoted our lives to Mathematices—and she is no easy mistress—let us serve her as effectively as we may. If we work best with an immediate practical problem in view, well and good. If mathematical fact comes to our mind, not as a chain of reasoning, built to answer a specific question, but as a whole body of learning, first seen as in a glass, darkly, then gaining substance and outline and logic, well and good also. The whole is greater than the parts, and in a lifetime of achievement, no one will care what particular question of practice was in the scholar's mind at such and such a moment."

This is of course the point of view of the German liberal scholar of the middle of the last century, imbued with the poetry of Schiller and Goethe, with the flaming intellectual passion of Heine, with the deep worship of the intellect of that magnificent line of philosophers from Kant to Schopenhauer. All modern professional scholarship is the heir of that Germany. I have sometimes wondered whether you fully understood it. There are many Germans of that period in your gallery of portraits, but they all seem to have a little taint of the grotesque, of the caricature of the professor in the college humorous journal rather than as he really is. They are friendly caricatures if you like, but at the bottom of it, we see the author laughing a little at anybody so remote from the normal outlook of humanity. "I'll be darned," he seems to say, "isn't he the funny little fellow." I do not find this same external attitude in your picture of mid-western scientists like Kettering. So too in Lewis' "Arrowsmith," Arrowsmith himself is understood, while Gottlieb and Sondelius are lay-figures arranged in interesting poses.

Thus I think that your sketches of scholars are likely to be misleading on account of a misplaced emphasis—an unavoidably misplaced emphasis. For all the wreck of the present Germany, the old German tradition of scholarship is the central tradition, abroad as well as in Germany. The scholars whom you portray best and clearest are a real but aberrant type, even within the scholarly world of this country. They are foreigners in the great international world of scholarship, which is as well represented here as abroad, whose bonds are personal as well as scientific, which is a great democracy with many tongues but only one habit of mind. It is much harder to bring this world home to the American lay public than to portray a character like Kettering, who is by your account the great American handy man writ large. I am not lacking in respect for that kind of ability, although I must deprecate the contempt for pure scholarship which you seem to attribute to Mr. Kettering. However, I do not think that that ability can attain its peak unless it is joined with some such intellectual subtlety as, let us say, Dean Bush possesses. I have no doubt that Mr. Kettering possesses some such subtlety, but you have not brought it out in your series of articles.

I question whether Mr. Kettering's laboratory stands in the class of those of the General Electric Company or The Bell Telephone Laboratories in its contributions to scholarship or to engineering. You certainly go too far in crediting Mr. Kettering with the distortionless telephone line. This was the discovery of Heaviside[11] about 1890. Heaviside, by the way, should be rich literary material for you.

In this connection, I should like to see a series of biographies of engineers from your hand. Other names could be Nicola Tesla and B. A. Behrend.[12] I should be glad to render you any possible help in the matter.

I trust that you will not take my remarks amiss. If they interest you in any way, I should be delighted to hear from you.

Very truly yours,
Norbert Wiener

Wiener Papers, Massachusetts Institute of Technology

1. DE KRUIF (1890–1971), originally a bacteriologist, had spent the years 1920–1922 at the Rockefeller Institute. To the dismay of Loeb and Flexner, he was the source of Sinclair Lewis's use of Loeb as a model for Gottleib in *Arrowsmith*. De Kruif had a very successful career as a popularizer of science. His *Microbe Hunters* (1926) was widely read.

2. In July, August, and September of 1933 a series on CHARLES FRANKLIN KETTERING (1876–1958) by de Kruif appeared in the *Saturday Evening Post*. A successful inventor and engineer, Kettering directed research at General Motors for 27 years. "Boss Kettering," the title of the series, represented a different tradition than the one Wiener extolled in this letter.

3. GEORG FRIEDRICH BERNHARD REIMANN (1826–1866) of Göttingen.

4. JOSEPH LOUIS LAGRANGE (1736–1818), the French mathematician.

5. BUSH (1890–1974), an electrical engineer, would become the most important individual in the emergence of a new science policy during and after World War II. He became president of CIW in 1939, serving until 1955. During the war Bush was the head of the Office of Scientific Research and Development. See his autobiography, *Pieces of the Action* (New York, 1970).

6. The great polymath GOTTFRIED W. B. LEIBNIZ (1646–1716).

7. An electrical engineer, SAMUEL HAWKS CALDWELL (1904–1960) toward the end of his career would invent a linotype machine for Chinese.

8. DIRK JAN STRUIK (1894–). JAN ARNOLDUS SCHOUTEN (1883–1971).

9. JESSE DOUGLAS (1897–1965).

10. EBERHARD HOPF (1902–), at Leipzig 1936–1944, Munich 1944–1949, since 1949 at Indiana University.

11. OLIVER HEAVISIDE (1850–1925) of Britain was self-taught. He was a mathematical physicist eminent enough to receive an honorary doctorate from Göttingen.

12. The electrical inventor NICOLA TESLA (1856–1943) was born in what is now Yugoslavia. B. A. BEHREND (1875–1932), a Swiss-born electrical engineer, had a notable career with several large American corporations.

Ernest Rutherford to Gilbert N. Lewis, October 30, 1933

Dear Lewis:

I have just got back from Brussels where I had the pleasure of meeting Lawrence and hearing all about his latest results. He is a broth of a boy, and has the enthusiasm which I remember from my own youth. We are expecting him at the Laboratory tomorrow.

The heavy water arrived safely and I will soon make use of it. Harteck[1] has had better luck than I anticipated in concentrating hydrogen. When I came back I found he has got a sufficient concentration to be useful for a good deal of our work. From now on we must be responsible for our own supply, but I would like again to thank you for your kindness in helping us out of your stock at the time when it was very precious.

With kind regards

Yours sincerely,
Rutherford

Lewis Papers, University of California, Berkeley

1. The physical chemist PAUL HARTECK (1902–), born in Austria, was a Rockefeller Fellow at the Cavendish Laboratory 1933–1934, then professor at the University of Hamburg 1934–1951; after that he was at Rensselaer Polytechnic Institute.

Percy W. Bridgman to Richard C. Tolman, September 9, 1934

Dear Tolman:

It was good to get your friendly letter. There seems to be a good deal of flutter recently on the subject of dimensions, and various echoes of it had reached me before your letter came. I had not been stimulated to doing any fresh thinking on the subject, however, so that I am afraid that I am in much the same position that you were and my reactions to your letter can signify only the present state of the precipitate left in my mind from a previous condition of activity.

In general, it seems to me that we are closer together than we have been before, and that I understand much better what your position is. In fact, I think that I would admit that you have a *right* to do nearly everything that you want to do, and that our differences have reduced largely to matters of taste. At the same time there do still seem to be real differences of taste, and I cannot for the life of me understand why you want to do some of the things that you apparently do.

I do not believe that there is as clean cut a separation between the two ways of looking at the subject as you would suggest. I suppose that you intend your description of the first way of looking at dimensional analysis, that is, the attitude which regards a dimensional formula as a compact way of summarizing certain operations which have been performed on numbers, as essentially a statement of my own attitude. But my attitude in the face of a dimensional formula contains much more than a consciousness of the way in which certain numbers were obtained—it contains also as a vital part of the background a consciousness of the physical operations involved in the measurements and also a realization, obtained from "the experience of all the ages," that it is useful to describe our experience in terms of these operations. In fact, it seems to me that in my background is embedded all that is to be found in your second way of looking at the matter, and more too. I will admit that if one wants as *comprehensive* a view as possible of all physics one may perhaps do well to adopt the classical division into geometry, kinetics, mechanics, electrodynamics, and thermodynamics, and therefore use five fundamental kinds of quantity in his definitions. But for the purposes for which dimensional analysis is used, one does not usually want to be reminded of the most comprehensive possible point of view, but instead one wants the most special one possible. For example, in treating a problem in ordinary elasticity, like the bending of a beam, one is saying more about nature and embodying a more extensive experience in defining force as a fourth kind of unit of its own kind, instead of defining it as mass times acceleration. It is saying more about nature, and it is more pertinent to say, that the problems with which we are confronted can be split up into subgroups, each capable of its own method of treatment, than it is to say that all phenomena can be treated under five grand subdivisions. And when you get right down to it, the five is not very significant, because we cannot yet describe biological and mental phenomena in terms of these five. In other words, in my experience the occasions are so rare on which I have found it profitable to be reminded that that selected subgroup of experience which we have chosen to call physics can be treated with complete generality under five main subdivisions, that I do not care to clutter my tool box by attaching special importance to this scheme of description, and I don't quite see why you do either. I think you will find this to be a fair description if you think over the uses made of dimensional analysis. I should class these as: changing units, checking equations, finding necessary relations in complicated experimental situations so as to reduce the number of necessary observations, as in model experiments, and perhaps in certain theoretical arguments as in the last chapter of my book. I think that all these purposes are served with a flexible system of definition as well if not better than with a system of five, and certainly the use in model experiments demands the maximum possible flexibility.

I think that I would be inclined to disagree with your position that your two ways of looking at dimensional analysis describe the attitude of actual physicists. It seems to me that there is still an awful lot of messy thinking about the whole thing, as in the classical statement of Eddington that a length of 10^{-41} cm must be the key to some essential physical structure.

I was glad to have you say what you did about the unattractiveness of and as fundamental kinds of quantity.[1] Leigh Page, who spends the summer next to me, has been writing an article with Adams,[2] which I hope will be published in the *American Physics Teacher,* in which he points out some disconcerting relations in this connection which are not ordinarily realized.

I too wish that we might have chances to talk together oftener. You will probably want to talk more about this, and I know that there are oodles of things about relativity theory that I am itching to get straight from you. Haven't you any inclination to spend another summer in the East? Randolph, only ten miles from the top of Mt. Washington, is very attractive in the summer, there are lots of nice cottages which you might rent, and we would be tickled to pieces to add you to our community.

Most sincerely,
[*Percy W. Bridgman*]

Bridgman Papers, Harvard University

1. This sentence had two blanks in the original as indicated here. The letter from Tolman to which this responds has not been found.
2. Perhaps Leigh Page and N. I. Adams, Jr., "Some Common Misconceptions in the Theory of Electricity," *American Physics Teacher* 3 (1935): 51–58.

Percy W. Bridgman to Richard C. Tolman, December 8, 1934

Dear Tolman:

I have been looking through your book on relativity. It seems to be just the sort of thing that I have been looking for for a long time, and I hope that sometime not too far away I can get down to really study it and think about it. It is a subject that I have always felt unhappy about, particularly in regard to the fundamentals. It looks to me as though you really have thought through a lot of the questions that are bothering me. From my point of view the ideal thing would be for you to spend a summer in the White Mountains as I suggested in my last letter, and then perhaps I might pump you while we were walking over the range. I suppose however that there is no very immediate prospect of anything so fine as I am going to anticipate by asking one question that has been exercising me this fall.

Imagine two gravitating masses held apart by cords attached to a frame as indicated. The thing is set in uniform translation of motion. Consider now the cords: They are under tension and in motion and there must be a flow of energy through them just as in a belt transmitting power from one pulley to another. This means that there is a continuous flow of energy into the one gravitating mass and out of the other. In the steady state energy must therefore get across from one mass to the other through empty space by a gravitational analogue of the Poynting vector, but no such vector exists. I have asked the two experts on gravitational theory in our neighborhood, Vallarta and Birkhoff,[1] and neither has ever thought of such a difficulty or can see what the solution is. In fact, Vallarta has now puzzled over it for several weeks with no success, yet there must be some simple answer and I hope you have it. One of the ambitions in the back of my head is to some day work out a general theory of relativity starting from the consideration of just such simple physical situations as this. It seems to me that something ought to be possible that would appeal much more to the blacksmith type of physicist like myself, and at the same time, make all the connections with experiment that gravitational theory now does, although perhaps not coinciding with present gravitational theory in higher order terms or being so elegant mathematically.

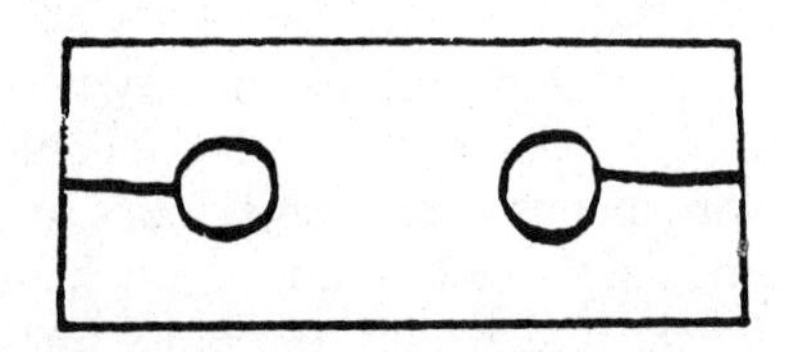

With most cordial greetings,
P. W. Bridgman

Bridgman Papers, Harvard University

1. G. D. Birkhoff. MANUEL SANDOVAL VALLARTA (1893–), a Mexican physicist, was then at M.I.T.

Richard C. Tolman to Percy W. Bridgman, December 15, 1934

Dear Bridgman,

It was very pleasant to hear from you. I am sometimes quite homesick for New England, although I suspect it is for the idealized country of my boyhood rather than for the reality that now exists. In addition, my blood has grown so thin here in Southern California, that I am afraid it would now seem very cold to me outdoors and stifling hot within, back in New England.

I will try to give some sort of answer to your question, where the numbers of equations and sections will correspond to my book.

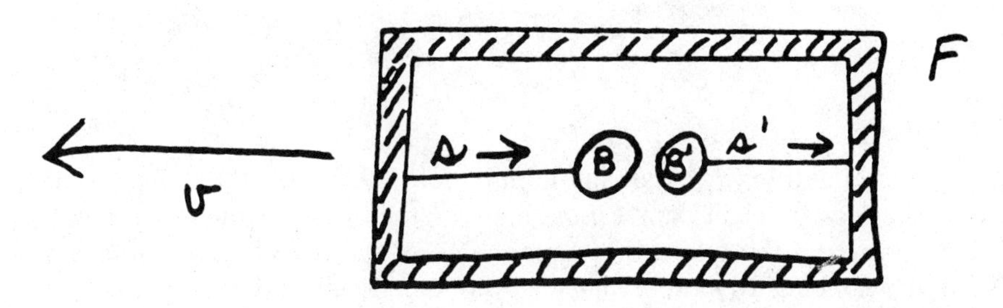

We consider a framework F, supporting two gravitationally attracting balls B and B′ with the help of cords, and then look at the system from a point of view such that the whole apparatus is moving with the uniform velocity v as shown in the figure.

We first try to interpret this physical situation with our older stock of physical notions as to the flow and conservation of energy. Since the two moving strings are under tension we see that there will be a flow of energy s, s′ along the two strings as indicated in the figure. And since the system is in a steady state, we then ask what becomes of the energy that flows into the ball B from its string, and where does the energy come from that flows out of ball B′ into the other string. In a general way we then answer this by saying that there must be some possibility for a flow of energy through space, out from ball B into its surroundings and into ball B′ from its surroundings. We realize, nevertheless, that our older stock of physical notions has given us no specific ideas as to the location and transfer of energy in a gravitational field and hence turn to the relativistic theory of gravitation for possible illumination at this point.

To investigate this we must first inquire into the possibility of expressing the principles of relativistic mechanics in a form which corresponds to those older ideas as to the transfer and conservation of energy which were made use of above, since I do not think that we have any a-priori right to demand the possibility of giving a relativistic answer to a problem which has been formulated in a pre-relativistic language. Fortunately for our present purposes, however, Einstein has shown that the principles of relativistic mechanics can be expressed in the non-tensor but nevertheless covariant form

$$\frac{\partial}{\partial x^{\nu}} (\mathcal{T}_{\mu}^{\nu} + t_{\mu}^{\nu}) = 0 \qquad (87.14)$$

which does correspond in general to our older principles of the conservation of energy (with $\mu = 4$) and of momentum (with $\nu = 1, 2, 3$), and indeed reduces at the limit of negligible gravitational field, in appropriate coordinates, to our usual expression

$$\frac{\partial T^{\nu}_{\mu}}{\partial x^{\nu}} = \frac{\partial T^{1}_{\mu}}{\partial x} + \frac{\partial T^{2}_{\mu}}{\partial y} + \frac{\partial T^{3}_{\mu}}{\partial z} + \frac{\partial T^{4}_{\mu}}{\partial t} = 0 \qquad (37.9)$$

for these principles.

The quantities τ^{ν}_{μ} occurring in equation (87.14) are the components of the tensor density corresponding to actual *material* energy and momentum and the t^{ν}_{μ} are the components of the pseudo-tensor density corresponding to potential *gravitational* energy and momentum. In relativistic mechanics, just as in Newtonian mechanics, the principle of the conservation of energy can only be maintained by introducing the idea of potential energy. In the Newtonian treatment this potential energy was regarded as somehow roughly associated with the gravitational field. In the relativistic treatment the t^{ν}_{μ}, as defined by equation (87.12), give us definite expressions for the densities of potential gravitational energy *and* momentum and for their densities of flow.

If we are interested in the special case of *energy* flow in regions *free from matter* and electromagnetic energy, equation (87.14) reduces to the form

$$\frac{\partial t^{1}_{4}}{\partial x} + \frac{\partial t^{2}_{4}}{\partial y} + \frac{\partial t^{3}_{4}}{\partial z} + \frac{\partial t^{4}_{4}}{\partial t} = 0$$

where x, y, z are space-like coordinates and t is the time-like coordinate. In agreement with this expression we may now interpret t^{4}_{4} as the density of our potential gravitational energy, and t^{1}_{4}, t^{2}_{4} and t^{3}_{4} as the components of the density of flow of this kind of energy. The general theory of relativity thus *does* provide something which is a sort of gravitational analogue to the Poynting vector.

I hope that this will make you feel somewhat more comfortable, especially when you note (see §§ 92, 97) that t^{4}_{4} is such a quantity as to make the total energy of a gravitating sphere of liquid reduce at the Newtonian level of approximation to the sum of the proper energy of the sphere plus the familiar expression for its potential energy.

I wouldn't want to make you feel too comfortable, however, as to the profits to be obtained by trying to treat gravitational phenomena in such a way as to make the maximum use of pre-relativistic notions. The components t^{ν}_{μ} of the pesudo-tensor density are defined in such a way as to make their calculation extremely complicated. Hence for practical computations we usually find it much easier to use the principles of relativistic mechanics in the form

$$\frac{\partial T^{\nu}_{\mu}}{\partial x^{\nu}} - \frac{1}{2} T^{\alpha\beta} \frac{\partial g_{\alpha\beta}}{\partial x^{\mu}} = 0 \qquad (84.7)$$

which makes no specific reference to the possibility of getting conservation laws by introducing the definition for a new quantity which can be regarded as related to potential energy and momentum. Furthermore, it is to be emphasized that t^{ν}_{μ} is only a *pseudo*-tensor density, whose components transform when you go from one system of coordinates to another in a non-linear and extremely complicated way. Indeed, at a given point in space-time the components may all be zero in one system of coordinates and not zero with respect to another system of coordinates, so that the localization that you give to your "invented" potential energy changes with the coordinate system in a specially drastic matter.

It may be very difficult to build up gravitational theory by the consideration of "simple" physical situations such as you suggest, since many of these situations may only remain simple so long as you apply the older stock of physical notions to them, and become very complicated as soon as you try to find the essentially new ideas which will be needed for the treatment of gravitation. If you consider the series

$$\left\{\begin{matrix}\text{Classical}\\ \text{Mechanics}\end{matrix}\right\} \underset{\longleftarrow}{\dashrightarrow} \left\{\begin{matrix}\text{Special}\\ \text{Relativity}\end{matrix}\right\} \underset{\longleftarrow}{\dashrightarrow} \left\{\begin{matrix}\text{General}\\ \text{Relativity}\end{matrix}\right\}$$

it takes a real process of generalization and invention to find out how to go from left to right, while the passage from right to left, after some great man like Einstein has given you what stands to the right, is one of strict implication.

Well, you probably are getting bored, and so I will merely tell you how much I enjoyed hearing from you and making a stab at an answer.

With all best wishes and season's greetings, sincerely your old fellow High School student.

Richard C. Tolman

Bridgman Papers, Harvard University

Ernest O. Lawrence to Howard Andrews Poillon,[1] December 29, 1934

Dear President Poillon:

In my report to you this year I have not given a detailed abstract of all our activities, as I understand that you are not particularly interested in many technical details. In the report I have endeavored to give a general impression of our activities, with particular emphasis on the discovery of artificial radioactivity. Appended to the report is a list of the publications during the year. This list does not include four papers now in process of publication. One of these is a detailed account by Sloan of the x-ray machine, and another is a report of my experiments on the transmutations of sodium by deutons.

Our progress this year has been extremely gratifying, and again I want to express to you personally my very deep appreciation of your interest and support.

With warmest New Year greetings and highest personal regards, I am

Sincerely yours,
Ernest O. Lawrence

Enclosure:
report

Report to the Research Corporation on the Activities during 1934 of the Radiation Laboratory, University of California

When we undertook the development of experimental methods for the study of the nucleus of the atom, we were entering a virgin field of investigation with little knowledge of what to expect; but we felt sure that important things would come from the study of the nucleus, as we knew at least that more than 90% of the atom's mass and energy was in the nucleus and that indeed the very nature of the atom was determined by the nucleus. During the past year, there have been several discoveries that have already demonstrated the great importance of nuclear physics, and in my report this year I wish to single out for mention the one which perhaps is the most important, namely, artificial radioactivity.

Early in the year Irene Curie, the daughter of Madame Curie, and her husband, F. Joliot,[2] discovered that some substances are rendered radioactive by bombardment with the alpha rays of radium. This discovery suggested that perhaps protons and neutrons, the high speed atomic projectiles produced by the apparatus in our laboratory, might be able to make various substances radioactive in a similar way. Following this lead, we bombarded various substances, and to our amazement discovered that almost every substance bombarded was rendered radioactive. We were able to manufacture in the laboratory radioactive atoms. This indeed was something new under the sun.

During the past six months we have been very busy with experiments, learning more about the many new radioactive substances. We have identified by chemical tests many of the new radioactive atoms such as radio-nitrogen, radio-oxygen, radio-fluorine, radio-sodium, radio-aluminum, and so forth.

Radio-sodium is of especial interest because of its applications to medicine. Not only are the rays given off from radio-sodium more powerful than those from radium, but sodium is not a poison. On the contrary, it is a wide-spread constituent of living matter, and therefore can be injected into the blood or taken internally. My medical colleagues tell me that the properties of radio-sodium are almost ideal for many medical applications, such as the treatment of cancer. Although of course our chief interest does not lie along medical lines, we appreciate the importance of this aspect of our work, and it is our intention to push forward the production of radio-sodium and others of these new radioactive substances in such quantities as will enable our medical colleagues to investigate their biological uses.

Our high voltage x-ray machine, installed in the University Hospital, has now been in daily operation almost a year. I am glad to report that the performance of the machine has already amply rewarded the expense and effort involved. Tests have shown that it is the most powerful generator of x-rays in existence, and for some medical treatments, is more effective than all the radium in the world. Through the months, hundreds of people afflicted with cancer have received treatments, and though it is too early to make any statements as to the percentage of cures, there can be no doubt that many have been benefitted. We are cooperating with Dr. Francis Carter Wood,[3] director of the Institute of Cancer Research, in the construction and installation of a similar x-ray machine at the Medical Center in New York. Most of the equipment has been built in New York, but some of the essential parts have been constructed in our laboratory here, and it seems likely that this second installation will be in operation soon.

Although in this report I have pointed out the medical applications of the results of our work, I do not wish to convey the impression that I regard this aspect as the most important. Our primary interest lies in the direction of furthering knowledge of the nucleus of the atom, knowledge that is fundamental to an understanding of all material things. A list of the publications of the Radiation Laboratory during 1934 is attached.

Again I wish to express my deep feeling of gratitude, and in this my colleagues join me, to the Research Corporation for the generous support which has made possible the work of the Radiation Laboratory.

Lawrence Papers, University of California, Berkeley

1. POILLON (1879–1954), an engineer, was president of the Research Corp.

2. The married couple IRÈNE CURIE (1897–1956) and F. JOLIOT (1900–1958) received a Nobel Prize in 1935.

3. WOOD (1869–1951), a pathologist. From 1912 to 1940 he was head of the Institute of Cancer Research at Columbia.

Ernest O. Lawrence to Ernest Rutherford, July 10, 1935

Dear Professor Rutherford:

Your very much appreciated letter was forwarded to me in New Haven, Connecticut, late in May. I was in the East about two months, engaged in my annual task of raising money for the support of our work in the radiation laboratory. I rather expected considerable difficulty in raising needed funds this year, and indeed was rather worried that we might have to restrict our work a great deal, but fortunately matters turned out otherwise. In this country medical research receives generous support, and it was the possible medical applications of the artificial radioactive substances and neutron radiation that made it possible for me to obtain adequate financial support. We are now able to produce several milli-curies activity of radiosodium. We are devoting a good deal of attention to the further development of the magnetic resonance accelerator for considerably larger currents and also higher voltages. It is reasonable to expect that it will not be very long before we will be producing ten times as much radioactive substance as at present. However, according to the medical people, at the present time we can provide enough radiosodium for beginning clinical investigations, and we have agreed to begin supplying the University Hospital here early this fall.

We have lately been making various tests of the performance of our apparatus with a view to the construction of an improved design. Perhaps the most interesting result is that the focusing action of the electric and magnetic fields is so nearly perfect that we can get just as large current of deuterons at 4.5 M.V. as at 2.5 M.V. At the present time the apparatus delivers several micro-amperes of deuterons having a range of 16.7 centimeters (about 4.5 M.V.). We have bombarded several substances, using these energetic deuterons, and it appears that almost the whole periodic table can be activated, the type of nuclear reaction involved being that in which the neutron of the deuteron is captured by the bombarded nucleus. We have found that gold can be activated in this way, a result which is very surprising. We shall do a good deal more work yet on these things before we can have confidence in the experimental results and theoretical interpretations.

We were all very much surprised to hear that Chadwick[1] is leaving you to be professor at Liverpool. I suppose it is a promotion for him, but I am sure that if I were he I would be very loathe to leave you and the Cavendish Laboratory.

Your newspaper articles regarding Kapitza[2] were widely read in this country, and we all hope that you were successful in influencing the Soviet government to allow Kapitza full liberty to spend as much of his time as he desires in the Mond Laboratory.

Doubtless Thornton[3] has written you that he is staying on here another year. I know he appreciated very much your willingness to allow him to come to the Laboratory.

I am very glad to hear that you are enjoying your usual good health, but you need hardly have told me that you are very busy with a multitude of things. That is obvious, even at this distance.

With best wishes and highest personal esteem, I am

Respectfully yours,
Ernest O. Lawrence
Professor of Physics

Lawrence Papers, University of California, Berkeley

1. JAMES CHADWICK (1891–1974), Nobel Prize–winner 1933. He was at Liverpool University 1935–1943, at Cambridge University 1944–1958.
2. The Russian physicist P. L. KAPITZA (1894–) was formerly at the Cavendish Laboratory and prevented by Stalin from returning.
3. ROBERT LYSTER THORNTON (1908–) is now emeritus professor of physics at University of California, Berkeley.

Warren Weaver to Percy W. Bridgman, August 8, 1935

Dear Dr. Bridgman:

Something over a year and half ago I took with me, for reading on the train from here to Seattle, a copy of Professor Planck's book "Where is Science Going."[1] I found the book very unsatisfying and I made a series of brief notes about difficulties. I took these notes on the boat with me when I went to Europe this spring and these notes, the sea air, or perhaps the slight feeling of nausea, led to the writing of the enclosed memorandum.[2]

There has been so much senseless and worthless "tripe" written about determinacy that I very much hesitate to add to the supply. Would you be willing to read this brief note and tell me entirely frankly whether any of the points involved seem to you to be a) wrong; b) meaningless; c) uninteresting. It is only fair to warn you that you should be very critical in your comment, for it you make any remarks that I can consider at all encouraging, I am quite likely to amplify the memorandum and attempt to publish it.

Very cordially,
Warren Weaver

Bridgman Papers, Harvard University

1. Published in translation in New York in 1932 with an introduction by Albert Einstein.
2. Not found.

Percy W. Bridgman to Warren Weaver, August 18, 1935

Dear Dr. Weaver:

Your memorandum on causality interested me very much. I too have found Planck most unsatisfying, and in fact I find myself inclined to sputter whenever a new batch of his writings on general topics comes out, as there recently has.

I wish you were hear to talk to; it is so difficult to be sure that one has caught the other fellow's meaning from his printed words, or to be sure that one has got one's own meaning across without the flash of the other fellow's response to guide him. But since you ask for my comment, I shall do my best by letter.

I think that my whole method of reaction to this difficult question of determinancy is fundamentally different from yours. It seems to me that most of our difficulties of this sort arise from the imperfections of the tools with which we think; in particular, our tools have a limited range of applicability, but we are not properly conscious of this and attempt to apply them out of their legitimate range. In this way arise many problems which disappear when adequately analyzed—"pseudo" problems as the Viennese school[1] calls them. It seems to me that the problem of determinancy as bearing on the freedom of our will and our course of action is a pseudo problem. Freedom of the will can have no objective meaning. If you maintain that you are free to take either the right or the left hand fork of the road, there is absolutely nothing that you can do to prove it; all you can do is to reiterate that that is the way you feel about it. If you analyze the reason for your feeling further I think that all you can possibly maintain is that all your past experience seems to indicate that your decisions have not been completely determined by what was in your consciousness. It seems to me that you can mean nothing more than this, and I think the most outspoken determinist would grant this. Now why it should be in the slightest degree disturbing to you to think that if the molecular structure of your entire brain were given, a correspondence would be found between your present decision and present molecular structure and between present molecular structure and past molecular structure, I could never understand. There seems to be a crying non sequitur here, which arises I think from the instinctive application to affairs of consciousness of modes of thought which are applicable only to the objective external worlds. Modes of thought which deal successfully with the external world cannot be taken over without modification to deal with consciousness.

So although I would not say that I disagree with your suggestion about statistical as opposed to individual behavior, I would not find such speculations comforting, because I had not felt the need for comfort.

With regard to the application of statistical ideas to the behavior of individual electrons and photons, I think there has been much loose talk by physicists as to what it is that they have actually done here. The probability of individual events at this level cannot be defined in the same way as probability at the level of ordinary experience. What we have is a purely mathematical concept, which determines the manipulations to which our mathematical model is amenable and which reduces very nearly to what we mean by probability on the large scale level.

With regard to the "reality" which you speak of in the first paragraph, it seems to me that a great deal of difficulty arises from not making an "operational" analysis of what we mean by "reality." I think that all reality can mean is something about the way in which we think about and handle our experience. Certain aspects of experience recur, and we have found out what to do in order to make them recur. Reality is a short hand restatement of what is involved in this complex. And having this feeling about reality, I of course would not say that the statement you ascribe to the scientist about reality is a statement of faith. This is another of those things that I don't get. I can see no reason for faith either in the scientist or in daily life. It seems to me that all that faith is my resolution to put through a certain program of action as a matter of practical convenience because I have not yet had time to think through a satisfactory analysis.

This I guess is about enough; you now can see how purely personal these reactions are, and you will be able to appraise them for what they are worth.

Most sincerely,
[*Percy W. Bridgman*]

Bridgman Papers, Harvard University

1. Of logical positivists.

Ernest O. Lawrence to James Chadwick, January 31, 1936

Dear Professor Chadwick:

I have been away for several weeks, which accounts in part for this delay in replying to your good letter.

Perhaps a satisfactory conception of the expenses involved in a cyclotron (slang for the magnetic resonance accelerator) may be obtained from the cost of the cyclotron being installed in the laboratory at Princeton. Their magnet and cyclotron vacuum chamber together, installed in the laboratory, is costing $8,000. The magnet has an iron core 40 inches in diameter with pole faces tapered to 35 inches diameter. It is confidently expected that the installation will be successful in producing many microamperes of deuterons having energies in excess of 10,000,000 volts. For the complete installation, including various accessories, motor generator, etc.—in other words, including all equipment needed for embarking on a program of nuclear investigations, they expect to spend a total of about $12,000. Practically the only significant expense of operation of the apparatus is that of power. For 5,000,000 volt deuterons about 20 kw. is put into the magnet and about an equal amount into the power oscillator. The power bill is per unit time of operation by no means insignificant, but on the other hand, when account is taken of the total time of actual operation of the cyclotron during the year, the power bill becomes a rather small part of one's whole laboratory budget.

I think Kinsey would be an awfully good man for you. He is a very hard worker and knows thoroughly high frequency technique and other matters involved in the cyclotron. He intends soon now to devote all of his time to work with the cyclotron here and is very eager to have an opportunity to help with the construction and use of one in England. I mentioned casually to him the possibility that you might be building one and that possibly he might have a chance to help you in the project. He said that nothing would please him more than an opportunity to assist you in such an undertaking. I feel confident that with Kinsey on your staff a cyclotron could be constructed and put into satisfactory operation in well within a year's time, and that you would not have to devote much attention to technical matters in connection with the installation.

There is, I know, a general impression that the cyclotron is a very tricky and difficult apparatus to operate, but this is really not so. Dr. Donald Cooksey[1] of Yale University came out here this fall on a leave of absence, and I suggested that he design a new cyclotron unit to be interchangeable with the present one (between the poles of the magnet.) He spent more than a month on the drawing-board, designing it carefully, incorporating certain improvements on the basis of our experience, and then supervised the construction in the shop. Finally, on completion of the construction of the new unit, it was placed between the poles of the magnet and a beam of 4 microamperes of 4½ MV deuterons was obtained immediately upon turning the switches. In other words this incident demonstrated that it is possible to design and construct a cyclotron which will operate from the beginning in an entirely predictable manner.

I am enclosing three views of this new cyclotron unit of Dr. Cooksey's design. It has lately been delivering 15 microamperes of 5.8 MV deuterons, and when it becomes desirable to do so, we will increase both the current and the voltage output. At the present time Dr. Cooksey is undertaking some modifications for the purpose of bringing the deuteron beam entirely out and away from the cyclotron chamber. There are many obvious advantages in bringing the beam entirely out of the magnetic field, and I think this will be accomplished in the next month or so. I am enclosing two photographs of the recoil protons in the cloud chamber filled with hydrogen at atmospheric pressure produced by the neutron radiation from a beryllium target bombarded by 8 microamperes of 5½ MV deuterons. The cloud chamber has a diameter of about 8 inches, and in the photograph where the tracks are, the chamber was placed at a distance of 4 feet from the bombarded target. In the other picture the cloud chamber was at a distance of 10 feet from the target. These pictures will give you a good idea of the tremendous intensity of the neutron emission when these high voltages are used.

Again let me say that in the event you undertake the construction of a cyclotron, you can count on our heartiest cooperation.

With kindest regards, I am

Sincerely yours,
Ernest O. Lawrence

P.S.—Perhaps it is worthy of mention that Dr. J. J. Livingood has now definitely established that bismuth bombarded with 5 MV deuterons results in the formation of radium E.

Lawrence Papers, University of California, Berkeley

1. COOKSEY (1892–1977) remained at the Radiation Laboratory for the remainder of his career.

Ernest O. Lawrence to Ernest Rutherford, March 21, 1936

Dear Professor Rutherford:

Thank you ever so much for your good letter. I should have known that you were responsible for the radium D-E-P, but I must confess that I didn't. As regards the yields of radium E by bombarding bismuth with five-million-volt deuterons, I must say that they are quite small. If I remember correctly, several hours bombardment with several microamperes gives, after a few weeks, something like thirty alpha-particles count per minute when the bismuth target is placed near the ionization chamber of the linear amplifier. Measurements on the range distribution of the alpha-particles from the bismuth indicate that the transmutation function is exceedingly steep (for nearly all of the alpha-particles have very near the full polonium alpha-particle range). It is probable, therefore, that at six million volts, which is the voltage we are now using, the radium E and polonium yield should be very much greater; and doubtless in the near future Dr. Livingood will continue experiments at this higher voltage.

We have recently made some alterations of the cyclotron which have made it possible to withdraw the beam completely from the vacuum chamber through a thin platinum window out into the air, and I assure you that we have got quite a thrill out of seeing the beam of six-million volt deutrons making a blue streak through the air for a distance of more than twenty-eight centimeters. Our purpose in bringing the beam out and away from the cyclotron chamber is twofold: partly to make it convenient to carry on scattering experiments, and partly to bring the beam to a target at a considerable distance from the vacuum chamber in order to get rid of the annoying neutron background produced by the circulating ions in the chamber striking various parts of the accelerating system. With this latest improvement in the design of the cyclotron, I think now we have an apparatus which closely approximates one's desires.

I believe in my last letter I mentioned that we have been carrying on experiments on the biological action of neutron rays. During the past two months such biological matters have taken a good share of my attention, because I feel that such matters, as well as nuclear physics, are of great importance. My brother, Dr. John H. Lawrence[1] of the medical faculty of Yale University, has been out here studying the effects of neutrons on a certain malignant tumor called "mouse sarcoma 180." He has compared the lethal effects of neutrons and X-rays on the tumor and on healthy mice and has very impressive evidence that this malignant tumor is relatively much more sensitive to neutron radiation than to X-radiation. If this is generally true for malignant tumors, we have here a very important possibility for cancer therapy. I am sure that it will not be long before neutrons will be used in the treatment of human cancer.

Kinsey has now got his lithium ion accelerator functioning beautifully. He obtains about thirty micro-amperes at a voltage slightly under one million, and is now getting ready to investigate possible nuclear reactions. He intends to wind up the work on lithium ions in the next month or two in order to devote his last six months here to the cyclotron. He feels that he should have first-hand experience with the cyclotron before returning to England. I think it is a pity that time will not permit him to carry on the lithium work, as now he has developed the apparatus and method which would enable him to go on up to several million volts, where nuclear reactions with the light elements should be observable. Kinsey has worked very industriously and intelligently during his period of time here and I regret that there is apparently so little to show for his work. However, I want to say to you personally that Kinsey has overcome some great experimental difficulties and he has really carried through an outstanding piece of experimental work. He is an exceedingly good man and should have every support.

I was interested to hear that you are beginning the new building for your two-million volt D.C. plant and that you are undertaking the construction of a large magnet.

I received the letter from Cockcroft and in the next few days will be sending him detailed information.

Several days ago I received an invitation to attend the meeting in September of the British Association for the Advancement of Science and I have written a tentative acceptance and I can arrange to be away from the laboratory at that time. I should like very much to come over to England to spend two weeks. In the event that you should decide to build a cyclotron, it is possible that I could be helpful by going over in detail with you matters of design.

I conveyed your greetings to Kinsey and I believe he intends to write you in the near future, if he has not done so already.

With best wishes and highest personal esteem, I am

Respectfully yours,
Ernest O. Lawrence

Lawrence Papers, University of California, Berkeley

1. JOHN H. LAWRENCE (1904–), Ernest O. Lawrence's physician brother, was working in medical physics.

Ernest O. Lawrence to John D. Cockcroft, November 16, 1938

Dear Cockcroft:

I was glad to get your letter and to hear that you have the first indications of a beam. The 35 KV between the dees is certainly too little for 10 MV deuterons. It is not even enough for 5 MV and I am sure that you have found that the beam current is a rapidly rising function of the voltage on the dees. For 5 MV you ought to have at least 50 KV. For 8 MV deuterons we are now regularly using about 90,000 volts although we can obtain a small beam at 60 KV, i.e., at 60 KV we get, let us say, 2 microamperes, while at 90 KV, 70 microamperes are obtained. With 40 KW input to your oscillators you have plenty of power provided that the cyclotron oscillatory circuit is efficient.

I presume you are using a coil and I would suggest that some time soon you replace the copper coil with two 3 inch diameter copper pipes extending more or less parallel out from the dee supports and joined together in a single loop. In this way the "Q" of the circuit will be greatly increased and you will have much higher voltage on the dees for a given power input.

I believe since you were here we changed over from using a coil made of half inch copper pipe to the 3 inch pipes and found that for the same voltage on the dees only one-third the power input was required.

The neutron treatments are going along nicely. They have been going long enough now so that we have some patients showing erythema effects on the skin and regressions of tumors, etc., but it is much too early yet to see whether the effects on the patients are definitely different than would be obtained with corresponding x-ray treatments. As far as I know, no one in the laboratory has ever suffered from over doses of neutrons. We regard a safe daily dose of neutrons as measured with an x-ray dosage meter as one-tenth that of x-rays.

You ask also whether we have had any special difficulties in pushing the energy up to 8 MV and beyond. No, we have had no greater difficulty in producing an 8 MV beam than lower voltages, as we have increased the voltage on the dees more rapidly than we have increased the output voltage. We can get a fairly good beam without any shims at all, but the current output is about doubled by using a set of symmetrical shims consisting of circular discs of sheet iron 11/1000″ thick of diameters to inches, increasing in inch steps. Such a pyramid of shims provides a slight curvature of the magnetic field all the way into the center, giving a better focusing of the ions. I am sure, however, the most effective thing you can do is to increase the voltage on the dees. If you get high enough voltage on the dees the beam will come through without any attention to shims at all.

I was interested to hear about your experiences in the war scare. It must be very difficult to work serenely on nuclear problems with such goings on. Although we can be glad that war was averted, it certainly is too bad that it is necessary to continue arming to the teeth.

The papers in this country are full of stories of every increasing persecutions of the Jews in Germany, an almost incredible thing in these modern times. There is much public indignation in this country, as I am sure there is in all democratic countries, and certainly such things cannot go on indefinitely.

Cordially,
Ernest O. Lawrence

Lawrence Papers, University of California, Berkeley

Ernest O. Lawrence to John D. Cockcroft, February 9, 1939

Dear Cockcroft,

Thank you very much for the invitation to participate in the British Association meeting at Dundee next September. I certainly would like to visit you; but, of course, it is yet too early to determine whether it will be possible. I am concerned about the international situation, which I personally think will be all right, but rather I think what I will be able to do next September will depend on the work in the Laboratory and my academic obligations here. September is almost in the middle of our fall semester, and it makes it quite difficult for me to be away at that time. I will write you later on in the spring as to prospects of my coming.

Also I am passing on your invitation to my brother. I haven't seen him today, but I am sure that he too greatly appreciates the invitation, and he will be eager to come over if he can manage to do so.

I was delighted to hear of your splendid success in getting 12 microamperes at 5 million volt protons, and I suppose by now you have 11 million volt deuterons. You certainly are to be congratulated.

The 37-inch cyclotron here is working long hours now everyday, giving steady beams of about 90 microamperes at 8 million volt deuterons. I believe I wrote you that since last September we have been treating cancer patients one day a week with neutrons. This work now has become almost as routine as the x-ray therapy, and the results the medical people are getting are encouraging, to say the least.

We are having right now a considerable flurry of excitement following Hahn's announcement of the splitting of uranium.[1] Within a day after having read about it in the newspapers, the heavy ionizing particles were observed here by using an ionization chamber and linear amplifier. A few days later we had pictures of the heavy particles in the cloud chamber. I believe the nicest work of all here along this line is being done by Abelson,[2] who has very clean evidence of iodine as being one of the so-called transuranic activities, as he has observed the iodine K x-rays in good intensity. He has also chemically identified several other of the activities including tellurium and, I believe, zenon. I believe there is some evidence that one of the activities that was thought to be element 93 is element 43, but I think this is still not certain. We are trying to find out whether neutrons are generally given off in the splitting of uranium; and if so, prospects for useful nuclear energy become very real.

Most of the parts of the medical cyclotron are now completed, and the assembly is proceeding quite rapidly, and we have good reasons to hope that we shall be looking for a beam early in April. We hope to start off with 15 million volt deuterons, though this may be a bit optimistic.

With kind regards,

Cordially,
[*Ernest O. Lawrence*]

Lawrence Papers, University of California, Berkeley

1. For the story of how this became known to U.S. physicists, see Richard G. Hewlett and Oscar E. Anderson, Jr., *The New World, 1939/1946* (University Park, Pa., 1962), pp. 10–11.

2. PHILLIP H. ABELSON (1913–) was a young physicist working in Lawrence's laboratory. He was director of CIW's Geophysical Laboratory 1953–1971, president of CIW 1971–1978, now editor of *Science*.

13
The Institute for Advanced Study

For Abraham Flexner, the launching of the Institute for Advanced Study was more than the culmination of his notable career. For him and others, the institute was both the end result of a long-term trend in the United States and the start of a new era. The idea for an institute was Flexner's who obtained financial support from Louis Bamberger and Mrs. Felix Fuld.[1] After being formally launched in 1930, the institute started operations at Princeton, New Jersey, in 1933 with Abraham Flexner its first head. He served until 1939. The institute very early gained a reputation for being a "paradise for scholars," a phrase Flexner was very fond of repeating. Its reputation influenced the spawning of other "think tanks" after World War II.

While Flexner merits considerable credit, the present nature of the institute is not quite what he originally intended. What finally emerged was a product of the interaction between Flexner and Oswald Veblen. A strong personality and a leading figure in his own right in the scientific community, Veblen was the leader of the institute's School of Mathematics, the most prestigious part of the organization. In 1924 Veblen had proposed an independent institute for mathematical research. In the following year he had unsuccessfully attempted to start one at Princeton with funds from the General Education Board. Almost from the start, he served on the Board of Trustees of the institute.

Flexner was not interested in simply creating a "paradise for scholars" in the sense of a great research center. That would make him a follower of Robert S. Woodward, the second president of the Carnegie Institution of Washington. On the contrary, Abraham Flexner was a follower of Daniel Coit Gilman, the first president of both the Johns Hopkins University and the Carnegie Institution. A graduate of Hopkins in its "golden age," Flexner throughout his life kept referring to that historical experience, and late in life he authored a biography of Gilman. Flexner conceived of the institute as the zenith of higher education in the United States—something beyond the graduate school. There is a faint, curious echo here of another incident from the previous century—G. Stanley Hall's unsuccessful attempt to have Clark University go beyond Gilman's creation.

Not only are the educational intentions quite clear in many of the documents, they are visible in print. The establishing letters of the founding donors barred undergraduate teaching but contemplated granting the Ph.D. In fact, the institute was authorized to grant that degree by the State of New Jersey in 1934, but it has never done so. Recognizing the adequacy of graduate facilities in the United States, Flexner took the route of G. Stanley Hall—education beyond the doctorate. For many years the institute's *Bulletin* made reference to tuition, a source of funds of little significance to the organization. Flexner wanted students, not surprising in a man who usually referred to himself as an educator. The comparison of the institute to Hale and Millikan's California Institute of Technology in the letter of March 4, 1932, is both revealing and significant. To Flexner, the Institute for Advanced Study was part of his campaign to deflate American higher education, to return it to the small scale of Gilman's Johns Hopkins and of his brother's Rockefeller Institute. His policy was greatly influenced by an almost mythic vision of the German university.[2]

Echoing his own experiences in medical education, Abraham Flexner called for full time research. By freeing the investigators from the distractions of routine teaching, public service, applied concerns, and the minutia of administration, Flexner expected more than increased contributions to knowledge. Elevating research to a full-time endeavor was more than an act of recognition of its great importance. It meant the recognition of the high status of the researchers and, consequently, of the deference due them in a proper national culture. The reported comments of Otto Warburg in the letter of March 4, 1932, plus Abraham Flexner's various statements to his brother Simon, clearly echo the sense of Siegbahn's words given earlier. Like other Americans calling for the honoring of knowledge and its creators, Flexner had an underlying nationalistic motive, the training of Americans and the honoring of an American Wissenschaft. No doubt Veblen shared many of these views. Yet, even in these generalities different nuances became apparent as well as pointed divergences on specific issues.

Hitler's rise to power speeded up the inevitable clash between the two men. Refugee intellectuals were a dilemma to Flexner. Much as he wanted to aid them, Flexner worried about the effects on his desire to train native intellectuals. For Veblen, the plight of the German refugees was an opportunity and a moral challenge. Wanting a great research center for pure mathematics and mathematical physics, Veblen moved quickly to add illustrious names to his School of Mathematics. Despite the depression, Veblen joined like-minded individuals in organizing efforts to help individual scientists and scholars to escape Germany and to relocate in the United States. He played a significant part in a great intellectual migration.

The differences with Flexner became acute once Veblen achieved his great School of Mathematics. There was the matter of the scope of the new institute. Both men agreed, more or less, on the humanities. Veblen and his colleagues differed markedly from Flexner on the social sciences. From the perspective of Flexner's observations of medicine, he wanted "clinical" practice as the basis for a science of economics. That argument had little chance of convincing the mathematicians. At one point Veblen came out for laboratory sciences, perhaps influenced by the presence of a Rockefeller Institute branch in Princeton, New Jersey, but Flexner had reservations.

Another source of contention revolved around the question of governance which, in turn, hinged on deep-rooted attitudes of Abraham Flexner. Following U.S. precedents, Flexner favored control by laymen and its corollary—insulation from the affairs of the world. The members of the institute had a role on the Board of Trustees and a modest voice in determining the scope of the program. Ostensibly, Flexner favored a lay director to free the researchers from the cares of everyday life. No doubt he was sincere in that position. But the implications were serious: (1) it went counter to the desire for self-government present among American and European academics alike; (2) it meant an insulation from the affairs of life, except on terms acceptable to the administration; (3) it implied a belief in the incapacity of scholars and scientists to form proper judgments outside their research fields.

An extreme consequence of Flexner's position is given in the documents about the dispute between the Einsteins and Flexner in the fall of 1933. Getting Albert Einstein was a great coup. A world figure, there was no way of avoiding publicity. While Einstein did not seek out the limelight, he apparently lacked Flexner's great squeamishness. Moreover, Einstein felt obligated to speak out on behalf of the victims of Nazism. All of this upset Flexner who was very conscious of the existence of anti-Semitic quotas in some American colleges and universities and feared a backlash. And like many of the leaders of the American scientific community, Flexner, a conservative, was very hostile to the New Deal (note the comment on Felix Frankfurter[3] in the letter of March 2, 1935). The result was the astounding incident of the director of the Institute for Advanced Study intercepting a White House invitation to the greatest scientist of the period. Flexner and Einstein resolved their differences in time, but the issues went beyond matters of personality.

In 1934 the question of governance became unexpectedly acute when Flexner moved to add two new schools (humanities, and economics and politics). Although aided by an anonymous gift of $1,000,000 for the latter, Flexner was forced to introduce differences in salary. This produced the clash with Felix Frankfurter, a trustee, given in the letters of February 6 and 21, 1934. In the next year Frankfurter was forced out of the trustees by Flexner.

The question of governance overlapped the issue of the institute's relationship with Princeton University. Initially, the institute was housed at the University. Flexner savored the relationship, implicitly viewing the institute as a postgraduate arm of the university. Veblen, a long-time faculty member of Princeton, wanted something both different and separate. He moved for the construction of a new building away from the Princeton campus, a move opposed by Flexner as late as February 1936. Veblen prevailed, and he selected the site of the present institute buildings.

Both Flexner and Einstein had involvements in public affairs before 1933. Both had functioned effectively within the frameworks of powerful institutions in the United States and in Germany. Unlike Flexner, before and after coming to this country Einstein displayed a willingness to go public on strongly felt issues, such as Zionism. Flexner reflected another tradition in which experts applied knowledge as proper, high-minded technicalities but not in controversial areas. In practice, the life of the mind was insulated from those practical affairs regarded as unsound by prevailing conservative viewpoints. Tacitly, many agreed that contemplation, at best, required a modern form of monasticism.

But just as Adolph Hitler helped make possible the greatness of the Institute for Advanced Study, so too did he kill the possibility of insulation from daily life. As the decade approached its end, Von Neumann began thinking of a commission in the Army Reserve. And when the peril of a German atom bomb arose, it was a letter signed by Albert Einstein which ultimately led to the Manhattan District, Alamogordo, Hiroshima, and Nagasaki, forever shattering the illusion of scientific theory isolated from the realities of human history.

1. The founding donors of the institute were BAMBERGER (1855–1944) and his sister, CAROLINE BAMBERGER FULD ([?]–1944). Their wealth came from a successful department store. Her late husband was one of Bamberger's partners.

2. Flexner's classic study, *Universities*, appeared in 1930. The first *Bulletin* of the institute had just appeared.

3. FRANKFURTER (1882–1965), a trustee of the institute, was then at the Harvard Law School and soon to go to the Supreme Court. Flexner's comments on Frankfurter in the letters that follow echo conservative wisdom of that day which saw Frankfurter as a kind of evil genius behind the New Deal.

Abraham Flexner to Simon Flexner, March 17, 1931

Dear Simon,

Your letter of February 19 missed us in America, but reached us here a few days ago. Eleanor's term has now ended and we are here in London for two or three days before going to the Continent. Our plans are not yet complete, but I suspect that we shall first go to Rome, where I want to talk with the mathematicians about the mathematical situation in America, this being the advice which I got at Princeton and elsewhere. Then we will proceed north and return to England for the spring term, after which I shall go back home while Anne and Eleanor remain here waiting for the season at Baireuth for which they have tickets.

Your letter was immensely interesting and stimulating, and I hope you will continue to discuss freely everything in connection with the proposed Institute. Nothing in the document which you read is really of a permanent character, and I mean to keep the thing fluid and flexible as long as I possibly can.

I have not lost sight of the administrative problems on which you touch, but I do not yet have to make up my mind about them. Sooner or later we shall want a Budget Committee, but we have no need of one as yet—indeed, the only Committees thus far appointed are the Executive Committee and the Finance Committee. Everything else is in abeyance.

I used the word "Genius" in a Pickwickian sense. Of course it might be restricted to persons like Shakespeare, Sir Isaac Newton and Faraday, but it may also be applied to persons of less fundamental gifts who have a peculiar genius for doing something extremely well. In this lesser sense I should say that Gildersleve[1] and Sylvester and Dr. Welch had genius, though none of them were geniuses.

You will also notice that I am not pinning my faith even to people who have a certain sort of genius, since in the same connection I mentioned "unusual talent" and "devotion." Whether we shall ever get hold of a genius is an absolute gamble—certainly we cannot count upon it. We ought to have a high proportion of talent and devotion.

On the economic side the situation in America is, in my opinion, critical. I do not believe that we appreciate the amount of rumbling discontent in our Universities, affecting the best and most pure-minded persons in them. There, in my judgment, a new precedent has simply got to be set. I see no reason why I, an ignorant manager, should be more highly remunerated than Tom Morgan, and yet I believe that for fifteen or twenty years my salary has been twice his. Surely I have had nothing in the way of comfort or opportunity for myself or the children which Morgan and his kind have not better deserved.

In England and on the Continent one hears everywhere the expression of amazement at the financial status of the American professor. Perhaps an example such as we may set may persuade administrators that they ought not to be putting up thirty millions' worth of buildings on the Yale and Harvard campus while there is not a single professor who is receiving a comfortable income.

How many subjects, or what subjects we can do on this basis I do not yet know, but it will be better to do a few subjects comfortably than to do many as they are now done. Unless the University Profession can be placed upon a sound economic basis, talent will be thrown into business and professional life even more strongly than to-day. I took good care to say that we should not try to compete with the highest business or professional income, but a professorial salary of $20,000 in the metropolitan area, awarded to a man of established reputation, is under existing circumstances in America sufficiently modest—though of course I am not committed to that or any other figure.

I am enclosing a copy of a letter from Chancellor Kirkland[2] which will greatly interest you. It appears that Robinson,[3] having himself left Vanderbilt, is now still further plucking the members of that faculty before they have had a chance to take root. I wonder whether you have heard that Rochester has frankly abandoned its full-time plan, and that McCann[4] and others are henceforth to be part-time men?

Our plans after leaving Rome are very vague. I shall feel my way along, hoping to be able to get back to America towards the end of May. Meanwhile write me in care of the Rockefeller Foundation, 20 rue de la Baume, Paris. I do hope that Helen has greatly improved and that you have maintained all your gains. Everyone in England asks for you both.

Saturday I went to tea at Lady Astor's, and there met Sir William Rothenstein. He fairly overflowed with praises of you and Helen and her kindness and hospitality. I met an interesting group—Mr. Baldwin, Tom Jones, Philip Kerr, and others.

With love to you both from us all,

Ever affectionately,
Abe

Simon Flexner Papers, American Philosophical Society

1. BASIL L. GILDERSLEEVE (1831–1924), a classical philologist, was one of the original faculty at Hopkins.
2. JAMES HAMPTON KIRKLAND (1859–1939) chancellor of Vanderbilt University 1893–1937.
3. GEORGE CANBY ROBINSON (1878–1960) had left Vanderbilt for Cornell's Medical School where he stayed from 1928 to 1935.
4. WILLIAM SHARP MC CANN (1889–1971) since 1924 was at University of Rochester's Medical School.

Abraham Flexner to Simon Flexner, March 4, 1932

Dear Simon:

Thank you for yours of the third. It is about two years now since my mind has been dwelling on the problem of how to make a start. My long experience at the General Education Board made me intimately familiar not only with the economic but with the mental situation of the American professoriate. On his last visit to America Otto Warburg,[1] talking to me here in the office on the subject and comparing our conditions with Germany's, even now, said that, if he could judge from the homes in which he had been entertained, the condition of the American professor and his family is "elend." Twenty-five years ago when Mall and I met in Munich and discussed the same problem, he emphasized the fact that Germany drew a larger share of first-rate brains partly because there were in the learned world a fair number of prizes not only economic but social, though the two hang closely together. At that time Waldeyer[2] told me that his income was 80,000 marks a year. Willstätter[3] told me—and the same is true of Weyl (the most eminent of living German mathematicians)—that he declined an American professorship simply because the salary offered would not permit him to continue the cultural side of his life—and it was the highest professorial salary paid in America at the time.

The prizes in America are reserved for university presidents, which to my way of thinking is quite wrong, for it removes the president socially and economically from easy intercourse with his faculty. I believe that Dr. Buttrick[4] was wise when within a short period after my entrance into the General Education Board he doubled my salary, making it equal to his, and never allowed it to sink below, despite the fact that he was president and I had only a limited sphere of responsibility. Wherever I have gone in America during these last twenty-five years and especially in the last two years and with whomsoever I have spoken on the subject of the future of American learning, the note of economic security has been almost invariably struck, often the first one struck. At my suggestion, our Committee on Site sent a letter to about forty-five leading American scholars and scientists asking them what in their opinion could be done by the new Institute to improve the academic situation in the United States. Freedom of speech and economic security were practically unanimous replies.

Like you, I am looking solely to the future, for my own connection with this enterprise will be a brief one, but it depends upon me to set the pace. It seems to me of small moment whether we do five subjects or fifteen but of infinite importance that whatever we do, we do as well as is humanly possible. I do not believe that the present depression will be permanent any more than the depression following 1873 was permanent. The resources of the country are too great. We shall ultimately have from the estate of Mr. Bamberger and his sister enough to do half a dozen subjects very well, if the men can be found—perhaps more. Hale and Millikan have made a distinguished institution with only a small group of allied subjects. I see no reason to think that the same cannot be done in the East, if sound judgment is exercised. What will happen as a result, I cannot prophesy, but if privately endowed institutions of learning are to be the pacesetters in this country, we must assume that in the future, as in the past, there will be other Mr. Bambergers just as there have previously been a Smithson, a Carnegie, a Rosenwald, etc. I have a profound conviction based upon heart-to-heart talks and intimate observation extending over a period of a quarter of century that the Director of this Institute ought not to be substantially better off than the most distinguished figures on its staff. On the other hand, I am equally convinced that hardship in the early years is helpful, not deterrent, and that men should be made to earn by hard work and genuine productivity such prizes as exist in the academic world. I am myself surely not too well endowed mentally nor have I gotten anything out of life in the way of books or travel or security for Anne and the children which is too good for a really distinguished professor. Nevertheless, there has never been a time when I could have lived the sort of life I have lived on the best university salary paid in the United States. This seems to be fundamentally wrong.

I do hope that you and Helen will have a fine vacation. When you get back, I shall welcome every possible opportunity for talk. It is criticism and deliberation, not commendation, that can be helpful to me.

With love to you all,

Ever affectionately,
Abe

P.S. I wonder if I told you that some weeks ago I had a two-hour talk with Taussig[5] about American and European economists, in the course of which he mentioned a number of European economists but was not willing to say of any American economist that he was "really first-rate." I asked him whether in his thirty years of experience in teaching graduates and undergraduates at Harvard he had ever had students of first-rate intellectual capacity.

"Oh yes," he replied, "quite a number."

"Where are they?" I asked.

"In law and in business," he answered, "how can anyone without independent means such as I possess continue in academic economics, law and business offering what they do?"

"If it is necessary," I replied, "to offer the financial inducements of law and business, we shall never have academic economics."

"That isn't requisite," he answered. "It is not even desirable, but we must do much better than we now do."

A. F.

Simon Flexner Papers, American Philosophical Society

1. OTTO H. WARBURG (1883–1970), a biochemist who was head of the cell physiology institute of the Kaiser Wilhelm Gesellschaft in Berlin. He received the Nobel Prize in 1931.

2. WILHELM VON WALDEYER (1836–1921).

3. The German chemist RICHARD WILLSTÄTTER (1872–1942) received the Nobel Prize in 1915.

4. WALLACE BUTTRICK (1853–1926), a clergyman, was chairman of the General Education Board.

5. The economist FRANK WILLIAM TAUSSIG (1859–1940) of Harvard was a founder of its School of Business Administration.

Abraham Flexner to Simon Flexner, March 12, 1932

Dear Simon:

Thank you very much for your two letters. I should have written you sooner, but I have been spending practically the entire morning day after day with the dentist and shall continue to have a daily session for the next couple of weeks to come.

Meanwhile, you will be glad to know that Boots has after further examination and conference with Swift outlined a course of procedure for Anne. He has completely won her confidence and has been very reassuring. She will, I think, be a thoroughly conscientious patient cooperating with him in every way. I am myself to have a talk with him within the next few days so that I too may understand precisely what it is he wants her to do. Unfortunately, she is not yet rid of the last traces of the cold which has hung about her for almost a month, and Mim has been laid up too with a cold which, however, seems to be simply of the ordinary nasal variety. I saw her yesterday afternoon, and she was very cheerful.

Both your letters have been extremely gratifying in so far as they have discussed the problem of the Institute. You are altogether too generous about me. I have not the least doubt that there are others, perhaps many others, who could do what I am trying to do and who are in point of years and training better equipped to do it.

I should think it very unfortunate if the salary became either too prominent an element or tempted any man to join the Institute who would not otherwise do so. In my conversation with Birkhoff it was the very last thing I mentioned, and in his letters to me he has never alluded to it. The thing that moves him primarily is the unique opportunity to free himself still further—though Harvard has done a lot for him—from routine that is perhaps inevitable in any large organization, though I think myself that there is more of it in American universities than even large organizations require. The salary alone would not have budged him. Of that I feel certain. On the other hand, the prospect of more adequate compensation during his active years, a more liberal retirement allowance, and a decent provision for his wife, if she survives, unquestionably make the opportunity even more attractive than it would otherwise be.

As to the president's or director's salary, in Europe, including England, there is such a thing as an academic status or caste. No professor or rector is much better off than anybody else in so far as his salary is concerned. In Germany there was up to 1920 a very marked distinction between the Ordinarii and the Extraordinarii, but this Becker[1] abolished with the result that the Extroardinarii, once an unhappy group, are now precisely the reverse. Unquestionably in America a director or president will have to do a few things that his faculty do not have to do. The fewer the better. He ought to be, I think, socially and economically in the same class as the professors. It was proposed originally that I should be paid more than a professor. I declined. Experience may show that a differential will be fair. If so, it can easily be provided. You are absolutely right in feeling that, if a man wants to do research, he will do it at whatever sacrifice. Claude Bernard[2] sacrificed his wife and children and lived in poverty. I cannot bring myself, however, to expect these men to get less out of life than I have gotten. Indeed they are the very people who ought to be enabled to enjoy art and music, the best medical, dental, and other attention, and to surround their children with cultural influences and to give them the best education attainable. The man who impressed that upon me first of all was Mall. He was not unhappy at Baltimore, as you know. He even insisted, when the first full-time gift of a million and a half was made, that no part of it should go to the laboratory men. On the other hand, it was he who first pointed out to me the abundance of "prizes" in the German universities—prizes, which were not easily earned, yet which were sufficient in number and adequate in size to make a university career as good a sporting chance as a career in law or medicine up to a certain financial limit.

Since I wrote you, I have had a two-hour lunch with Walter Stewart.[3] Stewart was born on a farm in Kansas, graduated at the University of Missouri, became an instructor there, later assistant professor at Michigan and Missouri, finally professor of economics at Amherst. Then, as he told me, the future of his two children began to worry him. He left Amherst to become Director of the Division of Research of the Federal Reserve Board, then for two years economic advisor to the Bank of England. Now he is Chairman of the Board of an investment security house in New York. He assured me that his style of living was still academic, but, he said, "I now am in position to give my children a better education than I got in the Middle West." We talked of American economics. He said it had been ruined by potboiling, law, and business, precisely what Taussig told me. I may add that Stewart was the American representative in the Basel conference on the question of German reparations a year ago.

This morning the mail brings me a letter from Evarts Greene,[4] a copy of which I am sending you under separate cover. It reflects the same state of mind which I have encountered in every university I have visited.

Mr. Gilman said in his Inaugural Address in 1876, "Almost every epoch requires a fresh start." He made a fresh start. Harper[5] made a fresh start. You made a fresh start. This new Institute ought to profit by these experiences and not hesitate to make a fresh start, to which other institutions may, if it proves successful, adapt themselves in years to come. But mind you, when I make this last statement, salary is only one item. There are others which are even more important, and of course in the matter of salary I am speaking only of men who have proved themselves and whose future, as far as one can judge, promises productivity and devotion. Finally, I have in mind no scale, and, should we ever have to employ young men, promising but still uncertain, I should feel that it is good for them to demonstrate their ability and their devotion, precisely as Germany always requires its Privatdocents to do.

With love and best wishes to you and Helen,

Ever affectionately,
Abe

Simon Flexner Papers, American Philosophical Society

1. CARL HEINRICH BECKER (1876–1933) was for many years the Prussian Kultusminister during the Weimar republic.
2. While it is true that BERNARD (1813–1878), the French physiologist, did struggle in his early years, he did not sacrifice wife and family for science. In fact, his financial struggles abated upon receiving his marriage dowry in 1845. Thereafter his conditions improved markedly.
3. WALTER W. STEWART (1885–1958), an economist. Stewart was a trustee of the institute from 1933 to 1941. He was at the institute 1938–1950 and from 1953 to 1955 was a member of the president's Council of Economic Advisors.
4. The historian EVARTS P. GREENE (1870–1947) of Columbia.
5. WILLIAM RAINEY HARPER (1856–1906) was the first president of the University of Chicago. A Yale Ph.D. in Semitic languages, Harper was one of the great university builders.

Abraham Flexner to Oswald Veblen, January 23, 1933

Dear Professor Veblen:

I have been giving a good deal of thought to your letter of January 18 and to the long conversation which we had on the day following. Let me lay before you the results of my cogitations.

1. As to the form of government. In Germany and France there is a governmental agency and in Germany a curator which relieve the faculty of an immense amount of administrative detail. In Oxford and Cambridge everything is done by the dons but with the result that in the last half century or so three times Royal Commissions have been appointed which have overhauled the universities. In addition, Curzon,[1] when he was Vice-Chancellor in Oxford, wrote a scathing criticism of the university and in the last year or two a voluntary committee has been formed, made up of Oxford men, only a few of whom are dons, for the purpose of criticising and improving the conduct of the university. It does not seem to me therefore that either the continental universities or the English universities can be quoted in favor of an arrangement for throwing everything upon the professors. Moreover, if I can judge from the amount of work, mostly useless, that passes over my desk, I do not see how a group of scientists can manage a going and growing institution without serious sacrifice of their work. At this moment, for example, in the Institute the members of the School of Mathematics would not only be thinking of mathematics and answering a lot of mail and inquiries with which they really do not need to be bothered, but they would be discussing what we should do next—classics, history, economics, or what not. The scheme under which we are proceeding aims to place the professors in positions of dignity and responsibility, to give them freedom from administrative duties, an adequate part in the development of the institution while at the same time utilizing laymen who can render important service, and someone whose main task it is to plan for future developments. I do not know whether this will work out or not, but it seems to me superior to the American or to the foreign organizations so far as I know them.

2. I cannot satisfy myself that it is right to give the name of associate to persons to whom we make grants in aid, for it does not seem to me to be fair. They belong to other institutions. They come to us primarily as workers. While the title of associates may improve their situation at the end of a year when they leave, I cannot help asking myself whether that is a thing to which we ought to be a party. Persons of this kind should, I think, be simply listed as receiving grants in aid. It will thus be understood that the rest of their means are derived from other sources.

Your letter does not state precisely what you have in mind regarding either Albert or Currier,[2] but I do not believe that I could recommend to the Board that we should simply make an appropriation—which is what it amounts to—to Harvard and the University of Chicago in order to enable them to send Currier and Albert to work with us at Princeton. These are both rich institutions, thoroughly able to give their men additional opportunities, of which they themselves will reap the benefit. The institutions and the men should between them make the sacrifice. We have undertaken to deal somewhat differently with van Kampen,[3] whether soundly or not I am not sure.

I feel strongly the importance of the point which I made as we were parting last week, namely, that we must proceed slowly in order to be as certain as we can that we set no precedents which will have to be undone or regretted.

With all good wishes and high regard,

Sincerely yours,
A. F.

Veblen Papers, Library of Congress

1. GEORGE NATHANIEL CURZON (1859–1925), chancellor of Oxford in 1907.

2. A. ADRIAN ALBERT (1905–1972) whose distinguished career was spent at Chicago. ALBERT E. CURRIER (1905–) joined the Naval Academy faculty in 1935 where he is now an emeritus professor.

3. EGBERT RUDOLPH VAN KAMPEN (1909–1942), a Netherlands mathematician, came to Hopkins in 1931.

Abraham Flexner to Oswald Veblen, March 17, 1933

Dear Professor Veblen:

I thank you for yours of the 16th. I shall be very happy indeed to bring Lefschetz and Tisdale[1] together as soon as I get in touch with Tisdale.

I have been giving a good deal of thought to our conversation of the other day, namely, as to the development of the mathematical group. It seems to me that we must leave open the question of expansion for the reasons which I gave you at that time. There is another consideration, however. We have got to fulfill two functions, namely—(1) enable our professors to continue their research under the most feasible possible circumstances; (2) enable them to train a few young men, mainly Americans. If we can not, in other words, make an American contribution to mathematics, the contribution which we can make by taking care here and there of a foreigner will be very slight. We have got to do what the Germans did during the 19th century, namely, make American Wissenschaft respectable. That is our prime and essential function. Hence we must look about among our 120 millions of Americans for young men who may be worthy of development.

Mrs. Flexner quite understood why you had to run away. We both hope we can very soon make our visit to Princeton.

I hope that Professor Einstein was not too trying today.

With all good wishes,

Very sincerely yours,
A. F.

March 18, 1933

P.S. Since dictating the above I have yours of the 17th regarding Dirac.[2] I do not feel inclined to make any move at the present moment. Our first task is, I am sure, to get the present mathematical nucleus to work. Dirac is young and will be available almost any time in the future.

Please let me know when we should remit to the American Mathematical Society, also when we should make a payment to Princeton University in connection with the Annals of Mathematics and whether a check for the whole amount, semi-annual, or quarterly payments.

A. F.

Veblen Papers, Library of Congress

1. The physicist WILBUR EARLE TISDALE (1885–1954) was at the National Research Council 1921–1926. From 1926 to 1929 he was in Europe, first representing the International Education Board (1926–1929), then the Rockefeller Foundation (1929–1940).

2. PAUL ADRIEN MAURICE DIRAC (1902–) was a British theoretical physicist at Cambridge. He won the Nobel Prize in 1933. Since 1971 he has been at Florida State University.

Oswald Veblen to Abraham Flexner, March 24, 1933

Dear Dr. Flexner:

I think I agree with you completely in your attitude with regard to American versus foreign appointments. The considerations which you adduce seem to be decisive. Also I appreciate your desire to be cautious in view of the present economic situation.

I have been expressing interest in Dirac in spite of the very small chance that he could be moved because he is a) young, b) extremely able, and c) interested in questions which are close to those being studied by Einstein, von Neumann and myself. I really think that the most we could hope to do would be to get him here for a single term once in a while.

I mention Wiener largely because of a desire to be fair. He seems to be now the most deserving American who is available and he would bring into our group an element which we lack (the sort of analysis of which Hardy is an exponent).

Should we wait with further junior appointments until the additional $5000 has been actually appropriated? I have about come to the conclusion that Mr. Bleick[1] should be given a $1000. scholarship or fellowship. He will receive the Ph.D. in Chemistry this year and wants to continue in mathematical physics. There would be no point in his taking a Ph.D. over again. Therefore, as he looks like an able man, I think he would be an appropriate student for the Institute and just about at a good point in his development to make use of the opportunities.

Yours sincerely,
[*Oswald Veblen*]

Veblen Papers, Library of Congress

1. WILLARD EVAN BLEICK (1907–), a mathematical physicist, has been at the Naval Postgraduate School since 1950.

Abraham Flexner to Oswald Veblen, March 27, 1933

Dear Professor Veblen:

I am very glad that you and I see eye to eye on this question of the form which our responsibility should take. Mr. Bamberger and Mrs. Fuld were very anxious from the outset that no distinction should be made as respects race, religion, nationality, etc., and of course I am in thorough sympathy with their point of view, but on the other hand if we do not develop America, who is going to do it, and the question arises how much we ought to do for others and how much to make sure that civilization in America advances. The matter has been very, very much on my mind, and I do not know that any two persons would solve it in exactly the same way. I can only say that I am glad to be assured that you realize that my mind is as wide open as it can possibly be, and in these days in view of the incredible things happening in Germany we do not wish to brand ourselves as nationalists in any way whatsoever. . . .[1]

Very sincerely yours,
A. F.

Veblen Papers, Library of Congress

1. Most of the remainder of this letter is on routine housekeeping matters.

Oswald Veblen to Simon Flexner, May 10, 1933

Dear Dr. Flexner:

Since our conversation in Washington about the problem as to what can be done to help the Jews and Liberals who are driven out of their positions in Germany, I have talked with a few of my colleagues here and some others. The idea which seems to receive most favor is that of having a committee for the natural sciences which should be composed in a large part of what the Germans would call aryan scientists, together with a few men of affairs who would know how to raise funds. The idea would be to distribute the German scientists who are helped in various countries in such a way as not to cause a undue concentration anywhere but so as to allow them to continue their scientific work. The scientific membership of the committee could be selected in such a way that the committee would possess first hand knowledge of the individuals who are to be helped.

No formal protest of any sort would be made but the existence of the committee and the nature of its membership would, I think, in the course of a year or two, have a good deal of practical value as a protest.

I went to see your brother about this matter day before yesterday and he sent me to Mr. Fosdick[1] whom I asked whether he would be willing to serve as chairman of the committee. He brought forward the difficulty which I had of course expected that he was so closely connected with the Rockefeller organizations and said that he would like to talk the matter over with Max Mason. Mason telephoned to me yesterday morning about the general question without saying anything about Fosdick. I put the question to him rather strongly by saying that I did not know where else to go in order to find the necessary leadership for such a committee. Mason promised to call me up later in the day but no message has yet come from him. I hope to hear from him today but I thought I would like to get this letter off to you so that you will prepared in case I try to consult you about the matter tomorrow or the next day.

I am hoping to leave for Maine some time next week but would like to see if someone capable of carrying out some sort of a relief undertaking for the German scientists can not be enlisted before I go. Do you suppose that Vincent might do it?

Yours sincerely,
[*Oswald Veblen*]
Princeton, New Jersey.

Veblen Papers, Library of Congress

1. RAYMOND BLAINE FOSDICK (1883–1972) was a lawyer and a very influential trustee of the Rockefeller Foundation of which he was president 1936–1948.

Oswald Veblen to John von Neumann, May 22, 1933

Dear Johnnie:

Please pardon me for dictating a reply to your letter of April 30th. We are starting off for Maine in our car tomorrow or the next day and after we get there I hope to write a better letter. So far as I can judge, our administration in Washington is not going to allow any more inflation than is absolutely forced upon it. The laws that are being passed all take the form of conferring on the President the power to do this and that. In several cases he has already announced his intention of not doing the things which he is empowered to do. That is really about all that anyone knows about the matter.

I believe that among the financial people there is a great deal of fear that further dishonest practices will be disclosed by the Congressional investigations. The comment is often being made that the capitalistic system may break down for lack of honest men to administer it. I should judge that the government means to go as far in the direction of state socialism as is necessary to preserve the present balance of power between the definite classes of the community.

There are a number of attempts being made to raise money to provide relief in this country for the Jews and Liberals who are being dispossessed in Germany. It would be a good idea to write me whatever you know in detail about the mathematicians and physicists who are in difficulties. What we lack here most of all is authentic news although I am not sure that we are any worse off than anyone else in this respect.

Yours sincerely,
[*Oswald Veblen*]

Veblen Papers, Library of Congress

Abraham Flexner to Simon Flexner, October 18, 1933

Dear Simon:

Dr. Berliner,[1] Editor of *Die Naturwissenschaften,* sends me the enclosed additional information about Dr. Bergmann. Please return the letter, as I have not answered it.

Yesterday was about the worst day I have had to endure for months. It was really no end of fun, but at the end of it I was absolutely done in, so much so that I stayed in bed up to lunch today partly of necessity and partly as a matter of precaution. The night before President Dodds[2] received a telephone message about ten o'clock from Mayor O'Brien[3] asking him to come to New York to welcome Einstein. He appealed to me, and I told him it wouldn't be necessary, so he declined. About eleven I had a telephone and a telegram from O'Brien to the same effect, to which I did not reply. Meanwhile, Felix Warburg[4] had communicated with Ben. None of us authorized the use of our names, though they did appear in yesterday's *Times*. O'Brien, Untermyer,[5] forty motor-cyclists, many plain-clothes-men, and prominent citizens, mostly Jews and Irish, assembled to meet the boat at eight o'clock in the rain. Meanwhile, Mr. Maass[6] left New York at five A.M. on a motorboat, picked up the Einsteins, and had them here at nine without any formalities whatsoever. I had a most courteous telegram from Secretary Hull,[7] informing me that they were delighted to cooperate with me. I called on the Einsteins about eleven and found them jubilant that they had escaped the excitement as well as the danger, thanks to your suggestion that I get in communication with the State Department. They stayed indoors all day except for a late afternoon walk to get the evening newspaper. All day long we were beseiged with long distance calls, the first from Untermyer who was as mad as a hornet because he had been kept waiting in the rain, the others from newspapers, all of which Mrs. Bailey attended to without worrying me. In the afternoon a swarm of newspapermen came to the Inn. I sent them away absolutely empty-handed, though in perfectly good humor. The one concession we made was done at Eisenhart's suggestion, for he said that, unless we permit Einstein to be photographed, the photographers would hang around the town and the campus and snap him unawares, so we agreed to give them all one shot at him at 11:30 this morning. When the hour came, he kicked like a steer because he and Veblen did not want to be interrupted. I do not myself know—and it is two P.M.—what they did.

Mrs. Bailey had occasion to take a telegram to Mrs. Einstein and was informed that they had never been happier than at the way the whole thing had been managed, for even in America these things can be done right if one wants to do so.

Now the reporters are beginning to camp on Weyl's tracks, but that will be easier. I am simply putting them off. They are under the impression that he is coming here, when as a matter of fact he is going first to Swarthmore. As long as they don't discover that, Weyl will not be annoyed. Einstein is very, very happy that Weyl is joining the group.

With love from Anne and me,

Ever affectionately,

Abe

P.S. You will be interested to know that Einstein and Veblen were quietly working at an algebraic problem as early as 9:30 this morning

Abe

Simon Flexner Papers, American Philosophical Society

1. ARNOLD BERLINER (1862–1942), a physicist.
2. HAROLD WILLIS DODDS (1889–1980) was Princeton's president from 1933 to 1957.
3. JOHN P. O'BRIEN (1873–1951).
4. FELIX M. WARBURG (1871–1937), the banker.
5. Perhaps the noted attorney SAMUEL UNTERMYER (1858–1940).

6. A New York City lawyer, HERBERT H. MAASS (1878–1957) was a trustee of the institute.
7. CORDELL HULL (1871–1955).

Abraham Flexner to Franklin D. Roosevelt, November 3, 1933

Dear Mr. President:
With genuine and profound reluctance, I felt myself compelled this afternoon to explain to your secretary, Mr. MacIntyre,[1] that Professor Einstein had come to Princeton for the purpose of carrying on his scientific work in seclusion and that it was absolutely impossible to make any exception which would inevitably bring him into public notice.

You are aware of the fact that there exists in New York an irresponsible group of Nazis. In addition, if the newspapers had access to him or if he accepted a single engagement or invitation that could possibly become public, it would be practically impossible for him to remain in the post which he has accepted in this Institute or in America at all. With his consent and at his desire I have declined in his behalf invitations from high officials and from scientific societies in whose work he is really interested.

I hope that you and your wife will appreciate the fact that in making this explanation to your secretary I do not forget that you are entitled to a degree of consideration wholly beyond anything that could be claimed or asked by any one else, but I am convinced that, unless Professor Einstein inflexibly adheres to the regime which we have with the utmost difficulty established during the last two weeks, his position will be an impossible one.

With great respect and very deep regret, I am

Very sincerely yours,
Abraham Flexner

Roosevelt Papers, Franklin D. Roosevelt Library

1. MARVIN H. MC INTYRE (1879–1943).

Abraham Flexner to Mrs. Albert Einstein, November 15, 1933

Dear Mrs. Einstein:
I thank you sincerely for your frank and candid letter of the 14th. It is quite clear to me and to President Dodds, with whom I have discussed the entire question, that, while your ideals and intentions and those of your husband are of the highest possible character, you

do not understand America. Every person of prominence is in this country subjected to pressure, precisely as your husband is. President Dodds has a secretary, who protects him from publicity which may be injurious to him or the University. In my own case, Mrs. Bailey[1] is absolutely inflexible, and she makes it impossible for any one to reach me unless, in her judgment, the matter is really worth my time and attention. Letters come to me daily requesting interviews, articles, etc., for which I would receive considerable sums which I should be happy to devote to the relief of the Jews in Europe. Mrs. Bailey declines them without even consulting me. You are apparently afraid that I wish to limit your husband's freedom. Precisely the contrary is the case, as I assured you at Caputh. I wish to protect his freedom in order that he may enjoy the advantages of his position in the Institute and in order that the dignity of the Institute and the dignity of Princeton University, with which we are so closely associated, may be preserved. You say in your letter that "Wenn man mit Menschen nett ist, so ist es hier in diesem Lande üblich, eine Pressenotiz darüber zu machen." This statement shows, as I said above, that you do not understand America. There are undoubtedly many people, whose names constantly appear in the newspapers, but they suffer in the judgment of their colleagues and of the best persons in the country for this very reason. Persons connected with institutions like the Rockefeller Institute never permit their names to appear in the newspapers except in connection with something directly associated with their work.

There is another consideration which I think you lose sight of. It is perfectly possible to create an anti-Semitic feeling in the United States. There is no danger that any such feeling would be created except by the Jews themselves. There are already signs which are unmistakable that anti-Semitism has increased in America. It is because I am myself a Jew and because I wish to help oppressed Jews in Germany—not only scholars but ordinary people—that my efforts, though continuous and in a measure successful, are absolutely quiet and anonymous.

There is no danger that there will be any breach of friendship between us. I write you in the kindest and most helpful spirit and only after conference with those who know America and Europe and who are as deeply concerned for your husband, for you, and for oppressed Jews as I am.

Let me once more assure you that you are entirely wrong if you think that I wish to make "irgendwelche Vorschriften in Bezug auf seine Lebensführung." I wish only to do for you husband what a first-class American secretary does with intelligence and inflexibility for her chief and thereby saves him from interruption and saves the institution, with which he is connected, from unfriendly criticism which can only harm him and the institution and the causes which he would like to help.

To show you how far I am from wishing to limit your real freedom, I may say that I have been asked by many Princetonians in the faculty and out of the faculty whether they were free to call on you. In every case I have said, "Yes," because I thought you and your husband would enjoy their company and that they would help to make you feel at home here. Let me repeat again and again that there is no question of freedom or "Vorschriften" involved. The questions involved are the dignity of your husband and the Institute according to the highest American standards and the most effective way of helping the Jewish race in America and Europe. Fortunately, not only in my judgment but in the judgment of others, Jews and Christians, whom I have consulted, all these good causes can be helped by the same course of action.

Very sincerely yours,
Abraham Flexner

Einstein Papers, Institute for Advanced Study

1. Esther S. Bailey was A. Flexner's secretary.

Albert Einstein to Eleanor Roosevelt, November 21, 1933

Dear Mrs. Roosevelt:—

Mr. Henry Morgenthau[1] wrote me that President Roosevelt extended an invitation to Mrs. Einstein and me. You can hardly imagine of what great interest it would have been for me to meet the man who is tackling with gigantic energy the greatest and most difficult problem of our time.

However, as a matter of fact, no invitation whatever has reached me. I only learned that such an invitation was intended but believed that the plan had been dropped.

I am writing this letter because it means a great deal to me to avoid the ugly impression that I had been negligent or discourteous in this matter.[2]

Very faithfully yours,
A. Einstein

Roosevelt Papers, Franklin D. Roosevelt Library

1. HENRY A. MORGENTHAU, JR. (1891–1967), Secretary of the Treasury.
2. Early in 1934 Einstein was so outraged by what he considered Flexner's misguided efforts to shield him that he prepared a letter to the trustees in which he suggested that he might leave the institute. He disclosed efforts by Flexner to prevent his participation in public meetings against the Nazis even before his arrival in America. The interception of the White House letter was only one of many of Flexner's such actions. Flexner also tried to prevent Einstein from aiding a benefit concert for refugees in New York in

December 1933. While Einstein's letter was never sent, the incidents only served to harden the sentiments of Veblen and others for a greater role in the conduct of the institute. Flexner could not conceive that others could honestly take exception to his paternalistic stance. The Einsteins, incidentally, did eventually lunch with the Roosevelts.

Abraham Flexner to Felix Frankfurter, February 6, 1934

Dear Felix:

I have your recent letter and I am glad you approve of broad differentiation. I agree with you that one cannot, within these categories, make small distinctions or offer unusual rewards, but I think it conceivable that a man may find himself in a position where it is necessary for an institution to do something unusual. That has happened, for example, in the case of Professor Weyl. As opposed to the other full professors, Einstein, Veblen, von Neumann and Alexander, who have no children and no encumbrances,[1] Weyl is a fugitive from Germany and has left everything behind; he has two children, and he and his wife have dependent parents. We therefore made an initial appropriation so as to free him from the burden of debt—it was only a question of a few thousand dollars—so that he might start from scratch. No exception has been taken to it by anyone.

I had an opportunity at the General Education Board to observe what rigid rules involved. Thus in the General Education Board in dealing with young men we never made any rule as to what allowance we would make when it came to sending them abroad for a year or giving them additional opportunities. Some of them were unmarried and without responsibility. Some of them were married and had children and dependents. I used to go into their circumstances carefully and figure out what they needed, and the number of persons involved was not so large that it was very difficult or impossible; nor were the sums so large that the purpose we had in view, viz.: that of enabling promising persons to go ahead, was ever misunderstood. At the same time the Rockefeller Foundation, the International Health Board, etc. all had absolutely fixed rules. The result was that they very often were unable to secure as fellows or workers the persons whom they most desired to secure. In G.E.B. we never failed once, and I have never heard, directly or indirectly, that anybody was offended by the irregularities in our grants-in-aid.

In the case you mention, for instance—Carl Becker[2]—if he were a young man I would not ask him "to base a claim." I would try to get him to be quite candid with me and figure out just what it was he needed in order to be able to be free of financial worries and to give his undivided attention to his scholarly work.

There are two considerations to bear in mind: (1) The German universities rose to their eminence by pursuing just such a policy as I have outlined above. There was in the best days a fixed basic salary, and then various allowances that varied from case to case, with the result that a German professor with a family did not have to resort to money making in order to keep afloat. (2) American professors are far too frequently compelled to do hack work, summer teaching, lecturing, and God knows what all, at terrible cost to their intellect and productivity, simply because their salaries are insufficient.

Simon has avoided this kind of thing at the Institute, which is much larger than our Institute will be for many years to come, and I believe we can avoid it too if we preserve the right spirit and grow gradually, thus maintaining the sort of intimate, confidential, personal relationship with which we have begun. But mind you, I absolutely agree with you in holding (1) that I would not make a financial bid for any man in competition with another institution; (2) I would shear away at once from anyone who is trying to make a good bargain.

"Serene collaboration" (your own phrase) is attainable, I believe, if the intellectual superiority is kept where it now is. If it falls below that, "serene collaboration" will fail, not only in the financial realm, but in the intellectual realm as well.

Since I dictated the preceding I have been called to the telephone by Dean Eisenhart of Princeton and Veblen of our own group. The three parties collaborate in a way which would be impossible if anyone were trying to get any advantage for himself or his own institution. Why should we not aim to keep this Institute a paradise for scholars and to create within it the conditions which make that a possibility, or even a probability? Once more, in its best days the German university did that. The Kaiser Wilhelm Institute at Berlin also did it. Arithmetical equality is not everything.[3]

Always sincerely,
Abraham Flexner

Frankfurter Papers, Library of Congress

1. Flexner was at least partly wrong. Both Einstein and von Neumann had children.
2. CARL L. BECKER (1873–1945) was a historian at Cornell University.
3. A two-page postscript has not been included since it is a further elaboration of the above arguments.

Felix Frankfurter to Abraham Flexner, February 21, 1934

Dear Abe:

Your letter of the 6th has come.

You are quite right. "The real point is: What are we aiming at?" Let us clear away confusion by eliminating what we are not aiming at.

1. For myself, I don't want to hear anything more about German Universities for a good long while anyhow. Partly I dare say this is the intolerance of an ignorant man, for I know very little and you know very much about German Universities. But I cannot bring myself to believe either that they have said the last word or that they furnish good criteria. After all, a university is something more than a means for contributing to knowledge. Especially in the social sciences, the test of knowledge lies in action. In successive moral crises not only have the universities of Germany failed to reveal the accumulated wisdom that we call civilization, but to a large extent they have been the centres of decivilization. I cannot believe that if one knew all there is to know one would not find something in the organization of German universities, in their relation to authority and in the relations of the professors *inter se* that would explain why in so large a measure German universities have been poisonous centres of anti-Semitism, militarism and Nazism. I can understand the coercive power of economic necessity and the consequences of subjecting the human spirit to such a tyranny as that which now rules Germany. But you'll have a very hard time to explain away the behavior of colleagues in letting a man like Mendelssohn-Bartholdy[1] be cast off as he has been withut so much as a word of decent human friendliness, however secretly imparted, from a single mother's son of them. This is not an isolated case. A society of scholars that has the moral account to settle which Redlich[2] can formulate with great particularity against the German professoriat during the last sixty years or so, had better not serve as a watershed for our enlightenment—except by way of dangers to avoid.

2. Nor do I think it very helpful to take too seriously the exuberant rhetoric of thinking of the Institute as a "paradise of scholars." For one thing, the natural history of paradise is none too encouraging as a precedent. Apparently it was an excellent place for one person, but it was fatal even for two—or at least for two when the snake entered, and the snake seems to be an early and congenial companion of man. Really, figures of speech are among the most fertile sources of intellectual confusion. Let's try to aim at something human, for we are dealing with humans and not with angels. I do not know by what right you may hope for a combination of greater disinterestedness and capacity than, say, the Harvard Law School is able to attract or, let us

say, than is now found on the Supreme Court, with five out of nine men of real size. I can assure you that neither of these institutions could be conducted on the assumption that it is a paradise. In both personal interactions play an important part; in both personal sensitiveness has not been wanting because of personal differentiations.

3. Equally irrelevant seems to me your account of grants-in-aid by the General Education Board or by the Institute to individual scholars. I am concerned with the Institute as a permanent group of scholars engaged in a permanent joint adventure. Individual benefactions for short periods are one thing; the building of a permanent institution quite another.

4. And so I have been brought, by way of denying what we are not aiming at, to what we are aiming at—a society of scholars. I quite agree with you that such a society cannot flourish best under financial pressure. I must add, however, that my stay here has not led me to conclude that great simplicity, even the need of thinking three times how a half crown is to be spent, is hostile to scholarship nor that the standards of commercial or financial income are relevant for a scholarly society. Let me also say that differentiations in income between the Colleges is not one of the most edifying aspects of Oxford nor making most for scholarship. But I wholeheartedly agree that the members of your Institute should be relieved of the necessity "to do hack work, summer teaching, lecturing and God knows what all, at terrible cost to their intellect and productivity,"[3]—circumstances which operate perhaps most unfairly against the wives and thereby greatly hamper in achieving a gracious society, which I deem an essential for a society of scholars. Therefore I think salaries should be ample. Put them as generously as you will—to me it seems that $10,000 for the younger men and $15,000 for older men is ample, considering the general economic level into which, I am sure, we are entering. I do not even have objection to a family allowance by way of so much for each child, in addition to a base salary. What I do insist on is that whatever classifications there be—and there ought to be a very few classes—they should be impersonal. This is what I meant by approving of "broad differentiation." It excludes all personal differentiation.

5. I need not repeat the grounds of my objection. But may I say that such a society of scholars as I envisage precludes an administrator who plays Lady Bountiful or, to keep my sex straight, Kris Kringle. Does it occur to you that the Carl Beckers might not want to discuss with you their intimate affairs? To be very blunt, he may feel that it is none of your business how he allocates his salary. Why should he have to tell you that he has a demented step-daughter? And why is it any of your concern whether, although I have no children, I have other ob-

ligations? The Institute's concern is so to fix salaries as to enable a man to live as a civilized gentleman in a world in which the family is the ordinary social unit. You seem to me to have a little bit too much the administrator's confidence in assuming (a) that you could spot the man "who is trying to make a good bargain" or (b) that you could plan the life of a man who is too shy or too proud to enter the realm of bargaining. And if you'll forgive me for saying so, you also have a little bit the optimism of the administrator who thinks his scheme "works perfectly" because evils have not as yet disclosed themselves, and particularly have not been disclosed to him.

6. From all of which you will gather that I feel very strongly about this. It is only one aspect of my conviction that a society of scholars implies a democratic aristocracy like unto the self-government by which, say, Balliol is conducted. This implies impersonal equality and self-government by the group. Those are the aims to which I am committed. I write thus frankly because you may think that, holding these views, I may not be a very useful member for your Board. If so, I'd better get off before I am on. In putting this to you, I am quite impersonal. It has nothing to do with our personal relations, and they would remain quite what they were before were you to tell me that perhaps it is just as well that I resign before I become active.

With warm regards,

Always yours,
[*Felix Frankfurter*]

Frankfurter Papers, Library of Congress

1. ALBRECHT MENDELSSOHN-BARTHOLDY (1874–1936) was in international law at the University of Hamburg until 1933, when he came to Balliol College at Oxford. At the time of this exchange with Abraham Flexner, Frankfurter was at Oxford for the year. Albrecht was the grandson of the composer, Felix Mendelssohn-Bartholdy.

2. JOSEPH REDLICH (1892–1936), originally from Austria, since 1926 was Frankfurter's colleague at the Harvard Law School.

3. When Flexner defended the German professoriate in a response of March 21, 1934, Frankfurter replied on April 24, 1934, with a litany of the money-making activities of its members. He would not let Flexner get away with myth making. See *Frankfurter Papers, Library of Congress*.

Abraham Flexner to Oswald Veblen, March 2, 1935

Dear Professor Veblen:

I have this morning yours of February 28. Let me suggest that you write Mrs. Manning how, in your opinion, she might most hopefully approach the Rockefeller Foundation. You can also say, as far as next year is concerned, there is no particular occasion to worry. If the

Rockefeller Foundation declines help towards the necessary endowment, we can put our heads together as to the best next step.

While I am writing, let me say a few things on which I have reflected a good deal since our conversation the other day in respect to additional facilities. This question must, and I think will, be viewed by the Trustees from the standpoint of the institution as a whole. You ought, it seems to me, on looking at it to forget that you are a member of the mathematical group, for your influence within the Board will in the long run depend upon your capacity to view the Institute as a developing concern which in the near future will take up economics and subsequently humanistic studies. This is an important point, because on it, I think, depends the extent to which what is erroneously called "faculty government" is likely to be inaugurated. In the precise sense of the words, "faculty government" exists nowhere on earth. In Germany the faculties deliberate and have always been subject to control by the education ministries. In Oxford the fellows of a college deliberate. The University is an anomalous affair, depending for its real progress upon a personality. Thus at Cambridge in recent years the moving force was Sir Hugh Anderson of Caius in cooperation with the University Grants Committee, made up mainly of lay-men. In addition, three times within our generation Parliament, finding that the colleges were not moving, has appointed statutory commissioners, mainly laymen, who have brought about far-reaching changes in the administraton of the Universities. Provincial universities in England are all governed by boards of trustees on which faculty members are in small minority. Against the judgment of pretty nearly every one whom I have consulted I am trying the experiment of conducting the Institute with a board made up of lay-men, outside scholars and scientists, and faculty members. How far faculty membership will be influential and how far it will extend in America are going to depend upon the ability of those chosen to forget the particular faculty, of which they are members, and to look at the Institute as a whole.

Frankfurter gave that movement a distinct setback, though, whenever any one has spoken to me about him as a "professor," I have always pointed out that it was unfair to saddle his conduct upon "professors" as a whole.

Complete faculty rule would mean that the mathematicians have got to think and devote time not only to teaching and investigation in mathematics but to searching in new fields like economics, humanism, and finding the persons as well as the money. The result at the present moment would be that the whole thing would be in the hands of mathematicians, and the mathematicians would do what I am doing, namely, devoting their entire time to things that do not bear upon mathematics.

This last week, for example, I have spent almost my entire time in conferences on the subjects of the humanities and economics. Why should a mathematical group want to do that or what particular capacity has it for doing it? Besides, our American institutions are so large that faculties are generally regarded as an inefficient instrument for purposes which they alone can discharge. We have in this country a definite tradition of lay management. I think this has not worked perfectly. We can't improve on it by adopting either the English or the German systems, neither of which was, however, a purely faculty system. Can we improve upon it by the device which we are now utilizing? That depends on the ability of the faculty members who come on the Board to look at the situation in its entirety.

This brings me to the subject of building. I do not know what the Board thinks, but I have had experience enough to make a pretty good guess. There are psychological, practical, and prudential reasons why the Institute should in the near future possess a building of its own. For the immediate present we can improvise. Riefler[1] yesterday told me he would be quite happy in a room adjoining ours at 20 Nassau Street. As far as mathematics is concerned, I wrote you somewhat fully on January 28. Originally, there was a question among some of the Trustees whether the combination in Fine Hall would work. I thought it would work, and it has worked, and I think nobody now has any doubts on the subject. The only questions I have discussed with you and Eisenhart is the possibility of a small wing or utilizing the basement of Fine Hall, the latter having now been ruled out as impracticable. I am sure that a suggestion to put a building on the Princeton Campus would probably be rejected by both Boards of Trustees, but you are perfectly free to make the suggestion in order to find out the sentiment of the Building Committee and the Board itself, for nothing that I say in this letter is anything else than my personal view with which I am acquainting you in order that you may think the thing over and thereafter advocate on your own responsibility any course that you think wise. Do not lose sight of the fact that the more money at this stage we put into buildings, the less income there is available for men.

It seems to me essential that within the next two years we should make a start in economics and in the humanities, and we ought to do it without cutting down what we are putting into mathematics. You see then that I am trying to view the problem as a whole over a period of, let us say, five years, during which our financial situation may have changed.

Do you recall saying that, "If we had pursued any other course than that which we had pursued, the Institute would have been a 'flop' "?

This I do not see. On the contrary, though our growth might have been and was, I imagine, expected by all of us to have been much slower, the six or seven professors with a few workers would have prevented its being a "flop." It can be stabilized at its present point or even limited in numbers without being a "flop." Pasteur, Claude Bernard, Rowland, and others whom you know better than I did not "flop," though they were mere individuals. I am much more afraid of "flopping" through bigness than through smallness. I agree with Justice Brandeis that size in itself is bad, and nowhere is it worse than in institutions of learning and research.

In all this (except the first paragraph) I am writing you as a trustee, not as a professor of mathematics, and I should not even write if I were not anxious that in the long run there should be more professors on the Board, but that must necessarily depend upon winning confidence—a task to which I have devoted myself with all the ingenuity I possess ever since I first met Mr. Bamberger. I have always been candid with him, as I have with the Board, but I have realized that every Board has got to be educated by experience to trust those upon whom the responsibility mainly falls. Hence "time" is not a controlling factor.

Always sincerely,
A. F.

Veblen Papers, Library of Congress

1. A Ph.D. in economics from Brookings, WINFIELD W. RIEFLER (1897–1974) was at the institute at this date. After the war he returned to the Federal Reserve Board where he remained from 1948 to 1959.

Abraham Flexner to Oswald Veblen, November 7, 1936

Dear Professor Veblen:

I have your note of November 5. I was surprised, as I told you in my reply, on reading your memorandum entitled "Building for School of Mathematics," to find that you had gone so far afield as to take in chemistry and biology without any previous communication to me in respect to this extension. Now to my further surprise I learn that your imagination has gone so far as to play a decisive part in your thinking not only about the future of the School of Mathematics but of the relationship of the Institute to the University. I feel that I should not have been presented with a problem in terms of space when, back of the problem of space, there were, unknown to me, implications of which I had had no intimation.

At the present moment the School of Mathematics is utilizing well nigh the entire income from the original endowment of the Institute.

The other two seeds which I was anxious to plant, though there was some difference of opinion as to the wisdom of my recommendations, are existing on a minimal basis and will naturally be entitled to prior consideration in the event of any substantial increase of income. I value very highly your fertile, fundamental thinking, but, as there is no immediate likelihood, even should space have been obtained, of any such expansion as you apparently have been thinking of, I should like to suggest that you commit your reflections and suggestions to paper in the form of a memorandum to which I and my successor can have access when the moment you contemplate arrives, and I hope it may before too long. As far as your colleagues are concerned, I should think it wise informally to impress upon them the importance of doing the utmost with the opportunities which you now possess, for, short of a surprise, these opportunities are not likely to be increased for some time to come; and the opportunities must really be very great if a man like G. H. Hardy asks me directly the question as to whether we really wish a monopoly on mathematics in Princeton.

I feel even more strongly as to you intimation that you have been thinking of the relationship of the Institute to the University and that you would like to consult your colleagues on that point also. I should regard any discussion on that subject as inopportune and ill-judged. Knowledge that such a discussion had taken place would almost inevitably spread and would do incalculable harm. The relations between the Institute and the University are very intimate, and they are important to each other in ways in which you and your mathematical associates do not and cannot possibly know. A discussion on the part of the mathematicians on that subject would be futile and might be harmful. It would be like pulling up a tender plant after a short period to find out whether it is growing. Should it ever become necessary, as I hope it may never become necessary, to have a faculty discussion on this point, the discussion could not be limited to the mathematical group. It would be called by me, and would be attended by all groups. I should preside and actively participate, for the very obvious reason that, leaving all else aside, I am far better informed than any one in any of the groups regarding the substance of the relationship. Any move that at this moment suggests that that relationship be modified, when it is the rock on which we now rest, and anything that could possibly interfere with the type of collaboration which we are trying to work out would be deplorable. In my opinion, therefore, and this is the result of very careful reflection, the whole subject should be dropped and the entire incident regarded as closed. Should the subject be mentioned by some one else, you can easily sidetrack it. At the moment we can not expand in any direction or for any purpose. Let us, therefore, be as productive as possible in our individual capacities and in addition do what we can for those who come here for inspiration and guidance.

I trust that you will not misunderstand this letter. You surely know that I set the highest value upon the services which you have rendered to me personally and to the Institute, but your memorandum and your letter have both disturbed me, and it seemed to me only right that I should put you quite candidly in full possession of every doubt that has crossed my mind since receiving them.

With all good wishes,

Sincerely yours,
A. F.

November 9, 1936

P.S. Since writing the above letter, I have had a call from one of the most important men in the Princeton faculty in reference to a specific topic, but without any intention upon his part or mine we drifted into a discussion of the mutual importance to each other of the University and the Institute. I confess I was amazed to find the enthusiasm with which he regarded the relationship and his anxiety to contribute everything in his power to its permanence and its interaction. I believe that this represents the general sentiment of the Princeton faculty, and it surely would not become any member of the Institute which has been a beneficiary of Princeton hospitality, to raise any question that affects the fundamental relations between the two institutions.

A. F.

Veblen Papers, Library of Congress

John von Neumann to Oswald Veblen, June 25, 1938

Dear Oswald,

Two months are gone since I sailed for Europe, and I feel by now, that there is an enormous amount of things about which I want to write to you. This is probably an illusion, and the number of these things will turn out to be very finite when I start writing them down. But in any case it is high time now for me to write to you.

I have been in Warsaw, the theoretical physics meeting there was quite interesting, although not very hot. I met a number of friends there: N. Bohr, Fowler, Wigner, Gamow, Møller.[1] I also got acquainted with Darwin, Langevin, Brillouin,[2] F. Perrin,[3] Bauer[4] (the 4 last ones being the Paris contingent). A fact of some note is, that Eddington spoke, and at long last got some contradiction—from all of us.

I also got acquainted with the man, who organises these meetings for the League of Nations. His name is Establier; he is a very pleasant man. In spite of his french name and habitat (his office is in Paris) he is a Spaniard from Alicante. He knows Flores, and told me, that Flores is now an artillery officer on the Loyalist side, commanding a battery on the Madrid front.

Bohr told me, that he intends to come to Princeton during the second term of 1938/39. He has a desideratum: He would like to bring Rosenfeld[5] with him. R. is a theoretical physicist of some merit, a pupil of Bohr, and presently professor in Louvain in Belgium. (He would, of course, return there.) The main reason for bringing R. along as Bohr's A.D.C. for his Princeton campaign is, that R. is very good in interpreting and expounding Bohr, when he has to commune with common mortals . . . Bohr is rather anxious to have him, and I have seen R. in action in Warsaw in this sense, and doing very well—so that I agree with Bohr about the desirability of this scheme. The finances of this matter are as follows: R. would get some support from his own university for this American trip, but Bohr estimates, that another $1,000 would be needed. Bohr asks us to raise this sum. I told him, that I thought, that the thing was not impossible, but rather uncertain—and hard to settle before fall. He asked me to call your attention to this matter. What is your reaction?

I also met some of the Warsaw mathematicians. (Knaster,[6] Kuratowski,[7] Mazurkiewicz,[8] Tarski[9]—Sierpinski[10] was in Budapest while I was in Warsaw, and gave a talk at the mathematical seminar there. At the same occasion Lebesgue[11] talked, too, whom I now met for the first time. He is on a tour collecting honorary degrees in Poland (from Cracow and Lwow).

The physics meeting was highly official, we were even given a lunch by the president of the polish republic. He is probably the best looking man I ever saw, and I don't think that he has much to say in running the country—the army takes care of that.

I am familiarized by now with the state of mind, the bellyaches and the illusions of this part of the world—such as they are since the annexation of Austria. The last item (illusions) is rather rare, the preceding one not at all . . . Hungary was well under way of being nazified by an *internal* process—which surprised me greatly—in March/April. The new government, which was formed in May, stopped this process, or slowed it down, but for how long, is not at all clear. As to the possibilities of war, you have probably the same information as people here, and guessing does not mean much. I think, that there will be war, although it may be at a distance of a half year or perhaps even one or two years yet. The general tension has unquestionably decreased since it became clear, that Checoslovakia would fight, that this would mean general war, and that Germany seems to hesitate to face this. The continued existence of a french government has a similar effect. But I still think, that these are only short-range blessings.

My private affairs move as scheduled, which—much to my regret—is not too fast. I expect to be married by the middle or end of October. Since there are a number of things to be done, which can only be done after I am married, it will be late in November in the best case, when I get back to Princeton.

Flexner wrote me from England. It seems, that he met Dirac in Cambridge, and had a "conversation" with him. He writes, that Dirac will not leave now Cambridge for good, particularly because Fowler has left, but that he may consider another one year invitation. He asks me to talk with Dirac about it when he comes to Budapest, which will probably happen in August. Do you approve of this, I mean of our return to the system of "visiting professors" in theoretical physics?

I will probably stay here until the middle of July, and then go for a "real" summer vacation to Switzerland or France. This will last until the middle of August. From then until the middle of September I plan to make a "scientific" voyage, probably to Copenhagen. After that I will return to Budapest, since I expect, that by then my presence here will be necessary.

Marina[12] is very well, and has really developed *very* considerably during the last year. And she seems to have a complete command of the situation . . .

There is nothing scientifical to be reported. I have been working steadily for the last 3 weeks, but it is only writing up or rewriting old things.

What news are there in Princeton, especially on the building-front? I hear that Robertson[13] got the Pasadena invitation and the promotion in Princeton, is this so?

I hope, that Elizabeth and you will enjoy soon—if not already—a cool and calm summer in Maine. And that I will hear from you soon.

Until then, with the best greetings,

I am Yours

John

P.S. Did I tell you, that I made "Curtesies and Customs of the Service" with 100 points?! I'll be a Master of Ceremonies yet!

I am sending my copy of "The Folklore of Capitalism" to your Princeton address, since I do not know, whether yours was found in my room, or not.

Veblen Papers, Library of Congress

1. The Danish physicist CHRISTIAN MØLLER (1904–[?]) at the University of Copenhagen.

2. LOUIS MARCEL BRILLOUIN (1854–1948).

3. FRANÇOIS PERRIN (1889–) was France's High Commissioner for Atomic Energy 1951–1970.

4. EDMOND HENRI BAUER (1880–1963) was then at the Collège de France.

5. LEON ROSENFELD (1904–), Belgium-born, since 1958 at the Nordic Institute for Theoretical Atomic Physics.

6. BRONISLAW KNASTER (1893–), at the faculty at Wrocław.

7. KASIMIERZ KURATOWSKI (1896–), at Warsaw University.

8. STEFAN MAZURKIEWICZ (1888–1945).

9. ALFRED TARSKI (1902–) joined the Berkeley faculty in 1942.

10. WACLAW SIERPINSKI (1882–1969) was at Warsaw University.

11. HENRI LEON LEBESGUE (1875–1941), of the Collège de France, is best known for his theory of integration.

12. Marina von Neumann Whitman, John von Neumann's daughter, is now an economist. She was a member of the President's Council of Economic Advisors for 1972 and 1973.

13. HOWARD PERCY ROBERTSON (1903–1961) was associate professor at Princeton 1936–1938, professor 1938–1947; after that he went to Cal. Tech. His interests were in mathematics and mathematical physics.

John von Neumann to Oswald Veblen, September 15, 1938

Dear Oswald,

I got your letter of Sept. 1. yesterday, and I hurry to answer it. The world has become so complex lately, that I don't see how this could be done, without an essential use of positive integers as ordinals.

Pro primo: Your political optimism cheered me up quite considerably. Things are so uncertain and so much in a liquid state—this is at least subjectively true at this moment, because it is 7:20 A.M., and I haven't seen this mornings Swedish newspapers yet—that I can imagine, that your diagnosis is right. Of course things have deteriorated considerably while your letter was in transit. I agree with you, that war at this moment is improbable, since neither side seems to want it just now—but the Sudeten-german-population seems to be very nearly out of control, so you can never tell. It also seems, as if Messrs. H. and M. were a little more emotional lately than rational, so you really cannot tell. So we may be much nearer to liquidation than it seemed 2 weeks ago. God knows what will happen—especially while this letter is in transit.

Pro Secundo: I don't know how to feel about it. There isn't much good in war in principle, of course. This particular war, on the other hand, might settle a thing or two, and settle them well. And finally, I have every personal reason to feel the way you indicated: Not to want it now. I need not tell you, that in my case the last mentioned feeling is by far the strongest one. Besides all these meditations are very futile.

Pro tertio: I was greatly interested by what you wrote me about the building question. I would like to have some more details, but I would *really* not want to bother you with that—could Miss Blake not send me a statement of the main quantitative and qualitative facts. (E.g.: How many rooms in tutto? How many for mathematics? Will there be space for all of us? [I infer from your letter, that there will be.] What shape will the building have? Will we have a full complement of assistants-rooms? Of what size will the various rooms be? How much of the library-budget is there, resp. what is the size of the library which has been approved? What "social" rooms will there be?)

Pro quarto: My military career. I have completed three courses: "Organization of the Army," "Military law," "Military discipline. Customs and courtesies of the Service," with ratings of 100, 75, 100 respectively. One more course is to be done ("Organization of the Ordnance Department"), but since the "Army extension courses" are vacationing from May 20. until Oct. 1., I can only do this after Oct. 1.—it seems, that I can do it while in Europe. So I may qualify for an officers commission after my return—end of November. The 75 grieves me badly, as you know, but then there is another 100 afterwards! Will I ever be a general? Well, that's that. (Need I point out to you, that the last theorem is capable of very wide generalizations and numerous applications in all walks of life, politics, etc.)

Pro quinto: My personal affairs develop accordingly to those—not too pleasant-rules, which govern such affairs: The "waiting period" is over on Sept. 23., the legal procedure lasts 3 to 6 weeks, after this getting married implies another week (or a little more) in formalities, and then getting a US visa and settling some other matters may take about 2 more weeks. So I expect to be able to leave, with all affairs settled, between Nov. 15. and something like Dec. 5. So much is alright, but whether one is entitled to count with periods of as super-geological length as these in present day Central Europe, is more than I can say. But I assume, that things will arrange themselves somehow.

Pro sexto: We decided, that I will only return to Budapest when the first part of these legal procedures is completed—a measure which, I think, is perhaps superfluous but can do no harm. So I spent a week in Copenhagen, where I saw much of the Bohrs, and other mathematicians, e.g. Neugebauer,[1] Nielsen,[2] Hecke[3] who was there on a 2 day visit, and also Hevesy.[4] I also gave 4 talks, and I had a very pleasant time.

I talked a lot about the "Zentralblatt" and I will write you in detail about this matter in a week, since I expect to get more technical details about its budget by then.[5] N. is decided to hold the fort as long as possible, but how long this will be possible, nobody knows. He is willing, to clarify the financial situation now, I'll report to you about this later (see above). By the way, it is clear to me, that if a research position in history of mathematics could be created for N. in USA, he would be glad to take it.

I am giving a talk here in Lund (visiting Marcell Riesz).[6] Then I'll go to Stockholm, and from there back to Copenhagen, where Niels Bohr asked me to stay with him for some days—for general conversations, and especially about ideas he has on biology. That's that.

Afterwards I'll go to Cambridge and London. (Of course, I must reserve the right of changing these dispositions in case of a world war.) I expect to be back to Budapest early in October.

The newspaper arrived now. So Mr. Chamberlain is flying to Germany. He seems to be very eager to have the next war postponed a bit, and I sympathize with this desire in the present special case.

I am finishing this letter, because I have to run to make a date with Riesz. I will continue it next week.

Please give my best greetings to Elizabeth. Hoping that everything will go well—on both sides—

as ever
John

P.S. I am glad to hear from you, what I also inferred from the newspapers, that things in America are on the upgrade again. By now I am convinced of it, too: It's obviously the "unbalancing" of the budget which does the trick in every case, and which, therefore, is perfectly justified. The progresses of Townsendism & its cousins is somewhat bewildering, but of all conceivable forms of political insanity, it is a relatively harmless one.

So it seems once more, that the Atlantic is a relatively clean cut between good and evil.

Veblen Papers, Library of Congress

1. OTTO NEUGEBAUER (1899–) was editing the *Zentralblatt* from Copenhagen while it was printed by Springer in Berlin. At this time he is at Brown University.

2. JAKOB NIELSEN (1890–[?]) at the University of Copenhagen.

3. ERICH HECKE (1887–1948), at Hamburg.

4. A Hungarian chemist, GEORGE CHARLES DE HEVESY (1885–1966), then at Copenhagen. He was noted for work with radioactive tracers; won the Nobel Prize in 1943.

5. The *Zentralblatt* was the leading abstract journal for mathematics. After the coming of Hitler, there was difficulty about noting the writings of non-Aryans. Neugebauer, its editor, went to Copenhagen. Veblen was moving toward founding *Mathematical Reviews* to fill the need no longer met by the German publication.

6. The Hungarian-born mathematician RIESZ (1886–[?]).

Albert Einstein[1] to Franklin D. Roosevelt, August 2, 1939

Sir:

Some recent work by E. Fermi and L. Szilard, which has been communicated to me in manuscript, leads me to expect that the element uranium may be turned into a new and important source of energy in the immediate future. Certain aspects of the situation which has arisen seem to call for watchfulness and, if necessary, quick action on the part of the Administration. I believe therefore that it is my duty to bring to your attention the following facts and recommendations:

In the course of the last four months it has been made probable—through the work of Joliot in France as well as Fermi and Szilard in America—that it may become possible to set up a nuclear chain reaction in a large mass of uranium, by which vast amounts of power and large quantities of new radium-like elements would be generated. Now it appears almost certain that this could be achieved in the immediate future.

This new phenomenon would also lead to the construction of bombs, and it is conceivable—though much less certain—that extremely powerful bombs of a new type may thus be constructed. A single bomb of this type, carried by boat and exploded in a port, might very well destroy the whole port together with some of the surrounding territory. However, such bombs might very well prove to be too heavy for transportation by air.

The United States has only very poor ores of uranium in moderate quantities. There is some good ore in Canada and the former Czechoslovakia, while the most important source of uranium is Belgian Congo.

In view of this situation you may think it desirable to have some permanent contact maintained between the Administration and the group of physicists working on chain reactions in America. One possible way of achieving this might be for you to entrust with this task a person who has your confidence and who could perhaps serve in an inofficial capacity. His task might comprise the following:

a) to approach Government Departments, keep them informed of the further development, and put forward recommendations for Government action, giving particular attention to the problem of securing a supply of uranium ore for the United States;

b) to speed up the experimental work, which is at present being carried on within the limits of the budgets of University laboratories, by providing funds, if such funds be required, through his contacts with private persons who are willing to make contributions for this cause, and perhaps also by obtaining the co-operation of industrial laboratories which have the necessary equipment.

I understand that Germany has actually stopped the sale of uranium from the Czechoslovakian mines which she has taken over. That she should have taken such early action might perhaps be understood on the ground that the son of the German Under-Secretary of State, von Weizsacker,[2] is attached to the Kaiser-Wilhelm-Institut in Berlin where some of the American work on uranium is now being repeated.

Yours very truly,
A. Einstein
(Albert Einstein)

Roosevelt Papers, Franklin D. Roosevelt Library

1. The origins of this letter are given in Richard G. Hewlett and Oscar E. Anderson, Jr., *The New World, 1939/1946* (University Park, Pa., 1962), vol. 1 of the history of the U.S. Atomic Energy Commission. The letter was not the first federal contact with the possibility of atomic energy arising from the results of the work of the German physicists Hahn, Strassmann, and Meitner. Its implications were noted in the United States and in France by Joliot. But the first contacts had not produced the action strongly sought by LEO SZILARD (1898–1964), a Hungarian-born physicist. ENRICO FERMI (1900–1954), a Nobel Laureate in physics, had left his native Italy and was at Columbia University. Szilard contacted Alexander Sachs, a New York banker with access to the White House who suggested getting Einstein's signature to counter the caution of Fermi and others. The letter was drafted by Szilard and Sachs. (An accompanying memorandum of a more technical nature is not given here.) Einstein readily agreed to sign. On receipt, Roosevelt started the process which would lead to the Manhattan Project, Alamagordo, and then to Hiroshima and Nagasaki. It was a new world, an ironic conclusion to the struggle of several generations of scientists interested in convincing their fellow citizens of the importance of supporting the disinterested search for truth.

2. CARL FRIEDRICH, FREIHERR VON WEIZSACKER (1912–1964) was a physicist with philosophic interests. In retrospect, this apprehension was exaggerated.

Appendix: Sources for Documents

Francis G. Benedict Papers. 10 boxes, ca. 1893–1957. Countway Library, Harvard Medical School.

John Shaw Billings Papers. 44 ft., 1861–1918. New York Public Library, New York.

Albert Francis Blakeslee Papers. Approx. 15,000 pieces, 1904–1954. Library, American Philosophical Society, Philadelphia.

Percy Williams Bridgman Papers. 20 ft., 1905–1961. Archives, Pusey Library, Harvard University.

James McKean Cattell Papers. Approx. 83,000 items, 1835–1945. Manuscript Division, Library of Congress.

Thomas C. Chamberlin Papers. 3 ft., ca. 1880–1928. Library, University of Chicago.

Arthur H. Compton Papers. 150 cu. ft., 1905–1962. Archives, Washington University.

Walter B. Cannon Papers. 164 boxes, 1890–1945. Countway Library, Harvard Medical School.

Carnegie Institution of Washington Archives. 180 ft., ca. 1900–date. [These figures are for the early historical materials kept at CIW headquarters, 1530 P St., N.W., Washington, D.C. 20005. Besides the trustees minutes, the former trustee files, the early general files, the executive committee minutes, and the departmental files were very useful.]

Frank Wigglesworth Clarke Papers. 0.2 cu. ft., 1873–1921. Archives, Smithsonian Institution.

E. G. Conklin Papers. 57 cartons and 10 file drawers, ca. 1863–1952. Seeley Mudd Manuscript Library, Princeton University.

Charles B. Davenport Papers. Approx. 20,000 pieces, 1874–1944. Library, American Philosophical Society.

Milislav Demerec Papers. 5,000 pieces, no date. Library, American Philosophical Society.

Albert Einstein Papers. Approx. 42,000 items, 1895–1955. Institute for Advanced Study, Princeton, N.J.

Joseph Erlanger Papers. 15 ft., 1890–1964. Library, Washington University School of Medicine.

Simon Flexner Papers. Approx. 200,000 pieces, 1891–1946. Library, American Philosophical Society.

Felix Frankfurter Papers. 69,800 items, 1864–1965. Manuscript Division, Library of Congress.

Genetics Department, University of California, Berkeley, Papers. 50 pieces, 1910–1947. Library, American Philosophical Society.

Richard Goldschmidt Papers. 5 boxes and 4 cartons, ca. 1900–1936. Bancroft Library, University of California, Berkeley.

George Ellery Hale Papers. 173 boxes, ca. 1891–1937. Archives, California Institute of Technology.

Hale-Goodwin Correspondence. Approx. 175 items, 1887–1937. Henry E. Huntington Library and Art Gallery, San Marino, Calif.

Edwin H. Hall Papers. Approx. 450 items, 1875–1938. Houghton Library, Harvard University.

Ross G. Harrison Papers. 69 ft., 1889–1959. Sterling Library, Yale University.

William James Papers. 16 boxes, ca. 1860–1920. Houghton Library, Harvard University.

Irving Langmuir Papers. Approx. 32,000 items, 1871–1957. Manuscript Division, Library of Congress.

Ernest Orlando Lawrence Papers. Approx. 90 ft., 1930–1958. Bancroft Library, University of California, Berkeley.

Gilbert Newton Lewis Papers. 1 ft., ca. 1912–1945. Archives, Bancroft Library, University of California, Berkeley.

Jacques Loeb Papers. Approx. 11,000 items, 1889–1924. Manuscript Division, Library of Congress.

Franklin P. Mall Papers. Approx. 1 ft., ca. 1900–1914. Department of Embryology, CIW, Baltimore.

Robert A. Millikan Papers. 40 ft., 1896–1953. Archives, California Institute of Technology.

National Research Council Record Series, 15 cu. ft., 1914–1918. Archives, National Academy of Sciences, Washington, D.C.

John William Strutt Rayleigh Papers. Approx. 2,300 items, 1862–1943. Air Force Geophysical Laboratory, L.G. Hanscom Field, Bedford, Mass.

Theodore William Richards Papers. Approx. 7 ft., 1889–1929. Archives, Pusey Library, Harvard University.

Owen Richardson Papers. Approx. 32,000 items, ca. 1899–1947. Humanities Research Center, University of Texas at Austin.

Rockefeller Foundation Archives. Project File. 1655 cu. ft., 1912–1975. Rockefeller Archive Center, Pocantico Hills, N.Y.

Franklin D. Roosevelt, Papers as President, 1933–1945. [From the president's secretary's file, 130 ft., and the president's personal file, 608 ft.] Franklin D. Roosevelt Library, Hyde Park, N.Y.

Henry Norris Russell Papers. 131 boxes, 1869–1957. Seeley Mudd Manuscript Library, Princeton University.

George Harrison Shull Papers. Approx. 20,000 pieces, 1904–1915, Library, American Philosophical Society.

Oswald Veblen Papers. Approx. 13,600 items, 1881–1960. Manuscript Division, Library of Congress.

A. G. Webster Papers. 0.6 cu. ft., 1892–1920. Archives, University of Illinois at Urbana-Champaign.

Norbert Wiener Papers. 30 ft., 1898–1964. Archives, Massachusetts Institute of Technology.

Robert M. Yerkes Papers. 54 ft., 1897–1956. Sterling Library, Yale University.

Index